Population and Human Development in Districts of India

Aalok Ranjan Chaurasia

Population and Human Development in Districts of India

Springer

Aalok Ranjan Chaurasia
Mewalal Chaurasia Foundation
and 'Shyam' Institute
Bhopal, Madhya Pradesh, India

ISBN 978-981-95-8168-9 ISBN 978-981-95-8169-6 (eBook)
https://doi.org/10.1007/978-981-95-8169-6

This Springer imprint is published by the registered company Springer Nature Singapore Pte Ltd.
The registered company address is: 152 Beach Road, #21-01/04 Gateway East, Singapore 189721, Singapore

Preface

The diversity of the country India is both mesmerising and perplexing. India is often christened as a country of countries. The social, cultural, economic and environmental diversity of the country, coupled with the vastness of its population has implications for economic growth, social equality and environmental sustainability that constitutes the core of the concept of sustainable development propagated by the United Nations in its 2030 agenda of sustainable development. There is evidence to suggest that diversity, in any form, influences the efficiency of population and development efforts in effectively meeting the diverse population and development needs of the people and, therefore, may perpetuate inequality in all dimensions of human well-being. The most commonly known income inequality is unacceptability high in India and appears to be increasing. The recent G20 report on income inequality reveals that the wealth of the richest 1 per cent Indians increased by more than 60 per cent between 2000 and 2023 while the bottom 50 per cent own only a small proportion of the total wealth of the country. The latest round of the National Family Health Survey, 2019–2021 suggests that availability of household assets in more than 70 per cent Scheduled Tribes households in the country is either low or very low. The Goal 10 of the United Nations 2030 Sustainable Development Agenda calls for reducing inequality of all forms among and within countries. Reducing inequality in India requires incorporating the dimension of diversity in population and development policy formulation and planning for population and development activities.

Concerns about the pervasive and persistent social, cultural, economic and environmental diversity in India in the population and development discourse is not new. In view of demographic and development implications of this diversity, the need for a decentralised district-based approach to population and development planning was emphasised even before the independence. After the independence, the First Five-year Plan recognised the need of decentralised population and development planning to effectively meet the diverse welfare and development needs of the people. The Constitution of India mandates creation of District Planning Committee (DPC) in each district for preparing and implementing District Development Plan. The erstwhile Planning Commission has prepared detailed guidelines for the preparation of District Development Plan.

Among a host of reasons that impede integrated district population and development planning, an important but perpetual one is the utter lack of district level data on population and its development needs that provide the evidence necessary for need-based planning. In the absence of SMART—specific, measurable, appropriate, reliable and timely—data, district population and development planning in India has either remained normative or analogical or anecdotal and, therefore, has largely remained inefficient in meeting the population and development needs of the people. The need for creating district population and development database to facilitate district population and development planning has long been emphasised in the policy discourse but there has been little progress. There are only two sources of district level data about people and their demographic and development attributes—the decennial population census and the National Family Health Survey. India had an unbroken series of decennial population census beginning 1881 but this continuity is now broken as the 2021 decennial population census could not be conducted. This means that whatever district level population and development data are available from the decennial population census, they are more than 15 years old and, therefore, are of little relevance to district population and development planning. The National Family Health Survey, on the other hand, was launched by the Government of India in 1990 to generate data related to health and family planning. The first three rounds of the survey did not provide district level data but the fourth round (2015–2016) and the fifth round (2019–2021) of the survey have provided data about selected demographic and development attributes of the people of all districts of the country as they existed at the time of the survey. There has hardly been any effort to create a system at the district level that generates data about the people of the district and their demographic and development attributes at regular intervals. The lack of district level data about the people and their demographic and development attributes is a major impediment to district level population and development planning in the country.

The present monograph is an attempt to characterise demography and development in the districts of India based on the data available from the decennial population census and the latest round of the National Family Health Survey. The monograph also highlights the pervasiveness of social, cultural, economic and environmental diversity in India as reflected through inter-district variation in different dimensions of demography and development. The monograph follows the population growth modelling approach to analyse population growth in districts since independence, to forecast district population up to the year 2041 and to estimate the likely time when population of different districts will stabilise. The monograph also characterises the demography of the district in terms of fertility and longevity, the two key determinants of population growth. On the other hand, the monograph follows the human development framework to characterise the level of development in the district by constructing the surface measure of human development that is an improvement over the conventional human development index. Finally, the monograph analyses the relationship between fertility and human development based on inter-district variation in fertility and inter-district variation in the surface measure of human development.

This analysis reveals that inter-district variation in human development is a weak predictor of inter-district variation in fertility.

The monograph will particularly be useful in understanding the district level population and development dynamics in India and may contribute significantly to the evidence-based district population and development planning so as to effectively address the diverse population and development needs of the people of different districts of the country. This understanding is crucial for the effectiveness of population and development planning in India.

Bhopal, India Aalok Ranjan Chaurasia

Competing Interests The author has no competing interests to declare that are relevant to the content of this manuscript.

Contents

About the Author

Aalok Ranjan Chaurasia holds an M.Sc. from Kanpur University, India, a Ph.D. from Jiwaji University, Gwalior, India, and a postgraduate diploma in Population and Development from Jawaharlal Nehru University, New Delhi, India. He participated in the United Nations Global Training Programme in Population and Development, 1990–1991 and has published more than 40 books and 160 research papers. He is a member of the International Union for the Scientific Study of Population and a life member of the Indian Association for the Study of Population. He is the editor of the *Indian Journal of Population and Development* and serves on the editorial board of the *Indian Journal of Human Development*. He has also served as a member of the editorial board of the journal *Demography India.*

Chapter 1
Population Growth, 1951–2011

Abstract This chapter analyses population growth trends in 640 districts of India, as they existed at the 2011 population census, during the 60 years period between 1951 and 2011. The analysis is based on the population of districts enumerated at different population censuses beginning 1951 through 2011 made available by the Registrar General and Census Commissioner of India. This chapter follows a modelling approach for analysing the population growth trend in each of the 640 districts of the country. The analysis reveals that the simple logistic growth model provides very good fit to population growth pattern during 1951–2011 in all but a few districts of the country. The annual estimates of the population of the district derived from the fitted logistic growth model have been used to analyse the trend in population growth in the district during the period 1951–2011 through the application of the joinpoint regression model. The trend analysis reveals that population growth trend in most of the districts changed at least five times during the 60 years between 1951 and 2011 and growth of population in different temporal segments has been different within the district and across the districts. The analysis also reveals that the average annual population growth rate during the period 1951–2011 has been more than 2 per cent in almost two-third of the districts of the country. There are, however, only five districts where estimated population growth rate has always been more than 2 per cent during this period.

Keywords India · Districts · Population growth · Modelling population growth

Introduction

A district is the third tier of the population and development administration system in India. Population and development policies and programmes in India are conceptualised and programmed at the national level, customised at state/Union Territory level but implemented at the district level. The district is also the lowest administrative unit for monitoring the implementation of different population and development programmes and interventions. The progress of population and development

A. R. Chaurasia, *Population and Human Development in Districts of India*,
https://doi.org/10.1007/978-981-95-8169-6_1

programmes and interventions at the state/Union Territory level and at the national level is simply an aggregation of the progress at the district level. The Constitution of India mandates formation of District Planning Committee in each district to prepare the development plan for the district in recognition of considerable inter-district diversity in all aspects of population and social and economic development. Since population transition and social and economic development processes are interdependent, there has also been emphasis on the integration of population factors in social and economic development planning process. However, integration of population factors with the social and economic development planning at the district level in India is hampered by the non-availability of necessary data. The only source of data about population at the district level in the country is the decennial population census. India had an enviable record of unbroken decennial population census since 1881 up to 2011. This continuity has now broken as the 2021 decennial population census in the country could not be conducted because of the COVID-19 pandemic.

Population and development planning may be defined as the planned allocation of resources to meet the development and welfare needs of the people. An understanding of the pattern of population growth and the factors that contribute to population growth is, therefore, an important component of social and economic development planning and programming process at the district level. There has, however, rarely been any attempt to analyse the pattern of population growth in the districts of the country. A major hindering factor is the change in the number of districts at every decennial population census. At the 1951 population census, the first population census after the independence, there were only 316 districts in the country. This number increased to 339 at the 1961 population census; 356 at the 1971 population census; 412 at the 1981 population census; 466 at the 1991 population census; 593 at the 2001 population census; and 640 at the 2011 population census. Although, there was no population census in the country after 2011 but the number of districts in the country increased to 751 in 2021 and 785 in 2024 (Government of India 2024) and this number continues to increase. The increase in the number of districts has resulted in frequent changes in the administrative boundaries of many districts which makes it difficult to analyse and explore how the population of many districts has evolved over time, especially, after the independence. The Government of India has now made available the population enumerated by sex at different decennial population censuses since 1901 in each of the 640 districts of the country as they existed at the time of the 2011 decennial population census. This database made available by the Government of India makes it possible, for the first time, to analyse the pattern of population growth in 640 districts of the country as they existed in 2011. This chapter uses the dataset made available by the Government of India to analyse the pattern of population growth in 640 districts of the country since the independence or more specifically, during the period 1951–2011.

This chapter is organised as follows. The next section of this chapter describes the data used for analysing the pattern of population growth in the districts. The analysis is based on the population enumerated at different decennial population censuses beginning 1951 through 2011 in 640 districts as they existed at the 2011 population census. The third section describes the approach adopted for analysing the pattern of population growth in the district. Since the enumerated population in the district is available at an interval of 10 years only, a modelling approach has been adopted to model the pattern of population growth in the district. It has been found that the simple logistic growth model provides very good fit to population growth during 1951–2011 in all but a few districts of the country. The simple logistic growth model has, therefore, been fitted to the population of each of the 640 districts enumerated at different decennial population censuses after the independence and the results of the modelling exercise are presented and discussed in section four of the paper. Section five of the chapter characterises the pattern of population growth in 640 districts in terms of the parameters of the logistic growth model. The modelling of district population growth reveals the diversity in the pattern of population growth across the districts during the period 1951–2011 which has implications for district level social and economic development planning and programming and for the future growth of population in the district. The variation in the pattern of population growth across the districts also justifies the need of integrating district-specific population factors in the district level social and economic development planning and programming process directed towards improving the quality of life of the people of the district.

The Data

The analysis is based on the population enumerated in the 640 districts of the country, as they existed at the time of the 2011 population census, at different decennial population censuses in India beginning 1951 through 2011. These data have been made available by the Registrar General and Census Commissioner of India. There are, however, 42 districts for which the population enumerated at the 1951 population census is not available whereas in 10 districts, population enumerated at 1951 and 1961 population censuses is not available. The population enumerated at the 1981 population census is also not available for all districts of Assam as the 1981 population census could not be conducted in the state. The analysis of population growth in 32 districts, therefore, has been carried out based on the population enumerated at different population censuses since 1961 through 2011 whereas, in 10 districts, the analysis has been carried out based on the population enumerated at different population censuses since 1971 through 2011. On the other hand, we have estimated population of the districts of Assam in 1981 as the average of the population enumerated at the 1971 population census and the population enumerated at the 1991 population census.

The population enumerated at the population census is known to be associated with several errors including errors of omission and duplication. The Registrar General and Census Commissioner of India undertakes post-enumeration survey after every decennial population census to estimate errors of omission and duplication in population enumeration. The post-enumeration survey conducted after the 2011 population census has revealed that there was a net omission of around 23 persons for every 1000 persons at the 2011 population census (Government of India 2014). There was an estimated undercount of 23.08 persons for every 1000 persons counted which was offset by an estimated duplication of 0.10 persons for every 1000 persons counted. The post-enumeration survey has also revealed that the net omission rate was markedly higher in the urban (29 persons per 1000 persons enumerated) as compared to the rural areas of the country (20 persons per 1000 persons enumerated). A comparison of the net omission rate at the 2011 population census with the net omission rate at the 2001 population census, however, suggests that there has been little change in the net omission rate in the two population censuses. The Registrar General and Census Commissioner of India, however, makes no adjustment in the enumerated population to consider the net omission rate in enumerating the population.

Estimates of the net omission rate at different population censuses are not available for the states/Union Territories and districts of the country. The Registrar General and Census Commissioner of India has, however, estimated the net omission rate in the census count at the 2011 population census in different zones of the country which reveals that the net omission rate varies across different zones. The net omission rate is estimated to be the lowest in the eastern zone of the country but the highest in the central zone (Government of India 2014). Although, estimates of the net omission rate are not available for the districts of the country, it may be argued that the net omission rate varies across the districts as well as in different population censuses since 1951 and, in the absence of any information about the net omission rate in census count and other errors associated with population enumeration, it is not possible to make any adjustment in the population enumerated in the districts at different population census. The present analysis has, therefore, been carried out without making any adjustment in the enumerated population in the 640 districts, as they existed at the 2011 population census, at 10 years interval beginning 1951 through 2011 as made available by the Registrar General and Census Commissioner of India. We have used population enumerated at 1951 through 2011 population censuses for analysing population growth in 598 districts and enumerated population at 1961 through 2011 population censuses for analysing population growth in 32 districts of the country. For the remaining 10 districts, we have used the population enumerated at 1971 through 2011 population censuses as enumerated population at 1951 and 1961 population censuses is not available in these districts.

Methodology

The database made available by the Registrar General and Census Commissioner of India gives the population enumerated at 10 years interval beginning 1901 in each of the 640 districts of the country, as they existed at the 2011 population census. Using these data, population growth is usually analysed in terms of average annual growth rate during the 10 years interval assuming that population grows exponentially within the interval. This approach also assumes that the trend in population growth changes only in that year when the population census is conducted and the time of change in the trend is the same for all districts. It is, however, not necessary that population grows exponentially between two population censuses. Similarly, it is not necessary that population growth trend changes in the census year only and the time of change is the same for all districts. A more appropriate approach is to first model population growth and then use model estimates to analyse population growth trend. The modelling of population growth is motivated by the demographic transition theory which classifies the growth of population in three distinct phases—initial slow growth phase, middle rapid growth phase and final phase in which population growth slows down to reach an upper limit. The demographic transition theory implies that the population growth rate follows a reverse V-shape growth trajectory. The population growth rate first increases, reaches a maximum, and then decreases. The reverse V-shape trajectory of population growth rate means that population cannot continue increasing indefinitely. Instead, growth of the population follows an *S*-shaped growth trajectory.

Different models have been proposed to model the S-shaped growth trajectory. These include, among others, the logistic growth model, the Gompertz model, the modified exponential model and the generalised logistic growth model. The logistic growth model is the simplest description of the *S*-shaped growth trajectory. It was first developed by Verhulst in 1838 (Verhlust 1838) and later re-invented independently by Pearl and Reed (Pearl and Reed 1920). The logistic growth model assumes exponential growth of the population under the constraint of an upper limit (Lotka 1956). Application of the logistic growth model to describe and forecast population growth has a long history. It was a popular method of describing and forecasting population growth in the past. Several studies have shown that the application of the logistic growth model can provide reasonably accurate description and forecast of population growth (Dorn 1950; Leach 1981). In recent years, there is a renewed interest in the logistic growth model to describe and forecast population growth (Hrytsiuk et al. 2023; Burg and Ausubel 2023; Mondol et al. 2018; Sharif Ullah et al. 2019). Bhat (1999) has used the logistic growth model to forecast the population of Delhi, the capital city of India when the data required for the application of the cohort-component method of population projection are not available.

If P denotes the population, U denotes the upper limit of population growth and r is the population growth rate, then the logistic growth model is defined by the differential equation:

$$\frac{dP}{dt} = rP\left(1 - \frac{P}{U}\right) \tag{1.1}$$

The cumulative function associated with Eq. (1.1) may be written as:

$$P_t = \frac{U - L}{1 + e^{-r(t - t_m)}} + L \tag{1.2}$$

where.

P_t = the population at time t,

L = lower asymptote of the model.
U = upper asymptote of the model.
r = intrinsic growth rate of the model.
t_m = the time of inflexion of the growth rate. In case of the logistic growth model, this time is also the time when the population reaches half of the upper asymptote of the model as the logistic growth model is symmetric about t_m.

The intrinsic growth rate r reflects the 'steepness' of the growth trajectory—the higher the r the steeper the growth trajectory. One approach frequently used to calculate r is based on the time required for the population to grow from 10 per cent to 90 per cent of the upper asymptote of the logistic growth model. This period is termed as the characteristic duration of growth and is denoted by Δt. If P_1 is 10 per cent of U at time t_1, then, we can write,

$$0.1 = \frac{P_1}{U} = \frac{1}{1 + e^{-r(t_1 - t_m)}} \; or \; 9 = \frac{1}{e^{-r(t_1 - t_m)}} \tag{1.3}$$

Similarly, if P_2 is 90 per cent of U at time t_2, then, we can write,

$$0.9 = \frac{P_2}{U} = \frac{1}{1 + e^{-r(t_2 - t_m)}} \; or \; 9 = e^{-r(t_2 - t_m)} \tag{1.4}$$

Combining Eqs. (1.2, 1.3), we get

$$9 \times 9 = 81 = \frac{e^{-r(t_2 - t_m)}}{e^{-r(t_1 - t_m)}} = e^{r(t_2 - t_1)} = e^{r \times \Delta t} \tag{1.5}$$

so that

$$r = \frac{\ln(81)}{\Delta t} \tag{1.6}$$

We have used the open source Loglet software package (Yung et al. 1999) to fit the logistic growth model to population growth during 1951–2011 in 640 districts of India as they existed at the 2011 population census. The software provides estimates of the lower asymptote d, upper asymptote K, the characteristics duration of population growth Δt and the time of inflexion of the rate of population growth, t_m. The software also provides goodness-of-fit statistics including mean absolute percentage error (MPAE), residual sum of squares (RSS), residual mean sum of squares (RMS) and R^2 which allow to test appropriateness of the logistic growth model to describe the pattern of district population growth. It may be pointed out that although, R^2 is not regarded as an appropriate measure for ascertaining the goodness-of-fit in case of nonlinear models as is the case with the logistic growth model (Spiess and Neumeyer 2010), yet R^2 can be used for testing the goodness-of-fit of the logistic growth model as the model can be transformed into a linear growth model through the Fisher-Pry transformation (Fisher and Pry 1971). We have used MAPE and R^2 to test the goodness-of-fit of the logistic growth model in describing district population growth.

The model is termed as highly accurate in describing the district population growth, if the MAPE is less than 10 per cent, good if MAPE is 10 per cent and more but less than 20 per cent, and reasonable if the MAPE ranges from 20 to 50 per cent. If the MAPE is 50 per cent and more than the model is termed as inaccurate in describing district population growth (Lewis 1982). On the other hand, the model is termed as appropriate for describing district population growth if the model explains at least 80 per cent of the variation in the original data and the higher this proportion the better the appropriateness of the model.

The next step in analysing district population growth involves constructing time series of annual estimates of population for each of the 640 districts based on the parameters of the logistic growth model and using the constructed time series to analyse population growth trend in each district. We have followed the joinpoint regression analysis approach for analysing the trend in population growth as the trend in population growth does not remain the same throughout the trend period (Kim et al. 2000). There are three steps in the joinpoint regression analysis. The first is to test whether the trend has changed during the period under reference or not and if the trend has changed, then how many times. If there is no change in the trend, then the trend analysis can be carried out by simply fitting a straight line (on a Log scale) to the observed time series and the annual per cent change can be estimated from the slope of the regression line. If the trend is found to have changed, then the second step involves identifying joinpoint(s) or the time(s) or year(s) when the trend has changed. Once the number of joinpoint(s) is identified, the last step in the joinpoint regression analysis involves estimating the regression function with identified joinpoint(s). If there are k joinpoints, then the entire trend period is divided into $k + 1$ temporal segments with different trends in population growth in different temporal segments. The population growth in different temporal segments may be measured in terms of the annual per cent change (APC) which provides complete characterisation of population growth in the temporal segment.

The number of times the trend in the growth of the population of a district has changed can be set in advance or can be determined statistically from the data. When the number of times the trend has changed is set in advance, the trend analysis can be carried out using piecewise or segmented regression modelling approach (Chaurasia 2020). On the other hand, there are different methods that have been proposed to determine, statistically, the number of times the trend in population growth has changed during the period under reference. These include permutation procedure (Kim et al. 2000); Bayesian Information Criterion (BIC) (Kim et al. 2009); Bayesian Information Method with a harsher penalty (BIC3) (Kim and Kim 2016); and the modified BIC method (Zhang and Siegmund 2007). A data-driven approach has also been proposed to identify the number of times the trend has changed in the time series (Kim et al. 2023). Identification of the number of times the trend has changed using the statistical, data-driven approach is preferred over arbitrarily setting the number of times the trend has changed in advance because statistical identification of the number of times the trend has changed is based on the real time data and, therefore, is free from the selection bias.

Let P_i denote the population in the year y_i such that $y_1 < y_2 < \ldots.. < y_n$ and $k_1 < k_2 < \ldots. < k_j$ are the years when the trend in population growth has changed or joinpoints. Then the joinpoint regression model is defined as:

$$\ln(P_t) = \alpha + \beta_1 * y_1 + \delta_1 * u_1 + \delta_2 * u_2 + \cdots + \delta_j * u_j + \varepsilon_i \tag{1.7}$$

where,

$$u_j = \begin{cases} (y_j - y_k), \ if \ y_j > k_j \\ 0, \text{otherwise} \end{cases} \tag{1.8}$$

When there is no change in the trend, the joinpoint regression analysis is the same as the linear regression analysis on a Log scale.

If P_0 is the population at the beginning and P_t is the population at the end of a temporal segment of length t years, then the annual per cent change (APC), p, in the temporal segment is calculated as

$$p = \frac{P_t - P_0}{P_0 \times t} \tag{1.9}$$

On the other hand, the annual population growth rate (AGR), r, in the temporal segment is calculated as

$$r = \frac{1}{t} \times \ln\left(\frac{P_t}{P_0}\right) \tag{1.10}$$

or

$$P_t = P_0 \times e^{(t \times r)} \tag{1.11}$$

Substituting from (9) into (7), we get

$$p = \frac{\left(P_0 \times e^{(t \times r)}\right) - P_0}{P_0 \times t} \tag{1.12}$$

which gives

$$r = \frac{1}{t} \times \ln(1 + (t \times p)) \tag{1.13}$$

The AGR in different temporal segments can be compared to examine the acceleration or deceleration in population growth. Moreover, the weighted average of AGR in different temporal segments with weights proportional to the length of the temporal segment gives the average annual growth rate (AAGR) of the population for the entire trend period. Thus, AAGR for the entire trend period (0, N) can be calculated as

$$AAGR = \sum_{i=1}^{N} w_i \times r_i, \quad w_i = \frac{t}{N} \tag{1.14}$$

The AAGR is argued to be a better approach to describe the long-term trend in population growth in situations when the trend changes over time than the commonly used approach in which a single regression line (on a Log scale) is fitted for the entire trend period and the average annual growth rate of the population is calculated from the slope of the regression equation (Clegg et al. 2009). The AAGR best summarises the trend in population growth that varies over time (Marrot 2010).

Following Kitagawa (1955), the difference in AAGR between two districts *A* and *B* can be decomposed as

$$\Delta r = r^A - r^B = \sum w_i^A \times r_i^A - \sum w_i^B \times r_i^B \tag{1.15}$$

$$\Delta r = \sum \left(w_i^A - w_i^B\right) \times \left(\frac{\left(r_i^A + r_i^B\right)}{2}\right) + \sum \left(r_i^A - r_i^B\right) \times \left(\frac{\left(w_i^A + w_i^B\right)}{2}\right) \tag{1.16}$$

Equation (1.16) suggests that the difference in AAGR between two districts is the sum of the difference in the length of different temporal segments and the difference in the rate of growth in different temporal segments. Eq. (1.16) provides a deeper understanding of the difference in the rate of growth of population between two districts as compared to the conventional approach which assumes that the rate of growth in population is linear throughout the trend period and there has been no change in the trend during the entire period.

We have used the Joinpoint Regression Program (National Institute of Health 2020) for analysing the trend in population growth in the districts during 1951–2011. The software requires, in advance, specification of maximum (>0) number of joinpoints. The Program starts with the minimum number of joinpoints (0) and tests whether more joinpoints are statistically significant and need to be added to the model. The grid search method is used to identify joinpoints (Lerman 1980). The grid search method allows a joinpoint to occur at the exact time *t*. A grid is created for all possible positions of joinpoint(s) or combination of joinpoint(s), the model is fitted for each position separately and that position is selected which minimises the sum of squared error (SSE). It may be pointed out that even if the final model has *k* joinpoint(s), the slope of all regression functions of $k + 1$ temporal segments may not be statistically significantly different from 0. Identification of *k* joinpoints means that the model with these joinpoint(s) provides the best fit to the observed data relative to all other models. If the slope of a regression function in a time segment is statistically insignificant then this means that there has been no change in the trend in the time segment. We have set the maximum number of joinpoints to 5 and the data-driven approach has been used to identify the joinpoints.

Joinpoint trend analysis has been used earlier in analysing the long-term trend in mortality and morbidity (Akinyede and Soyemi 2016; Chatenoud et al. 2015; Doucet and Rochette 2016; John and Hanke 2015; Mogos et al. 2016; Missikpode et al.

2015; Puzo et al. 2016; Qiu et al. 2009; Rea et al. 2017; Tyczynski and Berkel 2005). It has also been used to estimate population parameters under changing population structure (Gillis and Edwards 2019). Joinpoint trend analysis has been recommended for analysing the trend in measures of population health, especially when the trend changes over time (Ingram et al. 2018). It has also been used to analyse long-term trend in infant mortality rate (Chaurasia 2020) and transition in mortality (Chaurasia 2023). The joinpoint trend analysis provides an easily interpretable characterisation of the nonlinear trend.

Modelling Population Growth.

We have fitted Eq. (1.2) to male and female population growth in 640 districts during 1951–2011 as revealed through the population enumerated at different population censuses since 1951 and added the estimates of the male and female population to estimate the population of the district. The MAPE and R^2 of the fitted model for each district are given in the table appended to this chapter while inter-district variation in MAPE and R^2 is summarised in Table 1.1. The modelling exercise suggests that model (2) has provided very good fit to population growth in all but a few districts. In all but 9 districts, the MAPE is estimated to be less than 10 per cent and there is only one district—district Kiphire in Nagaland—where MAPE is estimated to be more than 20 per cent. In this district, the population enumerated at the 1961 population census was only around 15.6 thousand which increased to more than 51.8 thousand at the 1991 population census. However, the population of the district nearly doubled between 1991 and 2001 population census to more than 106 thousand, but decreased sharply to around 74 thousand at the 2011 population census. Model (2) did not provide good fit to population growth in district New Delhi and Central district of the National Capital Territory of Delhi. The enumerated population of these districts at the 2011 population census decreased, instead increased, relative to 1961 population censuses. On the other hand, model (2) explained at least 95 per cent of the variation in population growth during 1951–2011 in 536 districts and between 90 and 95 per cent in 78 districts. There are only 22 districts where model (2) explained less than 90 per cent of the variation in population growth during 1951–2011, and only one district where it explained less than 80 per cent of the variation.

Results of the population growth modelling exercise thus suggest that in all but a few districts of the country, population growth during the post-independence period may be characterised by an S-shaped growth trajectory which can be very satisfactorily modelled by the simple logistic growth model. This implies that the variation in the pattern of population growth during 1951–2011 in the districts can be explained in terms of the variation across in the parameters of the simple logistic growth model. The table appended to this chapter presents estimates of three parameters of the simple logistic growth model for each district: (1) characteristic duration of population growth; (2) the intrinsic population growth rate; and (3) the time of inflexion of the population growth rate. The variation in the three parameters of the simple logistic growth model across districts is summarised in Table 1.2. In all but a few districts, the characteristic duration of population growth or the duration during which the population of the district increases from 10 per cent to 90 per cent of the upper asymptote of the model varies between 56 and 60 years, but the range is very wide, from around

Table 1.1 Results of fitting of the logistic growth model to population growth in districts, 1951–2011. Variation in MAPE and R^2 across districts

MAPE			R^2		
Range	Males	Females	Range	Males	Females
<0.02	206	202	≥0.95	558	520
0.02–0.04	315	305	0.90–0.95	62	91
0.04–0.06	80	94	0.85–0.90	10	19
0.06–0.08	21	22	0.80–0.85	5	4
0.08–0.10	8	7	<0.80	3	4
≥0.10	10	8			
No data	2	2	No data	2	2
Total	640	640	Total	640	640

Source Author

10 years in district Kiphire in Nagaland to around 107 years in district Karim Nagar and district Medak in Andhra Pradesh. In 112 districts of the country, the characteristic duration of population growth is less than 56 years whereas in 23 districts, it is more than 60 years. In all districts of Sikkim, Arunachal Pradesh, Nagaland and Mizoram, the characteristics duration of population growth is less than 52 years.

Similarly, in most of the districts of the country, the intrinsic growth rate, r, varies between 5–10 per cent. There are only 33 districts where the intrinsic growth rate has been 10 per cent and more. The intrinsic growth rate is inversely related to the characteristic duration of population growth—the higher the intrinsic growth rate, the steeper the population growth trajectory and the shorter the characteristic duration of population growth and vice versa. The intrinsic growth rate has been the most rapid in district Kiphire of Nagaland so that the characteristic duration of population growth or the time during which the population of the district increases from 10 per

Table 1.2 Inter-district variation in the parameters of the logistic growth model

Characteristic growth period		Intrinsic growth rate		Year of inflexion of the growth trajectory	
Number of years	Districts	Rate	Districts	Year	Districts
< 52	70	<0.050	14	Before 1980	52
52–54	10	0.050–0.075	253	1980–1990	236
54–56	32	0.075–0.100	338	1990–2000	292
56–58	140	0.100–0.150	25	2000–2010	50
58–60	363	0.150–0.200	6	2010–2020	3
≥60	23	≥0.200	2	2020 and after	5
No data	2	No data	2	No data	2
Total	640		640		640

Source Author

cent to 90 per cent of the upper asymptote of the logistic growth model is estimated to be only around 10 years, the shortest among all districts of the country. On the other hand, there are 14 districts where the intrinsic growth rate is estimated to be less than 5 per cent so that the characteristic duration of population growth in these districts is very long. The population of these districts is likely to take very long time to increase from 10 per cent to 90 per cent of the upper asymptote of the simple logistic growth model. The intrinsic population growth rate is found to be the lowest in district Karim Nagar and district Medak of Andhra Pradesh. The very low intrinsic population growth rate means that the characteristic duration of population growth or the time taken by the population of these districts to increase from 10 per cent to 90 per cent of the upper asymptote of the logistic growth model is the longest among the districts of the country. The population of these two districts is likely to increase from 10 per cent to 90 per cent of the upper asymptote of the logistic growth model or the upper limit of population growth in a period of around 107 years.

The population growth modelling exercise also suggests that in most of the districts of the country, population growth is now flattening as the time of inflexion in the population growth rate trajectory has already been crossed. There are only four districts where the population growth rate still appeared to be accelerating around the year 2011. In 538 of the 640 districts, the inflexion point in the population growth rate trajectory was reached sometimes during the period 1980–2000. There are only 52 districts, in which the inflexion point in the population growth rate trajectory was reached before 1980. Similarly, there are only in 53 districts, the inflexion point in the population growth rate trajectory was reached sometimes after 2000 but before 2020. There is only one district where the inflexion point in the population growth rate trajectory was reached around the year 2020. The four districts, where the inflexion point in the population growth rate trajectory is yet to reach the inflexion point are district Kurung Kumey in Assam, district Rangareddy in Andhra Pradesh, district Yanam in Puducherry and district Daman in Daman and Diu. In district Kurung Kumey of Assam, the inflexion point in the population growth rate trajectory is expected to be reached by the year 2047. In the Rangareddy district of Andhra Pradesh, the inflexion point in the population growth rate trajectory is expected to be reached only by the year 2043. In district Yanam, of Puducherry, the inflexion point in the population growth rate trajectory is the most likely to be reached by the year 2037. Lastly, in district Daman, in Daman and Diu, the inflexion point in the population growth rate trajectory is expected to be reached by the year 2028. The variation in the inflexion point in the population growth rate trajectory is expected as different districts of the country are at different stages of population transition.

Figure 1.1 compares the population enumerated at the 2011 population census in 640 districts with the population estimated in the year 2011 based on the simple logistic growth model. The figure confirms that the simple logistic growth model provides very good fit to population growth in the districts of the country during the period 1951–2011. The sum of the population estimates of 640 districts for the year 2011 derived from the simple logistic growth model gives an estimate of 1193 million of the population of the country in 2011 which is very close to 1206 million people enumerated in the country at the 2011 population census. There is an error of just around 1 per cent. This means that the annual estimates of the population of each of the 640 districts of the country for the year 1951–2011 may be derived from the fitted logistic growth model for the district based on the population of the district enumerated at different decennial population censuses in the country beginning 1951. We have used the annual estimates of the population of a district derived from the logistic growth model to characterise the population growth trend in each of the 640 districts through the application of the joinpoint regression model. We have assumed that the trend in population growth in each district has changed at the most 5 times during the 60 years period between 1951 and 2011. This means that the period 1951 through 2011 can be divided into 6 time segments and the population growth in each time segment may be different. The years when the trend in population growth has changed have been determined through the data-driven approach and has not been identified in advance which means that the length of the 6 time segments may not be the same in the same district and across districts. This also means that the years when the trend in population growth has changed in a district may not be the same as the years of the enumeration of the population in the district. In other words, the analysis of the trend in population growth in the districts assumes that in each district, the rate of population growth is different in different time segments and the length of different time segments is not the same. The growth of population in a district during the period 1951–2011 depends upon both the population growth rate in different time segments and the length of different time segments which is different in different districts.

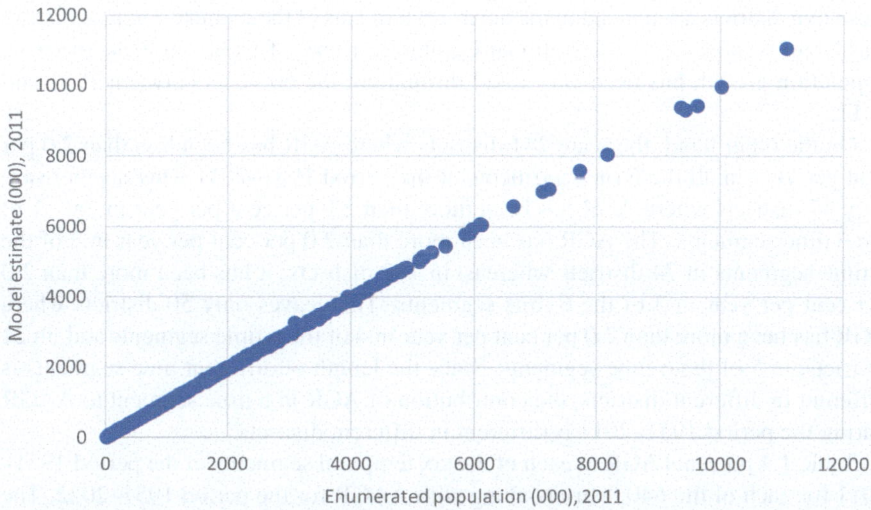

Fig. 1.1 Population (000) enumerated at the 2011 population census and population (000) estimated from the model in districts of India, 2011. *Source* Author

Trend in District Population Growth

Results of the analysis of the population growth trend during 1951–2011 in each of the 640 districts are presented in Table 1.3 which gives length of each of the 6 time segments, annual population growth rate (AGR) in each time segment and average annual growth rate (AAGR) during the 60 years period between 1951 and 2011. Districts vary in terms of the length of the time segments, in terms of the AGR in each time segment and hence in terms of AAGR. Among the 640 districts of the country, the average annual population growth rate (AAGR) has been the slowest in district Satara of Maharashtra where the AAGR during the period 1951–2011 was 0.392 per cent per year only but the highest in district East in the National Capital Territory of Delhi where the population increased at an average annual rate of 8.025 per cent per year during this period. There are 18 districts where AAGR of population has been less than 1.0 per cent per year. On the other hand, there are 37 districts where the population increased, on average, at a rate of at least 2.5 per cent per year (Fig. 1.2). In majority of the districts, the AAGR of population ranged between 1.5–2.0 per cent per year. There are 145 districts where population increased at an average annual rate of 1.0–1.5 per cent per year whereas, in 91 districts, it increased at an average annual rate of 2.0–2.5 per cent per year. There are, however, only 5 districts where the AGR has been more than 2 per cent per year in all the 6 time segments identified through the joinpoint regression analysis. These districts are Faridabad in Haryana, West Kameng in Arunachal Pradesh, Zunheboto in Nagaland, South Garo Hilla in Meghalaya and South Andaman in Andaman and Nicobar Islands. Three of these five districts are located in the north-eastern part of the country whereas district Faridabad is adjacent to the National Capital Territory of Delhi. In these districts, population growth has been very rapid throughout the 60 years between 1951 and 2011.

On the other hand, there are 241 districts where AGR has been less than 2.0 per cent per year in all the 6 time segments of the period 1951–2011 whereas there are only 67 districts where AGR has been more than 2.0 per cent per year in only 1 of the 6 time segments. The AGR has been more than 2.0 per cent per year in 2 of the 6 time segments in 74 districts whereas, in 182 districts, it has been more than 2.0 per cent per year in 3 of the 6 time segments. This leaves only 50 districts where AGR has been more than 2.0 per cent per year in 4 of the 6 time segments and, in 21 districts, in 5 of the 6 time segments. Since the length of different time segments is different in different districts, the contribution of AGR in a time segment to AAGR during the period 1951–2011 is different in different districts.

Table 1.4 presents AGR in each of the six temporal segments of the period 1951–2011 for each of the 640 districts along with AAGR for the period 1951–2022. The six temporal segments are different for different districts. The mean of the model-based AGR across the 640 districts increased from 1.419 ± 3.369 per cent in the first temporal segment to 2.149 ± 1.348 per cent in the third temporal segment and then decreased to 1.195 ± 0.578 per cent in the sixth temporal segment. The range of AGR across districts, however, decreased rapidly from temporal segment one to

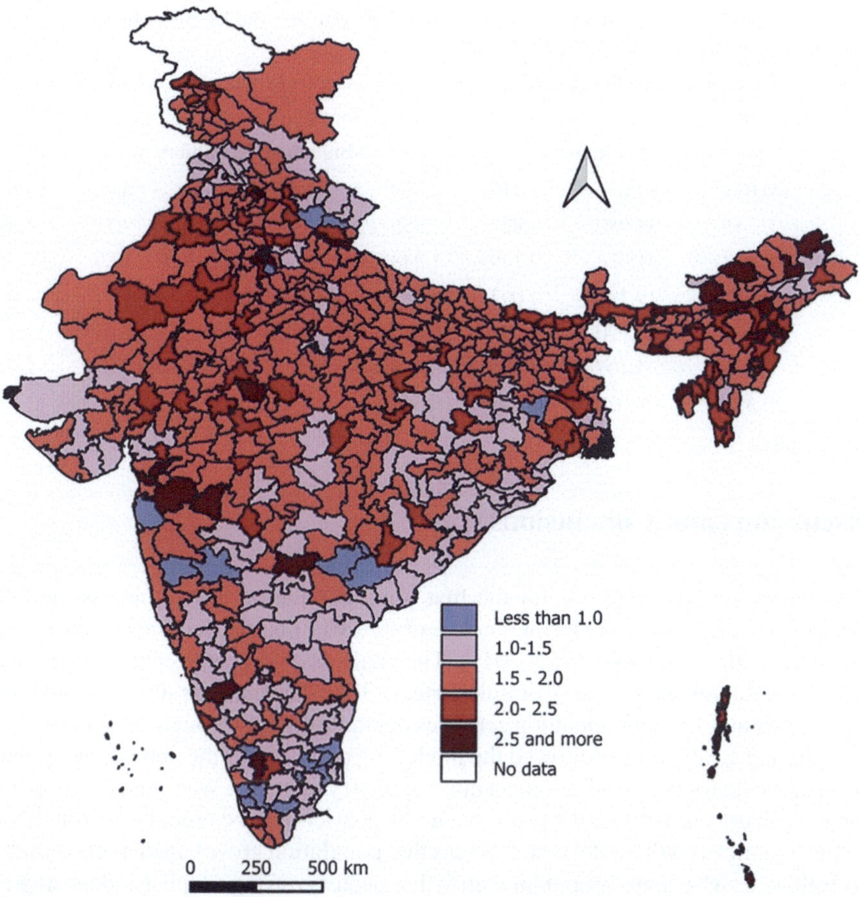

Fig. 1.2 Average annual population growth rate during 1951–2011 in 640 districts of India. *Source* Author

temporal segment six which indicates that AGR has converged across the districts during the 60 years under reference. On the other hand, the mean length of the temporal segments across districts has increased up to the temporal segment four and decreased thereafter. The mean length of the temporal segment was 8.4 ± 2.6 years which increased to 13.1 ± 6.5 years in the fourth temporal segment but decreased to 8.6 ± 2.7 years in the sixth temporal segment. There was very little change in the mean length of second, third, fourth and fifth temporal segments. The mean length of temporal segments indicates the duration for which a given population growth rate prevailed in a district. The variation in the mean length suggests that the duration for which a given population growth rate prevailed has also been different in different districts. This variation has also influenced the growth of population in the district during the period 1951–2011.

Table 1.3 Distribution of districts by population growth rate and the length of time segment in different time segments of the period 1951–2011

Time segment	Population growth rate (per cent year)				Length of time segment (years)			
	Range		Average	SD	Range		Average	SD
	Minimum	Maximum			Minimum	Maximum		
1	0.041	84.656	1.419	3.369	2	26	8.4	2.6
2	0.245	34.080	1.827	1.455	3	23	8.7	3.4
3	0.458	31.026	2.149	1.348	3	22	12.7	6.3
4	0.352	19.196	2.097	0.987	3	23	13.1	6.5
5	0.240	5.990	1.724	0.655	4	23	8.6	3.1
6	0.028	5.107	1.195	0.578	5	44	8.6	2.7
All	0.392	8.025	1.773	0.546				

Source Author

Discussions and Conclusions

The present chapter analyses, for the first time, population growth across the 640 districts of India as they existed at the time of the 2011 population census, during the post-independence period (1951–2011). The analysis could be possible because the Registrar General and Census Commissioner of India has made available population enumerated at different population censuses beginning 1901 through 2011 in the 640 districts. The analysis has followed the modelling approach as the enumerated population of the districts is available at an interval of 10 years only. We found that population growth during 1951–2011 in all, but a few districts can be modelled through the simple logistic growth model which means that population growth in districts of India has followed an S-shaped population growth trajectory during the post-independence period. This observation implies that the assumption that population between two successive population censuses is linear or the population changes at a constant rate between two successive population censuses is not the right approach to analyse the trend in population growth and for interpolating the population of districts for years between successive population censuses. We have ated annual estimates of population of districts derived by fitting the simple logistic growth model in each district.

The present analysis reveals that population growth trend in 640 districts of the country during the post-independence period has varied widely. This is expected as population transition and social and economic development has varied widely across the districts of the country. The diversity in population growth trend across the districts may be judged from the observation that the time taken by the population of a district to increase from 10 per cent to 90 per cent of the upper limit of population growth of the district varies from just around 10 years to more than 100 years, and the shorter this time the rapid the population growth. The analysis also reveals that in all but a few districts of the country, population growth rate trajectory has crossed

Table 1.4 Average annual growth rate (AAGR) and an ual growth rate (AGR) in different temporal segments of the period 1951–2011 in the districts of India

District Code	Year (1951–2011): AGR, segment 1	Segment 2	Segment 3	Segment 4	Segment 5	Segment 6	AAGR
001	0.891	1.467	2.060	2.655	2.666	2.472	2.159
002	1.126	1.690	2.252	2.360	2.256	1.725	2.004
003	0.795	1.443	2.131	2.350	2.058	1.368	1.784
004	0.673	1.354	2.056	2.403	2.180	1.513	1.770
005	0.700	1.169	1.678	2.143	2.203	2.021	1.747
006	0.827	1.285	1.783	2.294	2.293	2.026	1.881
007	1.130	1.603	2.068	2.132	1.896	1.292	1.784
008	1.150	1.690	2.198	2.320	2.197	1.615	1.964
009	1.134	1.766	2.344	2.470	2.483	1.946	2.100
010	1.012	1.548	2.095	2.237	2.097	1.578	1.867
011	0.896	1.418	2.003	2.514	2.488	2.297	2.051
012	0.916	1.472	2.007	2.210	2.162	1.658	1.821
013	0.965	1.515	2.075	2.269	2.217	1.781	1.889
014	0.784	1.312	1.881	2.426	2.515	2.474	2.004
015	0.877	1.565	2.328	2.639	2.152	1.143	1.920
016	0.991	1.550	2.115	2.297	2.242	1.750	1.904
017	0.891	1.513	2.144	2.371	2.356	1.898	1.913
018	1.004	1.514	2.031	2.152	2.004	1.473	1.787
019	1.146	1.694	2.220	2.327	2.160	1.570	1.958
020	1.052	1.612	2.184	2.302	2.144	1.560	1.908
021	1.367	2.072	2.371	2.314	1.751	1.196	1.959
022	1.642	2.175	2.331	2.237	1.666	1.102	1.965
023	1.130	1.674	1.915	1.810	1.399	0.943	1.559
024	1.127	1.568	1.765	1.616	1.179	0.764	1.422
025	3.998	3.152	2.140	1.161	0.454	0.087	1.156
026	0.937	1.397	1.853	2.015	1.835	1.322	1.661
027	1.612	2.046	2.126	1.867	1.278	0.728	1.666
028	1.125	1.472	1.582	1.386	1.017	0.638	1.273
029	0.759	1.143	1.536	1.715	1.514	1.075	1.385
030	1.409	1.846	1.960	1.727	1.225	0.759	1.569
031	0.944	1.482	2.057	2.207	1.975	1.380	1.781
032	1.160	1.715	1.966	1.868	1.484	1.068	1.643
033	1.128	1.607	1.843	1.703	1.294	0.877	1.501
034	1.239	1.608	1.736	1.463	0.969	0.534	1.315
035	1.145	1.617	1.823	1.654	1.190	0.759	1.456
036	1.335	1.770	1.904	1.703	1.233	0.731	1.510

(continued)

Table 1.4 (continued)

District Code	Year																																																												AAGR
	1951	1952	1953	1954	1955	1956	1957	1958	1959	1960	1961	1962	1963	1964	1965	1966	1967	1968	1969	1970	1971	1972	1973	1974	1975	1976	1977	1978	1979	1980	1981	1982	1983	1984	1985	1986	1987	1988	1989	1990	1991	1992	1993	1994	1995	1996	1997	1998	1999	2000	2001	2002	2003	2004	2005	2006	2007	2008	2009	2010	2011	
037	0.981									1.451									1.695																					1.594							1.230							0.863								1.388
038	0.973								1.357								1.532																				1.363								0.992								0.640									1.221
039	0.733									1.242								1.549																		1.293							0.859								0.443											1.080
040	1.459							1.895							1.985																				1.755									1.256								0.786										1.602
041	1.358									2.066									2.367																					2.309							1.748							1.192								1.955
042	1.099								1.552								1.747																					1.545								1.144							0.753									1.405
043	1.036							1.452						1.849							1.947																						1.665									1.109										1.612
044	1.209								1.721									1.931																					1.801							1.396							0.943									1.594
045	1.379								1.905								2.064																				1.902								1.393								0.904									1.686
046	1.408								1.943								2.113																					1.963								1.489							1.003									1.755
047	1.339							1.800								1.960																				1.779								1.302								0.813										1.582
048	1.559							2.054								2.161																					1.999								1.485								0.979									1.807
049	0.760									1.215								1.657								1.841																				1.678								1.248								1.480
050	0.558									0.936									1.336								1.519																				1.382							1.047								1.192
051	1.070								1.538									1.771																				1.611								1.212							0.813									1.428
052	1.162									1.817								2.463									2.625																					2.659							2.159							2.231
053	1.141								1.633									1.873																				1.728								1.307							0.881									1.522
054	1.181								1.716									1.975																				1.806								1.345							0.889									1.589
055	4.852																4.938										4.395								3.486								2.520								1.511											3.756
056	1.136										1.775									2.036																					1.897							1.471						1.025								1.656
057	1.098							1.587								1.834																				1.558								1.063								0.612										1.392
058	0.903									1.399								1.653																				1.422							1.033							0.621										1.253
059	0.811									1.356								1.737																			1.435							0.912								0.478										1.213
060	1.153								1.744								2.330								2.452																					2.380							1.822									2.085
061	0.983								1.388																		1.108						0.732						0.390									0.138														0.820
062	1.029						1.323							1.450																			1.226									0.828									0.470											1.103
063	1.357									1.650																				1.493							1.117							0.738								0.415										1.219
064	0.876						1.268					1.488															1.054								0.508										0.141																	0.828
065	1.929						2.376						2.344																					2.117									1.471									0.887										1.922
066	1.295							1.837							2.396								2.472																						2.346								1.724									2.134
067	2.659							3.306								3.168																					3.300								2.601								1.820									2.889
068	1.073									1.690								2.269								2.396																					2.350							1.824								2.014
069	1.258									2.232							3.145							3.275																				2.988								1.854										2.584
070	0.999							1.404						1.789							1.909																						1.645									1.097										1.577
071	1.558								2.181									2.391																					2.304								1.767							1.247								2.031
072	1.633							2.150								2.253																				2.093								1.560								0.991										1.864
073	2.497								2.568																					2.563							2.047							1.452								0.887										2.096

(continued)

Table 1.4 (continued)

District Code	Year (1951–2011)						AAGR
074	1.647	2.221	2.346	2.256	1.698	1.136	1.977
075	1.263	1.900	2.529	2.597	2.511	1.922	2.227
076	1.219	1.823	2.066	1.946	1.494	0.995	1.679
077	1.753	2.188	2.251	2.058	1.449	0.888	1.840
078	2.571	2.779	2.830	2.218	1.600	0.980	2.281
079	2.561	2.698	2.671	2.096	1.530	0.993	2.240
080	1.630	2.142	2.265	2.148	1.607	1.068	1.912
081	1.553	2.058	2.206	2.077	1.542	1.016	1.846
082	1.170	1.685	1.920	1.746	1.306	0.868	1.550
083	1.420	1.888	2.042	1.890	1.401	0.890	1.670
084	1.297	1.851	2.062	1.938	1.502	1.013	1.709
085	1.041	1.544	1.997	2.087	1.839	1.284	1.729
086	0.957	1.589	2.289	2.975	3.585	3.241	2.576
087	1.105	1.693	2.288	2.432	2.338	1.771	2.047
088	3.310	4.562	4.328	4.602	3.405	2.079	3.869
089	0.081	[illegible].245	0.469	0.715	0.965	1.156	0.592
090	5.282	7.250	6.466	4.997	3.452	1.872	4.997
091	4.340	5.503	4.551	3.542	2.458	1.436	3.814
092	1.761	4.401	7.572	3.930	0.589	0.080	1.479
093	84.656	34.080	31.026	19.196	5.990	0.028	8.025
094	2.827	3.861	3.833	3.871	2.772	1.618	3.251
095	1.662	2.619	3.535	3.511	3.443	2.415	3.008
096	2.363	3.502	3.696	3.719	2.757	1.740	3.122
097	2.356	2.575	2.659	2.169	1.620	1.060	2.198
098	3.116	2.971	3.016	2.429	1.752	1.085	2.485
099	1.718	2.315	2.810	2.709	2.591	1.901	2.440
100	1.536	2.147	2.344	2.307	1.809	1.284	2.017
101	1.535	2.066	2.246	2.103	1.537	0.993	1.858
102	1.152	1.686	2.188	2.317	2.225	1.647	1.968
103	1.252	1.758	2.246	2.281	2.078	1.441	1.939
104	1.048	1.641	2.195	2.316	2.183	1.654	1.925
105	1.124	1.644	2.13	2.266	2.163	1.599	1.921
106	1.207	1.704	2.183	2.245	2.047	1.416	1.900
107	1.052	1.649	2.208	2.336	2.281	1.770	1.962
108	1.477	2.070	2.655	2.696	2.570	1.878	2.347
109	1.436	2.075	2.303	2.296	1.803	1.288	1.961
110	1.343	1.858	2.319	2.360	2.139	1.462	2.018

(continued)

Table 1.4 (continued)

District Code	Year						AAGR
111	1.377	1.956	2.538	2.628	2.491	1.799	2.261
112	1.333	2.052	2.726	2.845	2.980	2.458	2.471
113	1.217	1.887	2.497	2.578	2.594	2.098	2.225
114	1.147	1.678	2.174	2.305	2.206	1.632	1.957
115	0.943	1.477	1.983	2.118	2.016	1.559	1.761
116	1.180	1.718	1.957	1.802	1.377	0.909	1.592
117	0.846	1.305	1.784	1.958	1.790	1.332	1.604
118	1.113	1.568	2.012	2.072	1.843	1.267	1.739
119	1.449	2.034	2.238	2.117	1.621	1.141	1.886
120	0.879	1.358	1.859	2.012	1.821	1.351	1.651
121	0.905	1.338	1.735	1.833	1.573	1.091	1.502
122	1.195	1.707	2.240	2.357	2.155	1.530	1.989
123	1.434	2.016	2.574	2.555	2.354	1.627	2.194
124	1.241	1.760	2.006	1.937	1.514	1.037	1.672
125	1.984	2.641	2.695	2.703	2.124	1.532	2.392
126	1.078	1.607	2.082	2.186	1.961	1.371	1.816
127	1.063	1.520	1.971	2.068	1.839	1.246	1.718
128	1.218	1.721	2.204	2.268	2.075	1.479	1.940
129	1.485	2.141	2.365	2.379	1.932	1.420	2.047
130	1.125	1.596	2.054	2.141	1.935	1.374	1.816
131	1.057	1.521	1.978	2.097	1.884	1.279	1.740
132	1.185	1.674	2.146	2.212	2.016	1.396	1.870
133	1.021	1.610	2.164	2.303	2.243	1.738	1.927
134	1.232	1.733	2.212	2.261	2.071	1.486	1.942
135	1.113	1.662	2.210	2.294	2.160	1.652	1.948
136	1.105	1.579	2.053	2.102	1.820	1.223	1.745
137	0.977	1.421	1.665	1.527	1.173	0.786	1.346
138	1.148	1.746	2.359	2.966	2.881	2.855	2.464
139	0.929	1.614	2.368	3.023	3.044	3.325	2.412
140	0.877	1.278	1.684	1.826	1.586	1.058	1.484
141	0.883	1.394	1.882	2.050	1.963	1.516	1.693
142	0.796	1.212	1.635	1.833	1.602	1.099	1.463
143	0.803	1.286	1.799	2.030	1.938	1.538	1.652
144	0.917	1.444	1.941	2.116	2.061	1.599	1.756
145	0.972	1.485	2.008	2.166	2.039	1.540	1.805
146	0.916	1.357	1.790	1.930	1.694	1.187	1.575
147	0.694	1.166	1.697	1.949	1.877	1.489	1.551

(continued)

Table 1.4 (continued)

District Code	Year						AAGR
148	0.898	1.432	1.941	2.136	2.086	1.612	1.764
149	1.177	1.666	2.139	2.201	1.986	1.409	1.874
150	0.737	1.264	1.818	2.035	1.985	1.608	1.623
151	0.860	1.418	2.031	2.251	2.261	1.840	1.844
152	0.758	1.258	1.811	2.063	2.016	1.606	1.657
153	0.788	1.256	1.707	1.904	1.821	1.403	1.556
154	0.880	1.329	1.743	1.883	1.626	1.112	1.527
155	0.969	1.542	2.130	2.362	2.354	1.887	1.960
156	0.770	1.231	1.725	1.943	1.827	1.400	1.562
157	0.844	1.313	1.805	1.975	1.791	1.332	1.614
158	1.044	1.491	1.930	2.036	1.822	1.244	1.693
159	0.896	1.301	1.710	1.841	1.597	1.066	1.500
160	1.056	1.613	1.911	1.790	1.394	0.983	1.564
161	0.937	1.326	1.704	1.827	1.537	1.009	1.492
162	1.308	1.877	2.111	2.003	1.547	1.038	1.747
163	1.033	1.422	1.782	1.868	1.625	1.110	1.566
164	1.340	1.903	2.130	1.992	1.509	1.053	1.774
165	0.979	1.542	2.114	2.315	2.283	1.788	1.916
166	1.134	1.633	1.859	1.682	1.261	0.838	1.499
167	1.032	1.446	1.840	1.938	1.664	1.117	1.608
168	1.143	1.576	1.977	2.029	1.760	1.196	1.709
169	0.916	1.467	2.004	2.171	2.065	1.578	1.785
170	0.913	1.326	1.747	1.866	1.604	1.066	1.519
171	0.813	1.292	1.753	1.922	1.760	1.315	1.557
172	0.773	1.302	1.897	2.133	2.035	1.602	1.704
173	1.093	1.613	2.114	2.234	2.073	1.514	1.876
174	0.673	1.173	1.709	1.978	1.961	1.581	1.558
175	0.711	1.218	1.698	1.936	1.818	1.356	1.523
176	0.849	1.320	1.814	1.996	1.830	1.359	1.632
177	0.779	1.250	1.756	1.974	1.861	1.429	1.589
178	0.771	1.235	1.703	2.173	2.217	1.991	1.786
179	0.546	0.992	1.521	2.030	2.160	2.121	1.609
180	0.598	1.010	1.462	1.911	2.039	1.855	1.577
181	0.597	1.014	1.477	1.912	2.008	1.796	1.562
182	0.537	0.941	1.399	1.885	2.031	1.917	1.520
183	0.720	1.163	1.642	1.876	1.767	1.353	1.499
184	0.608	1.140	1.709	1.986	1.901	1.500	1.508

(continued)

Table 1.4 (continued)

District Code	Year (1951–2011)						AAGR
185	0.839	1.373	1.953	2.172	2.129	1.711	1.768
186	0.843	1.301	1.780	1.941	1.755	1.304	1.589
187	0.816	1.353	1.943	2.180	2.131	1.699	1.762
188	0.883	1.322	1.760	1.927	1.734	1.243	1.578
189	0.882	1.349	1.832	1.995	1.838	1.377	1.646
190	0.915	1.441	1.938	2.106	1.984	1.500	1.729
191	0.793	1.225	1.680	1.849	1.662	1.231	1.505
192	0.939	1.398	1.813	1.943	1.717	1.196	1.597
193	0.815	1.332	1.847	2.053	1.875	1.373	1.641
194	0.998	1.513	2.040	2.150	1.972	1.484	1.795
195	1.111	1.574	2.023	2.111	1.903	1.352	1.789
196	1.153	1.757	2.401	2.510	2.170	1.342	2.005
197	1.022	1.557	2.100	2.239	2.110	1.551	1.858
198	1.532	2.121	2.667	2.617	2.484	1.822	2.312
199	0.821	1.261	1.717	1.881	1.729	1.260	1.532
200	0.768	1.190	1.631	1.819	1.662	1.242	1.482
201	0.824	1.397	1.986	2.205	2.212	1.806	1.782
202	0.698	1.170	1.686	2.164	2.205	1.989	1.747
203	0.587	1.030	1.537	2.074	2.227	2.205	1.679
204	0.635	1.086	1.588	2.092	2.234	2.057	1.723
205	0.755	1.287	1.842	2.070	2.051	1.668	1.657
206	1.413	1.952	2.461	2.426	2.226	1.562	2.097
207	1.295	1.895	2.452	2.531	2.422	1.786	2.168
208	1.270	1.846	2.376	2.454	2.346	1.740	2.105
209	1.173	1.778	2.391	2.464	2.315	1.748	2.087
210	1.077	1.690	2.257	2.399	2.311	1.763	2.001
211	0.814	1.407	2.031	2.281	2.306	1.870	1.829
212	0.872	1.440	2.057	2.297	2.285	1.832	1.871
213	0.912	1.435	1.931	2.109	2.055	1.593	1.749
214	0.788	1.348	1.931	2.173	2.189	1.782	1.745
215	0.833	1.330	1.810	1.994	1.898	1.459	1.633
216	0.737	1.230	1.783	2.011	1.909	1.509	1.606
217	0.759	1.230	1.741	2.002	1.910	1.501	1.613
218	0.837	1.418	2.013	2.227	2.232	1.822	1.802
219	0.754	1.278	1.870	2.139	2.075	1.634	1.704
220	0.873	1.431	2.037	2.238	2.174	1.738	1.825
221	0.794	1.230	1.709	2.189	2.214	1.962	1.810

(continued)

Table 1.4 (continued)

District Code	Year						AAGR
222	0.884	1.418	1.935	2.121	2.029	1.550	1.739
223	0.845	1.342	1.863	2.095	2.031	1.618	1.717
224	0.803	1.239	1.695	1.855	1.678	1.210	1.502
225	0.755	1.225	1.737	1.990	1.881	1.475	1.600
226	0.749	1.200	1.683	1.904	1.780	1.405	1.539
227	0.912	1.355	1.738	1.951	1.735	1.220	1.591
228	0.945	1.490	2.004	2.174	2.124	1.646	1.808
229	0.772	1.241	1.745	1.981	1.881	1.441	1.591
230	0.770	1.239	1.745	1.986	1.888	1.492	1.607
231	0.800	1.233	1.700	2.170	2.170	1.894	1.782
232	0.990	1.468	1.942	2.090	1.939	1.410	1.739
233	0.839	1.412	1.996	2.208	2.225	1.824	1.792
234	0.832	1.388	2.003	2.245	2.194	1.745	1.812
235	0.852	1.393	1.983	2.189	2.126	1.701	1.782
236	0.859	1.441	2.032	2.222	2.215	1.815	1.808
237	0.760	1.285	1.823	2.046	1.984	1.592	1.631
238	0.801	1.271	1.723	1.914	1.824	1.402	1.566
239	2.334	3.576	3.766	3.255	1.816	0.665	2.599
240	0.390	1.093	1.948	2.423	1.804	0.873	1.469
241	0.477	1.244	2.183	2.704	2.088	1.029	1.707
242	0.611	1.445	2.461	2.913	2.305	1.192	1.942
243	0.518	1.071	1.688	2.323	2.468	2.140	1.772
244	0.629	1.369	2.171	2.617	2.104	1.108	1.748
245	0.392	0.805	1.310	1.855	2.026	1.656	1.438
246	2.841	3.804	4.624	4.035	4.276	3.034	3.847
247	0.483	1.008	1.601	2.212	2.392	2.103	1.706
248	0.973	1.764	2.265	1.992	1.286	0.661	1.593
249	0.953	2.039	2.676	2.390	1.571	0.854	1.859
250	2.151	2.653	2.474	1.738	1.051	0.499	1.877
251	3.234	4.047	3.706	3.576	2.426	1.310	3.082
252	0.954	1.676	2.077	1.887	1.338	0.803	1.551
253	0.390	0.951	1.669	2.425	2.684	2.701	1.763
254	0.233	0.887	1.799	2.892	4.046	5.107	1.885
255	2.363	3.470	3.932	2.769	1.094	0.184	1.744
256	2.610	4.309	4.455	4.095	2.481	1.001	3.247
257	4.172	3.998	4.424	3.450	2.420	1.389	3.432
258	0.628	1.048	1.310	1.190	0.924	0.654	1.031

(continued)

Table 1.4 (continued)

District Code	Year (1951–2011)						AAGR
259	0.213	1.713	3.818	5.069	3.335	0.803	2.132
260	0.474	1.597	2.989	3.640	2.191	0.563	1.907
261	0.359	1.328	2.559	3.289	2.089	0.718	1.764
262	0.331	1.748	3.810	4.827	3.635	1.281	2.461
263	4.889	6.037	5.491	4.400	3.157	2.005	4.536
264	0.957	1.972	3.019	3.317	2.618	1.244	2.326
265	0.536	1.522	2.811	3.478	2.721	1.267	2.167
266	0.147	4.003	8.745	6.852	2.660	0.178	2.375
267	0.041	2.377	7.451	10.877	5.753	0.327	2.413
268	1.649	2.478	3.281	3.246	2.874	1.760	2.708
269	0.771	2.304	4.275	4.869	3.661	1.208	2.934
270	1.493	2.136	2.819	3.435	3.188	3.359	2.880
271	1.135	1.771	2.355	2.472	2.387	1.829	2.076
272	2.138	2.713	2.667	2.622	1.986	1.333	2.336
273	2.203	2.403	2.362	1.863	1.298	0.806	1.945
274	1.614	2.153	2.319	2.207	1.630	1.067	1.941
275	1.602	2.091	2.182	2.017	1.483	0.930	1.801
276	2.425	2.609	2.651	2.114	1.495	0.941	2.158
277	1.108	1.710	2.294	2.451	2.457	1.953	2.070
278	1.208	2.008	2.800	2.959	2.925	2.173	2.453
279	1.428	2.562	3.108	2.666	1.493	0.522	2.009
280	0.371	1.186	2.168	2.752	2.487	1.683	1.711
281	0.781	1.879	3.219	3.568	2.840	1.474	2.426
282	0.391	1.102	1.895	2.432	1.958	1.152	1.545
283	0.136	0.482	0.921	1.459	1.824	1.226	1.017
284	0.432	1.127	1.981	2.525	1.971	0.979	1.585
285	0.467	1.355	2.491	3.102	2.814	1.900	2.038
286	1.011	2.044	3.171	3.483	2.820	1.399	2.461
287	2.536	2.627	2.624	2.078	1.457	0.877	2.132
288	3.233	3.464	2.772	2.033	1.342	0.730	2.397
289	4.688	4.641	3.858	2.860	1.848	0.954	3.216
290	1.595	2.104	2.262	2.169	1.620	1.077	1.907
291	1.180	1.753	2.321	2.386	2.264	1.736	2.040
292	1.374	1.985	2.540	2.571	2.462	1.832	2.226
293	0.806	1.326	1.897	2.115	2.044	1.632	1.711
294	1.394	2.012	2.570	2.604	2.533	1.955	2.284
295	2.090	2.722	3.348	3.083	3.112	2.359	2.881

(continued)

Table 1.4 (continued)

District Code	Year (1951–2011)						AAGR
296	1.183	1.671	2.177	2.270	2.060	1.473	1.923
297	1.450	2.131	2.781	2.767	2.728	2.126	2.429
298	3.136	3.105	2.894	2.112	1.308	0.587	2.191
299	1.374	1.989	2.220	2.197	1.775	1.301	1.904
300	1.683	2.281	2.390	2.275	1.699	1.124	2.005
301	1.285	1.826	2.035	1.913	1.485	1.004	1.688
302	1.443	2.019	2.212	2.084	1.596	1.125	1.865
303	1.488	2.072	2.263	2.131	1.612	1.081	1.883
304	1.594	2.077	2.155	1.978	1.457	0.915	1.778
305	2.315	2.467	2.487	1.979	1.405	0.858	2.009
306	4.204	4.301	3.584	2.671	1.818	1.038	3.086
307	1.508	1.990	2.095	1.901	1.386	0.862	1.709
308	1.382	1.738	1.822	1.592	1.110	0.650	1.432
309	1.132	1.530	1.702	1.519	1.107	0.691	1.357
310	0.988	1.375	1.589	1.396	0.993	0.600	1.237
311	1.780	2.171	2.149	1.923	1.358	0.807	1.738
312	3.786	3.263	3.201	2.509	1.722	0.997	2.588
313	2.560	2.744	2.881	2.309	1.712	1.129	2.346
314	1.033	1.428	1.794	1.899	1.663	1.130	1.588
315	0.978	1.435	1.845	1.940	1.703	1.197	1.607
316	0.981	1.460	1.893	1.996	1.752	1.178	1.627
317	1.772	2.304	2.398	2.298	1.742	1.131	2.022
318	3.967	3.846	3.010	2.033	1.125	0.434	2.285
319	1.201	1.662	1.859	1.715	1.298	0.878	1.530
320	3.921	3.502	3.769	3.124	2.296	1.506	3.115
321	1.942	2.086	1.920	1.447	0.986	0.575	1.599
322	3.058	3.132	2.508	1.849	1.185	0.604	2.130
323	1.235	1.849	2.144	2.084	1.646	1.177	1.794
324	3.719	3.496	2.768	1.977	1.212	0.561	2.228
325	1.567	2.124	2.241	2.113	1.633	1.118	1.898
326	1.790	2.291	2.314	2.104	1.534	0.974	1.914
327	2.483	2.467	2.374	1.788	1.221	0.707	1.906
328	1.764	2.419	2.557	2.569	2.041	1.467	2.245
329	1.725	2.223	2.280	2.075	1.505	0.950	1.875
330	1.244	1.750	2.223	2.290	2.007	1.333	1.920
331	1.269	1.741	2.166	2.203	1.966	1.350	1.880
332	1.181	1.697	1.945	1.801	1.359	0.913	1.581

(continued)

Table 1.4 (continued)

District Code	Year (1951–2011)						AAGR
333	1.844	2.114	2.034	1.572	1.117	0.710	1.698
334	2.155	2.427	2.476	1.941	1.405	0.907	2.013
335	2.117	2.582	2.507	2.324	1.642	1.004	2.093
336	1.932	2.174	2.107	1.648	1.145	0.680	1.707
337	1.172	1.574	1.728	1.540	1.128	0.707	1.384
338	0.779	1.122	1.469	1.597	1.365	0.914	1.294
339	1.257	1.729	1.884	1.710	1.299	0.877	1.554
340	0.997	1.125	0.905	0.624	0.361	0.151	0.682
341	1.399	1.960	2.183	2.082	1.596	1.129	1.840
342	1.419	1.875	1.986	1.772	1.281	0.787	1.605
343	1.331	1.872	2.059	1.876	1.349	0.857	1.658
344	1.109	1.689	2.271	2.414	2.332	1.773	2.038
345	1.042	1.634	2.230	2.415	2.387	1.867	2.010
346	0.799	1.248	1.744	2.243	2.287	2.056	1.864
347	0.936	1.493	2.067	2.297	2.268	1.768	1.886
348	0.811	1.381	1.970	2.204	2.224	1.814	1.776
349	0.691	1.177	1.686	1.916	1.880	1.528	1.524
350	0.831	1.316	1.824	2.037	1.948	1.507	1.656
351	0.679	1.061	1.489	1.928	1.993	1.716	1.601
352	1.448	2.012	2.165	1.991	1.503	1.003	1.794
353	2.041	2.601	2.554	2.430	1.834	1.180	2.181
354	0.609	1.021	1.471	1.918	2.047	1.865	1.586
355	1.217	1.806	2.041	1.935	1.496	1.006	1.668
356	1.012	1.592	2.139	2.271	2.194	1.696	1.898
357	0.811	1.350	1.950	2.189	2.127	1.689	1.763
358	1.188	1.699	2.231	2.357	2.160	1.536	1.988
359	2.354	2.626	2.664	2.072	1.486	0.949	2.169
360	0.581	1.011	1.438	1.640	1.517	1.186	1.282
361	0.630	1.092	1.591	1.835	1.776	1.429	1.440
362	1.064	1.592	2.107	2.256	2.088	1.504	1.877
363	0.443	0.721	1.017	1.328	1.439	1.185	1.106
364	0.519	0.919	1.362	1.633	1.573	1.259	1.256
365	0.558	0.863	1.189	1.340	1.161	0.831	1.060
366	0.478	0.844	1.251	1.492	1.400	1.116	1.143
367	0.690	1.160	1.692	1.943	1.858	1.467	1.543
368	1.123	1.502	1.661	1.453	1.039	0.634	1.315
369	0.843	1.203	1.559	1.704	1.412	0.919	1.375

(continued)

Table 1.4 (continued)

District Code	Year (1951–2011): period 1	period 2	period 3	period 4	period 5	period 6	AAGR
370	1.370	1.787	1.894	1.664	1.187	0.741	1.519
371	1.434	1.880	1.968	1.755	1.280	0.796	1.599
372	2.061	2.345	2.240	1.751	1.215	0.723	1.813
373	1.224	1.693	1.878	1.733	1.271	0.829	1.526
374	0.726	1.080	1.405	1.554	1.337	0.926	1.254
375	1.409	1.962	2.140	1.985	1.446	0.931	1.744
376	1.346	1.865	2.026	1.846	1.392	0.930	1.670
377	1.337	1.716	1.840	1.591	1.074	0.605	1.417
378	1.243	1.633	1.804	1.585	1.082	0.598	1.390
379	1.204	1.663	1.854	1.636	1.167	0.707	1.460
380	1.150	1.658	1.881	1.696	1.270	0.844	1.516
381	1.368	1.728	1.824	1.592	1.101	0.663	1.450
382	1.348	1.801	1.971	1.817	1.339	0.876	1.625
383	1.249	1.553	1.644	1.427	0.982	0.569	1.285
384	1.235	1.776	2.313	2.350	2.075	1.385	1.963
385	1.129	1.576	1.757	1.568	1.174	0.783	1.426
386	0.978	1.460	1.682	1.529	1.167	0.805	1.363
387	0.815	1.232	1.466	1.333	1.017	0.677	1.160
388	0.773	1.159	1.512	1.666	1.444	1.000	1.348
389	0.724	1.118	1.536	1.707	1.533	1.103	1.371
390	0.997	1.442	1.692	1.572	1.211	0.816	1.375
391	0.699	1.045	1.369	1.496	1.246	0.850	1.199
392	1.147	1.609	1.781	1.581	1.138	0.724	1.420
393	0.751	1.126	1.506	1.699	1.524	1.093	1.376
394	1.049	1.557	1.795	1.686	1.305	0.883	1.456
395	1.276	1.862	2.100	2.047	1.596	1.132	1.763
396	0.801	1.185	1.532	1.666	1.448	1.011	1.359
397	2.971	2.856	2.853	2.225	1.570	0.973	2.331
398	1.804	2.265	2.332	2.130	1.489	0.901	1.900
399	1.045	1.492	1.963	2.104	1.889	1.338	1.756
400	1.087	1.488	1.694	1.518	1.096	0.677	1.340
401	0.781	1.171	1.568	1.736	1.528	1.089	1.406
402	1.690	2.326	2.464	2.360	1.829	1.299	2.116
403	0.476	0.853	1.254	1.675	1.852	1.655	1.370
404	0.839	1.343	1.875	2.111	2.038	1.570	1.711
405	0.952	1.498	2.056	2.250	2.188	1.703	1.855
406	0.877	1.320	1.731	1.864	1.600	1.095	1.511

(continued)

Table 1.4 (continued)

District Code	Year (1951–2011)						AAGR
407	1.595	2.128	2.290	2.161	1.603	1.015	1.887
408	1.071	1.630	2.197	2.325	2.199	1.612	1.936
409	0.986	1.434	1.673	1.527	1.170	0.780	1.351
410	0.977	1.354	1.710	1.806	1.542	1.039	1.498
411	1.732	2.170	2.236	2.026	1.416	0.860	1.817
412	1.206	1.764	1.979	1.871	1.451	1.019	1.644
413	1.233	1.698	2.118	2.172	1.936	1.325	1.846
414	1.486	1.840	1.872	1.616	1.112	0.641	1.473
415	1.522	2.100	2.285	2.065	1.462	0.872	1.820
416	1.263	1.785	2.316	2.421	2.257	1.626	2.067
417	1.170	1.729	2.223	2.304	2.097	1.479	1.935
418	1.028	1.444	1.838	1.948	1.674	1.114	1.611
419	1.198	1.693	2.198	2.310	2.135	1.540	1.964
420	1.043	1.492	1.935	2.040	1.822	1.244	1.695
421	1.018	1.593	2.130	2.257	2.123	1.612	1.873
422	1.143	1.625	2.126	2.245	2.040	1.453	1.892
423	0.982	1.512	2.060	2.206	2.052	1.490	1.815
424	1.288	1.903	2.182	2.151	1.716	1.242	1.846
425	1.333	1.949	2.152	2.051	1.590	1.114	1.800
426	1.345	1.891	2.101	1.969	1.504	1.059	1.760
427	1.218	1.717	2.201	2.245	2.026	1.395	1.898
428	1.082	1.581	2.058	2.166	1.999	1.466	1.823
429	1.184	1.674	2.179	2.288	2.093	1.500	1.939
430	1.507	2.004	2.169	2.039	1.507	0.989	1.807
431	1.468	2.006	2.156	2.021	1.547	1.052	1.808
432	1.156	1.638	2.113	2.155	1.901	1.292	1.806
433	1.384	1.939	2.164	2.063	1.582	1.120	1.823
434	1.115	1.573	2.020	2.073	1.833	1.255	1.738
435	1.530	2.159	2.387	2.372	1.851	1.310	2.049
436	1.106	1.671	2.241	2.348	2.218	1.632	1.965
437	1.264	1.945	2.559	2.610	2.614	2.056	2.248
438	1.278	1.786	2.275	2.279	2.050	1.420	1.941
439	1.033	1.619	2.167	2.313	2.277	1.771	1.941
440	0.939	1.477	1.985	2.134	1.996	1.508	1.756
441	1.094	1.567	2.032	2.126	1.901	1.293	1.771
442	2.748	3.410	3.262	3.449	2.730	1.923	2.997
443	1.387	1.893	2.341	2.338	2.109	1.455	2.016

(continued)

Table 1.4 (continued)

District Code	Year (1951–2011)						AAGR
444	1.448	2.077	2.276	2.247	1.770	1.267	1.942
445	1.582	2.056	2.152	2.001	1.480	0.936	1.782
446	1.144	1.619	2.079	2.150	1.933	1.370	1.826
447	1.275	1.882	2.119	2.034	1.580	1.070	1.744
448	1.008	1.500	1.946	2.047	1.797	1.248	1.689
449	1.727	2.210	2.251	2.051	1.499	0.956	1.860
450	1.231	1.753	1.990	1.899	1.480	1.007	1.652
451	1.408	1.926	2.078	1.938	1.435	0.944	1.713
452	1.039	1.549	1.810	1.698	1.341	0.962	1.496
453	1.365	1.869	2.021	1.866	1.425	0.968	1.682
454	1.116	1.685	1.953	1.836	1.394	0.956	1.592
455	1.026	1.453	1.641	1.442	1.028	0.647	1.293
456	1.254	1.827	2.357	2.441	2.338	1.788	2.110
457	1.130	1.654	2.121	2.171	1.923	1.351	1.820
458	1.384	1.944	2.146	2.007	1.537	1.044	1.777
459	1.442	2.076	2.301	2.102	1.536	0.990	1.866
460	1.134	1.690	2.242	2.329	2.208	1.640	1.965
461	1.523	2.184	2.763	2.776	2.769	2.096	2.446
462	1.248	1.811	2.327	2.424	2.338	1.742	2.080
463	1.201	1.692	2.198	2.278	2.061	1.475	1.934
464	1.270	1.876	2.136	2.081	1.608	1.128	1.782
465	1.742	2.283	2.380	2.251	1.688	1.081	1.988
466	0.818	1.352	1.937	2.183	2.156	1.726	1.768
467	1.272	1.784	2.306	2.366	2.157	1.552	2.023
468	1.403	1.778	1.869	1.620	1.112	0.638	1.456
469	1.463	1.807	1.836	1.581	1.090	0.632	1.445
470	1.536	1.986	2.049	1.865	1.383	0.877	1.692
471	1.558	2.066	2.187	2.018	1.494	0.943	1.796
472	1.566	2.242	2.419	2.400	1.890	1.351	2.076
473	1.203	1.763	1.996	1.913	1.483	1.045	1.662
474	1.406	2.022	2.218	2.164	1.700	1.212	1.883
475	1.646	2.050	2.086	1.889	1.334	0.772	1.680
476	1.734	1.884	1.715	1.294	0.889	0.503	1.412
477	1.844	2.166	2.044	1.587	1.142	0.706	1.676
478	1.813	1.984	1.801	1.354	0.862	0.428	1.416
479	1.444	2.004	2.174	2.015	1.521	1.017	1.802
480	1.243	1.593	1.725	1.520	1.106	0.662	1.374

(continued)

Table 1.4 (continued)

District Code	Year						AAGR
481	1.412	1.784	1.877	1.636	1.127	0.650	1.466
482	1.163	1.741	2.030	1.964	1.504	1.047	1.673
483	1.173	1.813	2.394	2.491	2.407	1.850	2.104
484	1.514	2.006	2.149	2.004	1.490	0.948	1.768
485	1.401	1.832	1.946	1.719	1.221	0.759	1.560
486	0.976	1.456	1.725	1.634	1.253	0.876	1.408
487	1.453	2.094	2.320	2.325	1.885	1.386	2.004
488	1.434	1.773	1.852	1.642	1.140	0.667	1.466
489	1.076	1.575	2.045	2.192	2.086	1.542	1.851
490	1.385	1.996	2.660	3.269	3.001	2.958	2.680
491	1.526	1.929	1.997	1.703	1.125	0.618	1.538
492	1.238	1.807	2.060	1.923	1.469	0.972	1.682
493	0.884	1.559	2.368	3.178	3.842	3.461	2.709
494	1.260	1.982	2.722	3.383	3.242	3.578	2.765
495	0.935	1.442	1.970	2.123	1.948	1.451	1.752
496	1.174	1.689	1.948	1.846	1.419	0.989	1.620
497	1.086	1.624	1.862	1.733	1.331	0.890	1.503
498	0.933	1.341	1.751	1.856	1.603	1.074	1.521
499	1.357	1.829	1.996	1.810	1.318	0.849	1.631
500	1.053	1.571	1.853	1.766	1.396	1.004	1.535
501	1.142	1.614	1.813	1.614	1.196	0.788	1.462
502	1.167	1.523	1.688	1.482	1.021	0.571	1.303
503	1.545	2.097	2.246	2.149	1.662	1.142	1.904
504	1.209	1.557	1.686	1.440	0.971	0.548	1.290
505	1.318	1.589	1.450	1.087	0.722	0.409	1.185
506	1.430	1.955	2.092	1.935	1.431	0.938	1.722
507	1.503	2.027	2.154	1.916	1.307	0.719	1.665
508	1.254	1.733	1.920	1.779	1.308	0.855	1.564
509	1.350	1.970	2.203	2.151	1.726	1.254	1.876
510	1.027	1.466	1.899	2.004	1.783	1.213	1.663
511	1.043	1.492	1.940	2.031	1.779	1.239	1.699
512	0.984	1.410	1.861	2.027	1.828	1.295	1.684
513	1.291	1.883	2.430	2.505	2.408	1.840	2.170
514	1.139	1.678	2.194	2.302	2.141	1.560	1.939
515	1.649	2.471	3.248	3.192	3.413	2.707	2.867
516	4.868	4.941	4.206	3.081	1.921	0.941	3.401
517	1.349	1.716	1.156	0.581	0.240	0.044	0.536

(continued)

Table 1.4 (continued)

District Code	Year (1951–2011)						AAGR
518	0.772	1.220	1.652	1.822	1.702	1.309	1.487
519	1.135	1.767	2.345	2.474	2.494	1.955	2.103
520	1.281	1.780	1.975	1.820	1.371	0.919	1.624
521	0.985	1.418	1.852	1.958	1.698	1.176	1.626
522	1.189	1.665	2.126	2.154	1.918	1.323	1.821
523	1.092	1.561	1.774	1.599	1.206	0.809	1.433
524	1.030	1.537	1.812	1.731	1.329	0.929	1.485
525	1.183	1.578	1.712	1.480	1.053	0.635	1.354
526	0.446	0.777	1.047	0.792	0.485	0.209	0.677
527	0.478	0.536	0.458	0.352	0.244	0.147	0.393
528	1.440	1.875	1.987	1.773	1.268	0.795	1.603
529	1.293	1.694	1.822	1.644	1.204	0.756	1.479
530	1.500	1.987	2.128	1.981	1.474	0.938	1.751
531	1.658	2.055	2.071	1.814	1.248	0.719	1.641
532	1.077	1.595	1.858	1.610	1.142	0.717	1.450
533	0.880	1.287	1.703	1.837	1.573	1.036	1.486
534	1.102	2.144	2.786	2.304	1.413	0.633	1.873
535	1.295	1.955	2.622	3.242	3.011	3.043	2.645
536	1.090	1.622	1.878	1.779	1.366	0.952	1.542
537	1.120	1.653	1.920	1.702	1.173	0.684	1.464
538	1.152	1.648	1.835	1.603	1.129	0.698	1.442
539	1.776	2.220	2.239	1.976	1.344	0.761	1.771
540	0.690	1.043	1.277	1.102	0.822	0.532	0.991
541	0.608	1.029	1.292	1.042	0.682	0.354	0.903
542	1.139	1.735	1.988	1.833	1.390	0.912	1.591
543	0.915	1.428	1.731	1.409	0.883	0.438	1.215
544	1.008	1.531	1.849	1.519	0.905	0.399	1.262
545	1.245	1.654	1.790	1.583	1.146	0.679	1.412
546	1.295	1.625	1.727	1.496	1.016	0.579	1.341
547	0.923	1.345	1.575	1.397	1.038	0.689	1.249
548	1.004	1.420	1.626	1.435	1.061	0.698	1.300
549	0.990	1.431	1.663	1.497	1.121	0.749	1.331
550	1.167	1.715	1.925	1.795	1.385	0.965	1.590
551	1.096	1.576	1.792	1.645	1.267	0.850	1.463
552	0.988	1.439	1.682	1.542	1.184	0.792	1.360
553	1.120	1.615	1.849	1.716	1.323	0.888	1.512
554	1.063	1.568	1.812	1.701	1.351	0.975	1.506

(continued)

Table 1.4 (continued)

District Code	Year						AAGR
555	0.783	1.217	1.677	1.850	1.668	1.195	1.488
556	1.151	1.693	1.937	1.852	1.434	1.009	1.605
557	1.318	1.941	2.203	2.168	1.681	1.185	1.847
558	1.105	1.559	1.999	2.078	1.870	1.289	1.745
559	1.091	1.453	1.606	1.404	1.008	0.619	1.274
560	1.513	1.974	2.066	1.885	1.388	0.873	1.697
561	1.963	2.027	1.828	1.331	0.839	0.446	1.460
562	1.143	1.614	1.789	1.576	1.128	0.712	1.419
563	1.179	1.670	2.149	2.199	1.966	1.344	1.849
564	1.289	1.711	1.836	1.614	1.162	0.714	1.470
565	1.853	2.178	2.078	1.618	1.157	0.689	1.719
566	2.780	2.568	2.329	1.671	1.038	0.475	1.748
567	1.207	1.554	1.385	0.993	0.628	0.322	1.088
568	2.276	2.151	1.661	1.177	0.734	0.351	1.445
569	1.289	1.626	1.703	1.434	0.960	0.518	1.285
570	1.475	2.191	2.907	2.974	3.157	2.608	2.645
571	1.920	2.086	1.864	1.301	0.760	0.340	1.391
572	1.784	1.932	1.737	1.252	0.780	0.407	1.371
573	1.426	1.776	1.862	1.636	1.122	0.646	1.462
574	2.128	1.953	1.529	1.062	0.622	0.290	1.277
575	1.342	1.853	2.011	1.843	1.349	0.874	1.638
576	1.166	1.520	1.646	1.371	0.912	0.472	1.223
577	0.939	1.406	1.833	1.957	1.708	1.179	1.602
578	0.695	1.223	1.753	1.959	1.828	1.417	1.541
579	0.932	1.348	1.600	1.490	1.149	0.776	1.298
580	1.273	1.634	1.769	1.563	1.137	0.679	1.410
581	0.893	1.281	1.670	1.794	1.562	1.053	1.468
582	1.055	1.389	1.527	1.246	0.817	0.414	1.118
583	1.250	1.678	1.806	1.566	1.114	0.645	1.408
584	1.834	2.103	2.039	1.593	1.107	0.660	1.647
585	1.305	1.922	2.200	1.885	1.256	0.695	1.665
586	2.042	2.188	2.039	1.565	1.084	0.642	1.697
587	2.267	2.263	1.961	1.359	0.822	0.390	1.512
588	3.193	3.140	2.484	1.813	1.144	0.571	2.154
589	1.940	2.011	1.817	1.324	0.874	0.492	1.482
590	1.656	2.151	2.216	2.048	1.513	0.952	1.837
591	1.229	1.551	1.662	1.418	0.951	0.534	1.276

(continued)

Table 1.4 (continued)

District Code	Year (1951 … 2011)						AAGR
592	1.607	1.710	1.527	1.108	0.702	0.362	1.197
593	1.840	1.835	1.636	1.206	0.781	0.398	1.279
594	4.327	3.572	2.693	1.709	0.880	0.286	1.811
595	1.306	1.326	1.112	0.787	0.487	0.227	0.860
596	1.564	1.401	1.044	0.712	0.430	0.189	0.917
597	1.857	1.525	1.085	0.677	0.341	0.113	0.828
598	2.111	1.962	1.505	1.017	0.602	0.283	1.285
599	1.997	1.896	1.463	1.037	0.676	0.354	1.334
600	1.089	1.661	2.230	2.354	2.387	1.997	2.010
601	2.615	2.438	2.261	1.704	1.133	0.602	1.768
602	0.891	1.325	1.776	2.217	2.204	2.031	1.852
603	0.979	1.358	1.562	1.440	1.112	0.801	1.296
604	0.759	1.092	1.293	1.163	0.907	0.623	1.042
605	0.710	0.983	1.252	1.353	1.170	0.823	1.116
606	0.966	1.376	1.567	1.467	1.154	0.830	1.303
607	0.672	0.942	1.233	1.387	1.200	0.851	1.132
608	1.015	1.340	1.492	1.353	1.021	0.672	1.212
609	2.177	2.210	1.856	1.257	0.685	0.261	1.346
610	0.818	1.109	1.264	1.128	0.869	0.607	1.032
611	0.474	0.708	0.955	1.093	0.936	0.694	0.873
612	1.100	1.429	1.556	1.395	1.003	0.625	1.244
613	0.810	1.150	1.352	1.266	1.028	0.769	1.130
614	0.882	1.113	1.225	1.035	0.746	0.479	0.973
615	1.033	1.342	1.478	1.317	0.977	0.630	1.195
616	0.873	1.127	1.228	1.035	0.743	0.476	0.968
617	1.002	1.224	1.102	0.852	0.612	0.399	0.952
618	1.046	1.283	1.345	1.138	0.800	0.485	1.053
619	1.166	1.476	1.597	1.379	0.986	0.623	1.280
620	0.832	1.105	1.234	1.072	0.793	0.507	0.981
621	1.222	1.560	1.654	1.458	1.064	0.690	1.340
622	1.267	1.488	1.355	1.036	0.706	0.412	1.125
623	1.029	1.340	1.482	1.344	1.017	0.692	1.225
624	1.346	1.588	1.481	1.154	0.808	0.493	1.220
625	0.848	1.045	1.144	0.973	0.676	0.401	0.885
626	0.761	1.073	1.242	1.144	0.903	0.653	1.025
627	1.656	1.662	1.493	1.123	0.747	0.412	1.201
628	1.375	1.760	1.865	1.737	1.330	0.883	1.558

(continued)

Table 1.4 (continued)

District Code	Year						AAGR
629	1.059	1.441	1.827	1.905	1.717	1.265	1.628
630	1.106	1.522	1.696	1.558	1.209	0.874	1.420
631	0.653	1.039	1.437	1.816	1.928	1.801	1.529
632	1.101	1.811	2.567	3.303	3.977	3.806	2.948
633	1.213	1.764	2.270	2.367	2.269	1.691	2.027
634	0.898	1.305	1.560	1.446	1.112	0.749	1.260
635	0.880	1.257	1.633	1.748	1.511	1.056	1.448
636	1.331	2.640	3.479	2.739	1.235	0.200	1.750
637	4.487	5.025	3.915	2.688	1.594	0.698	3.193
638	4.562	4.956	4.199	3.272	2.249	1.293	3.624
639	1.090	1.660	2.255	2.823	2.725	2.678	2.333
640	2.992	3.848	3.683	4.051	3.358	2.521	3.491

Source Author

the inflexion point which implies that population growth is now slowing down in all, but a few districts of the country or population growth trajectory is flattening. This suggests that projecting or forecasting district population beyond 2011, the year when the last population census was conducted in India, should not be based on the constant rate of growth estimated as the annual population growth rate between 2001 and 2011 population censuses but should be based on the simple logistic growth model derived from population growth during the period 1951–2011. The simple logistic growth model may also serve as the basis for forecasting population of districts beyond the year 2011 and the forecasted district population may be added to obtain population forecast for the country to constitute the bottom-up approach of population forecasting.

The balancing equation of population growth (Preston et al. 2001) informs us that the change in population in a time interval is the result of the four and only four factors—number of births during the time interval which is determined by the level of fertility in the population; number of deaths during the time interval which is determined by the level of mortality in the population; the number of in-migrants; and the number of out-migrants during the time interval. Estimates of fertility and mortality, and information about the number of in-migrants and the number of out-migrants are not available at the district level in India. It is, therefore, not possible to analyse the contribution of the change in fertility and mortality, and net migration to the population growth in the districts of the country. At the national and state/Union Territory levels, population growth is primarily be attributed to the change in the difference between the birth rate and the death rate or the natural population growth. Net migration constitutes a small proportion of population growth at country and state/Union Territory levels. However, at the district level, migration, in and out of the district, may often be the dominant factor in deciding population growth.

The decadal annual population growth rate estimated from the population enumerated at different population censuses depicts a discontinuous trajectory of population growth. It means that population increases linearly or at a constant rate of growth throughout the 10 years' interval between two successive population censuses and this constant rate of increase is different for different ten years intervals. The decadal annual population growth rate based on successive population censuses ten years apart masks the variation in the growth of population within the ten years interval. This chapter follows the modelling approach to address the issue of discontinuity in the population growth trajectory based on the data 10 years apart available from decennial population census. Using the annual estimates of the population of the district derived from the population growth model, it is straightforward to obtain annual estimates of the population of the country and states/Union Territories of the country. An advantage of the modelling approach adopted in this chapter is that the district population growth model may be used to forecast district population at least in the near future and district population forecast may be used to forecast population of the country. This is important as the 2021 population census in India has been deferred and projecting district population using the conventional cohort-component projection method is not possible because of the non-availability of the necessary data at the district level. Modelling district population growth also leads to the bottom-up approach of forecasting the population of the country.

Appendix

See Table 1.5.

Table 1.5 Parameters of the simple logistic model of population growth in districts of india during 1951–2011

District code	State/Union Territory/ District	Results of fitting of the logistic model				
		MAPE	R^2	Δt	*r*	t_m
Jammu and Kashmir						
001	Kupwara	0.011	0.999	57.9	0.076	2005
002	Badgam	0.016	0.999	59.6	0.074	1997
003	Leh (Ladakh)	0.007	1.000	47.2	0.093	1992
004	Kargil	0.009	0.999	50.9	0.086	1996
005	Punch	0.018	0.993	57.8	0.076	2002
006	Rajouri	0.042	0.959	58.6	0.075	2002
007	Kathua	0.025	0.994	59.5	0.074	1992
008	Baramula	0.015	0.997	58.0	0.076	1996
009	Bandipore	0.020	0.997	55.8	0.079	1998
010	Srinagar	0.019	0.995	58.0	0.076	1996
011	Ganderbal	0.024	0.996	55.9	0.079	2002
012	Pulwama	0.027	0.993	59.5	0.074	1998
013	Shupiyan	0.020	0.996	58.7	0.075	1999
014	Anantnag	0.036	0.966	59.2	0.074	2007
015	Kulgam	0.029	0.992	44.6	0.099	1990
016	Doda	0.033	0.979	58.8	0.075	1999
017	Ramban	0.039	0.964	58.5	0.075	2001
018	Kishtwar	0.030	0.981	56.3	0.078	1995
019	Udhampur	0.016	0.997	58.7	0.075	1996
020	Reasi	0.032	0.993	57.3	0.077	1995
021	Jammu	0.023	0.995	56.7	0.078	1991
022	Samba	0.031	0.992	58.5	0.075	1988

(continued)

Table 1.5 (continued)

District code	State/Union Territory/ District	Results of fitting of the logistic model				
		MAPE	R^2	Δt	r	t_m
Himachal Pradesh						
023	Chamba	0.016	0.991	59.6	0.074	1988
024	Kangra	0.012	0.996	58.8	0.075	1985
025	Lahul & Spiti	0.024	0.816	42.6	0.103	1953
026	Kullu	0.015	0.994	58.2	0.076	1994
027	Mandi	0.016	0.997	58.7	0.075	1982
028	Hamirpur	0.011	0.995	58.6	0.075	1982
029	Una	0.019	0.979	58.1	0.076	1992
030	Bilaspur	0.020	0.994	58.7	0.075	1984
031	Solan	0.016	0.996	57.4	0.077	1994
032	Sirmaur	0.018	0.994	58.4	0.075	1990
033	Shimla	0.019	0.992	57.5	0.076	1987
034	Kinnaur	0.011	0.997	55.5	0.079	1980
Punjab						
035	Gurdaspur	0.016	0.997	58.4	0.075	1985
036	Kapurthala	0.008	0.999	55.7	0.079	1983
037	Jalandhar	0.022	0.984	57.3	0.077	1988
038	Hoshiarpur	0.008	0.999	57.9	0.076	1983
039	Shahid	0.005	0.999	50.8	0.087	1981
040	Fatehgarh Sahib	0.026	0.991	58.5	0.075	1984
041	Ludhiana	0.017	0.996	57.5	0.076	1991
042	Moga	0.013	0.996	58.0	0.076	1985
043	Firozpur	0.026	0.978	59.2	0.074	1990
044	Muktsar	0.047	0.913	59.1	0.074	1988
045	Faridkot	0.031	0.972	56.6	0.078	1987
046	Bathinda	0.040	0.969	57.2	0.077	1987
047	Mansa	0.035	0.980	58.9	0.075	1985
048	Patiala	0.046	0.973	59.7	0.074	1987
049	Amritsar	0.025	0.965	57.9	0.076	1994
050	Tarn Taran	0.037	0.953	57.2	0.077	1993
051	Rupnagar	0.020	0.983	59.0	0.075	1986
052	Sahibzada Ajit Singh Nagar	0.028	0.960	59.2	0.074	2002
053	Sangrur	0.021	0.991	58.9	0.075	1987
054	Barnala	0.027	0.983	58.9	0.075	1987

(continued)

Table 1.5 (continued)

District code	State/Union Territory/ District	Results of fitting of the logistic model				
		MAPE	R^2	Δt	r	t_m
Chandigarh						
055	Chandigarh	0.112	0.977	55.8	0.079	1989
Uttarakhand						
056	Uttarkashi	0.006	1.000	59.0	0.075	1990
057	Chamoli	0.012	0.996	52.4	0.084	1983
058	Rudraprayag	0.009	0.998	50.9	0.086	1983
059	Tehri Garhwal	0.014	0.991	47.3	0.093	1982
060	Dehradun	0.041	0.979	58.7	0.075	1999
061	Garhwal	0.008	0.923	41.7	0.105	1971
062	Pithoragarh	0.012	0.995	58.7	0.075	1979
063	Bageshwar	0.005	1.000	57.3	0.077	1977
064	Almora	0.010	0.965	38.5	0.114	1971
065	Champawat	0.029	0.993	59.5	0.074	1984
066	Nainital	0.040	0.929	58.7	0.075	1997
067	Udham Singh Nagar	0.101	0.873	59.1	0.074	1995
068	Hardwar	0.045	0.959	58.0	0.076	1999
Haryana						
069	Panchkula	0.039	0.966	45.4	0.097	1995
070	Ambala	0.026	0.976	58.4	0.075	1990
071	Yamunanagar	0.037	0.978	59.6	0.074	1991
072	Kurukshetra	0.068	0.927	59.5	0.074	1986
073	Kaithal	0.033	0.993	58.7	0.075	1983
074	Karnal	0.033	0.986	59.1	0.074	1989
075	Panipat	0.019	0.991	59.6	0.074	1999
076	Sonipat	0.023	0.988	59.4	0.074	1989
077	Jind	0.029	0.991	58.1	0.076	1985
078	Fatehabad	0.062	0.978	59.1	0.074	1985
079	Sirsa	0.051	0.984	59.4	0.074	1985
080	Hisar	0.036	0.983	58.7	0.075	1987
081	Bhiwani	0.026	0.993	59.2	0.074	1987
082	Rohtak	0.034	0.977	57.2	0.077	1986
083	Jhajjar	0.026	0.991	57.8	0.076	1985
084	Mahendragarh	0.014	0.997	59.3	0.074	1988
085	Rewari	0.023	0.983	59.1	0.074	1993
086	Gurgaon	0.086	0.988	59.2	0.074	2020
087	Mewat	0.047	0.879	59.1	0.074	2007
088	Faridabad	0.028	0.999	54.9	0.080	2000
089	Palwal	0.040	0.939	58.3	0.075	1998

(continued)

Table 1.5 (continued)

District code	State/Union Territory/ District	Results of fitting of the logistic model				
		MAPE	R^2	Δt	r	t_m
National Capital Territory of Delhi						
090	Northwest	0.018	1.000	46.1	0.095	1996
091	North	0.016	0.936	48.4	0.091	2007
092	Northeast	0.075	0.998	37.7	0.116	1994
093	East	0.016	1.000	47.0	0.093	1990
094	New Delhi	0.665	NA	NA	NA	NA
095	Central	0.973	NA	NA	NA	NA
096	West	0.065	0.990	49.3	0.089	1992
097	Southwest	0.061	0.972	48.7	0.090	1999
098	South	0.042	0.996	49.2	0.089	1993
Rajasthan						
099	Ganganagar	0.045	0.986	58.7	0.075	1986
100	Hanumangarh	0.046	0.991	56.7	0.078	1985
101	Bikaner	0.012	0.998	59.2	0.074	1997
102	Churu	0.021	0.995	58.7	0.075	1991
103	Jhunjhunun	0.006	1.000	59.1	0.074	1987
104	Alwar	0.031	0.957	58.6	0.075	1996
105	Bharatpur	0.039	0.948	58.3	0.075	1993
106	Dhaulpur	0.025	0.976	59.1	0.074	1997
107	Karauli	0.022	0.989	57.3	0.077	1995
108	Sawai Madhopur	0.032	0.978	59.0	0.074	1993
109	Dausa	0.017	0.995	59.4	0.074	1998
110	Jaipur	0.020	0.994	58.6	0.075	1997
111	Sikar	0.009	0.999	58.0	0.076	1991
112	Nagaur	0.020	0.995	59.5	0.074	1994
113	Jodhpur	0.038	0.977	58.7	0.075	1997
114	Jaisalmer	0.024	0.988	58.1	0.076	2003
115	Barmer	0.049	0.948	58.4	0.075	2001
116	Jalor	0.039	0.940	57.8	0.076	1996
117	Sirohi	0.035	0.942	57.8	0.076	1997
118	Pali	0.022	0.992	57.6	0.076	1987
119	Ajmer	0.027	0.952	57.7	0.076	1995
120	Tonk	0.025	0.987	59.0	0.075	1992
121	Bundi	0.019	0.997	59.1	0.074	1990
122	Bhilwara	0.026	0.960	59.1	0.074	1995
123	Rajsamand	0.028	0.974	58.3	0.075	1992

(continued)

Table 1.5 (continued)

District code	State/Union Territory/ District	Results of fitting of the logistic model				
		MAPE	R^2	Δt	r	t_m
124	Dungarpur	0.050	0.911	55.2	0.080	1994
125	Banswara	0.042	0.970	58.6	0.075	1995
126	Chittaurgarh	0.025	0.991	58.3	0.075	1989
127	Kota	0.027	0.996	57.7	0.076	1993
128	Baran	0.025	0.983	57.9	0.076	1994
129	Jhalawar	0.031	0.975	59.1	0.074	1993
130	Udaipur	0.036	0.977	58.5	0.075	1994
131	Pratapgarh	0.038	0.979	57.3	0.077	1993
Uttar Pradesh						
132	Saharanpur	0.020	0.991	59.8	0.074	1994
133	Muzaffarnagar	0.016	0.995	58.6	0.075	1992
134	Bijnor	0.018	0.994	58.8	0.075	1993
135	Moradabad	0.023	0.973	56.8	0.077	1997
136	Rampur	0.030	0.981	58.4	0.075	1994
137	Jyotiba Phule Nagar	0.020	0.988	59.5	0.074	1997
138	Meerut	0.014	0.995	59.4	0.074	1992
139	Baghpat	0.011	0.996	59.2	0.074	1986
140	Ghaziabad	0.025	0.981	58.1	0.076	2007
141	Gautam Buddha Nagar	0.021	0.999	58.8	0.075	2013
142	Bulandshahr	0.018	0.992	57.2	0.077	1990
143	Aligarh	0.016	0.991	59.1	0.074	1996
144	Mahamaya Nagar	0.017	0.990	58.5	0.075	1992
145	Mathura	0.027	0.960	59.2	0.074	1997
146	Agra	0.038	0.924	57.9	0.076	1997
147	Firozabad	0.036	0.933	58.8	0.075	1997
148	Mainpuri	0.023	0.978	59.0	0.075	1993
149	Budaun	0.014	0.984	59.5	0.074	1998
150	Bareilly	0.022	0.983	59.0	0.075	1997
151	Pilibhit	0.016	0.994	58.9	0.075	1993
152	Shahjahanpur	0.019	0.993	57.2	0.077	1998
153	Kheri	0.026	0.962	58.6	0.075	2000
154	Sitapur	0.029	0.982	56.9	0.077	1997
155	Hardoi	0.028	0.960	55.3	0.080	1995
156	Unnao	0.018	0.988	58.1	0.076	1992
157	Lucknow	0.019	0.991	57.7	0.076	1999
158	Rae Bareli	0.014	0.989	57.9	0.076	1996

(continued)

Table 1.5 (continued)

District code	State/Union Territory/ District	Results of fitting of the logistic model				
		MAPE	R^2	Δt	r	t_m
159	Farrukhabad	0.021	0.977	59.6	0.074	1995
160	Kannauj	0.019	0.992	59.6	0.074	1992
161	Etawah	0.031	0.965	58.6	0.075	1991
162	Auraiya	0.028	0.976	58.6	0.075	1989
163	Kanpur Dehat	0.028	0.973	55.5	0.079	1989
164	Kanpur Nagar	0.026	0.991	57.7	0.076	1989
165	Jalaun	0.024	0.981	56.9	0.077	1990
166	Jhansi	0.018	0.994	59.4	0.074	1989
167	Lalitpur	0.018	0.983	57.5	0.076	1999
168	Hamirpur	0.019	0.992	57.7	0.076	1986
169	Mahoba	0.020	0.992	59.4	0.074	1991
170	Banda	0.029	0.987	59.6	0.074	1991
171	Chitrakoot	0.028	0.976	59.3	0.074	1997
172	Fatehpur	0.020	0.989	57.6	0.076	1990
173	Pratapgarh	0.012	0.996	59.0	0.075	1995
174	Kaushambi	0.021	0.974	58.9	0.075	1999
175	Allahabad	0.015	0.993	59.0	0.075	1995
176	Bara Banki	0.019	0.962	57.2	0.077	1999
177	Faizabad	0.008	0.997	57.9	0.076	1997
178	Ambedkar Nagar	0.011	0.997	58.6	0.075	1996
179	Sultanpur	0.006	0.999	57.8	0.076	1996
180	Bahraich	0.020	0.981	59.0	0.075	2003
181	Shrawasti	0.014	0.960	58.4	0.075	2006
182	Balrampur	0.018	0.984	58.4	0.075	2001
183	Gonda	0.011	0.997	59.1	0.074	2002
184	Siddharthnagar	0.011	0.995	58.8	0.075	2005
185	Basti	0.022	0.967	59.5	0.074	1996
186	Sant Kabir Nagar	0.009	0.995	59.0	0.075	1999
187	Mahrajganj	0.021	0.982	59.2	0.074	1999
188	Gorakhpur	0.012	0.994	58.4	0.075	1994
189	Kushinagar	0.013	0.994	58.3	0.075	1999
190	Deoria	0.007	0.999	58.6	0.075	1993
191	Azamgarh	0.010	0.996	57.4	0.077	1995
192	Mau	0.016	0.985	59.2	0.074	1997
193	Ballia	0.014	0.995	57.9	0.076	1994
194	Jaunpur	0.006	0.999	58.2	0.076	1992

(continued)

Table 1.5 (continued)

District code	State/Union Territory/ District	Results of fitting of the logistic model				
		MAPE	R^2	Δt	*r*	t_m
195	Ghazipur	0.016	0.992	58.8	0.075	1996
196	Chandauli	0.020	0.982	57.8	0.076	1995
197	Varanasi	0.011	0.995	58.5	0.075	1993
198	Sant Ravidas Nagar (Bhadohi)	0.012	0.998	51.3	0.086	1991
199	Mirzapur	0.013	0.993	57.7	0.076	1996
200	Sonbhadra	0.024	0.988	59.2	0.074	1997
201	Etah	0.021	0.979	58.7	0.075	1994
202	Kanshiram Nagar	0.025	0.945	58.8	0.075	1995
Bihar						
203	Pashchim Champaran	0.052	0.930	59.3	0.074	2000
204	Purba Champaran	0.026	0.956	59.3	0.074	2003
205	Sheohar	0.023	0.970	56.6	0.078	2004
206	Sitamarhi	0.025	0.965	59.0	0.075	2003
207	Madhubani	0.032	0.917	58.3	0.075	1999
208	Supaul	0.082	0.872	59.1	0.074	1994
209	Araria	0.070	0.928	59.6	0.074	1998
210	Kishanganj	0.069	0.951	58.6	0.075	1996
211	Purnia	0.062	0.896	59.3	0.074	1999
212	Katihar	0.049	0.917	59.1	0.074	1999
213	Madhepura	0.043	0.902	59.6	0.074	2002
214	Saharsa	0.030	0.931	59.0	0.075	2000
215	Darbhanga	0.020	0.980	57.9	0.076	1997
216	Muzaffarpur	0.027	0.926	58.2	0.076	2001
217	Gopalganj	0.013	0.995	58.9	0.075	1996
218	Siwan	0.025	0.970	59.3	0.074	1998
219	Saran	0.023	0.974	57.8	0.076	1997
220	Vaishali	0.029	0.893	58.8	0.075	2001
221	Samastipur	0.022	0.958	58.5	0.075	1999
222	Begusarai	0.033	0.939	57.2	0.077	1998
223	Khagaria	0.036	0.888	58.3	0.075	2002
224	Bhagalpur	0.041	0.922	57.7	0.076	1997
225	Banka	0.032	0.891	57.4	0.077	1998
226	Munger	0.034	0.944	58.0	0.076	1994
227	Lakhisarai	0.034	0.970	58.3	0.075	1997
228	Sheikhpura	0.032	0.957	58.5	0.075	1997

(continued)

Table 1.5 (continued)

District code	State/Union Territory/ District	Results of fitting of the logistic model				
		MAPE	R^2	Δt	r	t_m
229	Nalanda	0.026	0.980	58.8	0.075	1993
230	Patna	0.029	0.973	58.7	0.075	1998
231	Bhojpur	0.029	0.923	58.7	0.075	1997
232	Buxar	0.036	0.956	59.5	0.074	1998
233	Kaimur (Bhabua)	0.041	0.923	59.5	0.074	2001
234	Rohtas	0.036	0.963	59.4	0.074	1995
235	Aurangabad	0.034	0.940	58.4	0.075	1999
236	Gaya	0.032	0.936	59.5	0.074	2000
237	Nawada	0.031	0.924	58.7	0.075	1999
238	Jamui	0.046	0.965	56.5	0.078	1998
239	Jehanabad	0.032	0.867	58.4	0.075	1999
240	Arwal	0.029	0.957	59.2	0.074	1996
Sikkim						
241	North District	0.057	0.970	38.6	0.114	1981
242	West District	0.009	0.999	37.2	0.118	1990
243	South District	0.026	0.994	37.8	0.116	1991
244	East District	0.043	0.986	39.4	0.111	1991
Arunachal Pradesh						
245	Tawang	0.029	0.932	47.9	0.092	2001
246	West Kameng	0.011	0.998	41.5	0.106	1992
247	East Kameng	0.061	0.947	48.3	0.091	1996
248	Papum Pare	0.043	0.995	49.1	0.090	2002
249	Upper Subansiri	0.093	0.870	47.3	0.093	2001
250	West Siang	0.011	0.999	45.4	0.097	1984
251	East Siang	0.046	0.974	42.9	0.102	1988
252	Upper Siang	0.102	0.952	49.1	0.090	1978
253	Changlang	0.056	0.990	48.8	0.090	1988
254	Tirap	0.027	0.984	46.9	0.094	1987
255	Lower Subansiri	0.064	0.841	48.6	0.090	2008
256	Kurung Kumey	0.097	1.000	47.0	0.093	2047
257	Dibang Valley	0.007	0.995	31.7	0.139	1968
258	Lower Dibang Valley	0.065	0.991	38.1	0.115	1987
259	Lohit	0.037	0.998	49.1	0.090	1989
260	Anjaw	0.035	0.930	48.5	0.091	1988

(continued)

Table 1.5 (continued)

District code	State/Union Territory/ District	Results of fitting of the logistic model				
		MAPE	R^2	Δt	*r*	t_m
Nagaland						
261	Mon	0.112	0.831	22.6	0.194	1991
262	Mokokchung	0.087	0.892	26.8	0.164	1986
263	Zunheboto	0.077	0.984	30.6	0.144	1989
264	Wokha	0.123	0.955	26.8	0.164	1993
265	Dimapur	0.080	0.995	45.9	0.096	1995
266	Phek	0.025	0.996	39.0	0.113	1990
267	Tuensang	0.062	0.949	33.6	0.131	1992
268	Longleng	0.059	1.000	13.7	0.320	1986
269	Kiphire	0.331	1.000	10.1	0.437	1991
270	Kohima	0.017	0.997	46.6	0.094	1994
271	Peren	0.046	0.990	29.0	0.152	1990
Manipur						
272	Senapati	0.113	0.909	59.0	0.075	2011
273	Tamenglong	0.010	1.000	58.9	0.075	1999
274	Churachandpur	0.048	0.981	58.2	0.076	1990
275	Bishnupur	0.023	0.995	58.2	0.076	1983
276	Thoubal	0.037	0.978	57.5	0.077	1988
277	Imphal West	0.024	0.993	59.4	0.074	1986
278	Imphal East	0.034	0.991	58.1	0.076	1985
279	Ukhrul	0.030	0.983	58.3	0.075	1999
280	Chandel	0.049	0.993	54.2	0.081	2000
Mizoram						
281	Mamit	0.080	0.804	36.6	0.120	1981
282	Kolasib	0.041	0.949	39.0	0.113	1998
283	Aizawl	0.027	0.996	38.1	0.115	1992
284	Champhai	0.015	0.996	38.4	0.115	1993
285	Serchhip	0.013	0.996	38.7	0.113	1998
286	Lunglei	0.035	0.979	36.8	0.119	1990
287	Lawngtlai	0.050	0.976	39.8	0.111	1998
288	Saiha	0.020	0.998	39.1	0.112	1991
Tripura						
289	West Tripura	0.054	0.980	58.1	0.076	1983
290	South Tripura	0.049	0.988	56.5	0.078	1979
291	Dhalai	0.054	0.984	54.9	0.080	1982
292	North Tripura	0.100	0.799	58.3	0.075	1988

(continued)

Table 1.5 (continued)

District code	State/Union Territory/ District	Results of fitting of the logistic model				
		MAPE	R^2	Δt	r	t_m
Meghalaya						
293	West Garo Hills	0.045	0.927	59.3	0.074	1998
294	East Garo Hills	0.023	0.984	59.0	0.074	1998
295	South Garo Hills	0.049	0.891	58.3	0.075	1998
296	West Khasi Hills	0.024	0.992	59.3	0.074	1999
297	Ribhoi	0.053	0.995	59.7	0.074	2001
298	East Khasi Hills	0.042	0.925	54.3	0.081	1993
299	Jaintia Hills	0.032	0.983	58.8	0.075	2000
Assam						
300	Kokrajhar	0.007	0.999	50.6	0.087	1977
301	Dhubri	0.053	0.964	59.3	0.074	1992
302	Goalpara	0.056	0.969	59.6	0.074	1989
303	Barpeta	0.055	0.944	58.6	0.075	1988
304	Morigaon	0.060	0.961	57.5	0.076	1988
305	Nagaon	0.054	0.958	58.9	0.075	1989
306	Sonitpur	0.033	0.986	59.4	0.074	1986
307	Lakhimpur	0.048	0.980	58.4	0.075	1983
308	Dhemaji	0.043	0.994	59.1	0.074	1984
309	Tinsukia	0.044	0.973	57.1	0.077	1985
310	Dibrugarh	0.033	0.976	58.0	0.076	1981
311	Sivasagar	0.022	0.987	59.0	0.075	1984
312	Jorhat	0.029	0.968	54.6	0.081	1982
313	Golaghat	0.018	0.994	55.7	0.079	1983
314	Karbi Anglong	0.055	0.987	57.2	0.077	1983
315	Dima Hasao	0.011	1.000	56.8	0.077	1987
316	Cachar	0.042	0.972	53.0	0.083	1988
317	Karimganj	0.045	0.901	58.9	0.075	1993
318	Hailakandi	0.047	0.953	58.1	0.076	1991
319	Bongaigaon	0.058	0.962	59.6	0.074	1988
320	Chirang	0.032	0.960	50.4	0.087	1972
321	Kamrup	0.041	0.978	58.7	0.075	1986
322	Kamrup Metropolitan	0.059	0.990	58.3	0.075	1991
323	Nalbari	0.040	0.970	58.2	0.076	1978
324	Baksa	0.018	0.991	57.8	0.076	1975
325	Darrang	0.046	0.979	59.1	0.074	1988
326	Udalguri	0.021	0.992	59.0	0.075	1973

(continued)

Table 1.5 (continued)

District code	State/Union Territory/ District	Results of fitting of the logistic model				
		MAPE	R^2	Δt	*r*	t_m
West Bengal						
327	Darjiling	0.030	0.982	57.6	0.076	1988
328	Jalpaiguri	0.043	0.975	58.5	0.075	1986
329	Koch Bihar	0.040	0.983	58.9	0.075	1979
330	Uttar Dinajpur	0.048	0.957	57.9	0.076	1993
331	Dakshin Dinajpur	0.016	0.998	59.0	0.075	1985
332	Maldah	0.034	0.976	59.8	0.074	1993
333	Murshidabad	0.040	0.952	58.7	0.075	1992
334	Birbhum	0.044	0.940	59.2	0.074	1987
335	Barddhaman	0.032	0.988	59.5	0.074	1982
336	Nadia	0.030	0.992	58.5	0.075	1984
337	North Twenty	0.031	0.992	59.2	0.074	1986
338	Hugli	0.028	0.992	58.5	0.075	1981
339	Bankura	0.032	0.975	57.0	0.077	1983
340	Puruliya	0.025	0.967	55.5	0.079	1989
341	Haora	0.023	0.989	59.5	0.074	1987
342	Kolkata	0.009	0.905	49.2	0.089	1968
343	South Twenty Four Parganas	0.036	0.975	58.7	0.075	1990
344	Paschim Medinipur	0.031	0.985	57.7	0.076	1984
345	Purba Medinipur	0.025	0.989	59.5	0.074	1986
Jharkhand						
346	Garhwa	0.030	0.973	58.1	0.076	1998
347	Chatra	0.034	0.955	57.5	0.076	1999
348	Kodarma	0.034	0.914	58.8	0.075	2002
349	Giridih	0.038	0.946	55.8	0.079	1997
350	Deoghar	0.028	0.967	56.1	0.078	2000
351	Godda	0.028	0.986	57.9	0.076	1997
352	Sahibganj	0.038	0.920	59.6	0.074	1998
353	Pakur	0.054	0.896	58.5	0.075	2000
354	Dhanbad	0.026	0.993	58.2	0.076	1987
355	Bokaro	0.011	0.999	58.0	0.076	1988
356	Lohardaga	0.025	0.927	59.2	0.074	2002
357	Purbi Singhbhum	0.037	0.976	57.7	0.076	1988
358	Palamu	0.032	0.955	59.4	0.074	1999

(continued)

Table 1.5 (continued)

District code	State/Union Territory/ District	Results of fitting of the logistic model				
		MAPE	R^2	Δt	*r*	t_m
359	Latehar	0.038	0.933	58.2	0.076	1998
360	Hazaribagh	0.032	0.982	59.6	0.074	1995
361	Ramgarh	0.012	0.999	56.2	0.078	1985
362	Dumka	0.027	0.961	57.8	0.076	1995
363	Jamtara	0.018	0.981	58.5	0.075	1997
364	Ranchi	0.040	0.960	58.6	0.075	1996
365	Khunti	0.031	0.841	59.1	0.074	1998
366	Gumla	0.042	0.932	56.0	0.079	1995
367	Simdega	0.038	0.937	58.0	0.076	1990
368	Pashchimi Singhbhum	0.040	0.883	56.7	0.078	1995
369	Saraikela-Kharsawan	0.031	0.940	58.4	0.075	1998
Odisha						
370	Bargarh	0.010	0.996	56.3	0.078	1982
371	Jharsuguda	0.013	0.996	53.2	0.083	1988
372	Sambalpur	0.021	0.993	59.5	0.074	1983
373	Debagarh	0.029	0.988	58.5	0.075	1984
374	Sundargarh	0.032	0.986	59.4	0.074	1981
375	Kendujhar	0.036	0.973	58.8	0.075	1986
376	Mayurbhanj	0.030	0.974	57.4	0.077	1989
377	Baleshwar	0.027	0.988	57.1	0.077	1987
378	Bhadrak	0.030	0.981	57.6	0.076	1987
379	Kendrapara	0.022	0.990	58.3	0.075	1981
380	Jagatsinghapur	0.006	0.999	57.7	0.076	1981
381	Cuttack	0.017	0.994	59.6	0.074	1983
382	Jajapur	0.022	0.989	58.2	0.076	1986
383	Dhenkanal	0.024	0.985	54.7	0.080	1982
384	Anugul	0.018	0.994	57.3	0.077	1986
385	Nayagarh	0.020	0.978	58.1	0.076	1979
386	Khordha	0.014	0.998	59.0	0.075	1993
387	Puri	0.020	0.992	57.3	0.077	1985
388	Ganjam	0.018	0.992	59.5	0.074	1987
389	Gajapati	0.023	0.990	55.8	0.079	1985
390	Kandhamal	0.017	0.983	56.5	0.078	1991
391	Baudh	0.027	0.972	56.4	0.078	1992
392	Subarnapur	0.009	0.998	58.7	0.075	1987
393	Balangir	0.042	0.938	59.0	0.075	1990

(continued)

Table 1.5 (continued)

District code	State/Union Territory/ District	Results of fitting of the logistic model				
		MAPE	R^2	Δt	*r*	t_m
394	Nuapada	0.029	0.982	58.8	0.075	1984
395	Kalahandi	0.029	0.943	57.8	0.076	1992
396	Rayagada	0.033	0.977	56.8	0.077	1987
397	Nabarangapur	0.035	0.976	59.1	0.074	1990
398	Koraput	0.018	0.995	58.6	0.075	1991
399	Malkangiri	0.067	0.978	58.8	0.075	1984
Chhattisgarh						
400	Koriya	0.020	0.996	58.9	0.075	1985
401	Surguja	0.033	0.956	56.0	0.078	1992
402	Jashpur	0.037	0.970	59.2	0.074	1984
403	Raigarh	0.033	0.890	59.6	0.074	1993
404	Korba	0.012	0.999	58.7	0.075	1991
405	Janjgir-Champa	0.018	0.987	57.1	0.077	2002
406	Bilaspur	0.051	0.945	58.2	0.076	1997
407	Kabeerdham	0.055	0.973	59.3	0.074	1997
408	Rajnandgaon	0.033	0.952	58.1	0.076	1991
409	Durg	0.034	0.986	58.7	0.075	1987
410	Raipur	0.057	0.904	57.4	0.077	1997
411	Mahasamund	0.029	0.975	59.2	0.074	1986
412	Dhamtari	0.024	0.983	58.4	0.075	1989
413	Uttar Bastar Kanker	0.039	0.983	56.5	0.078	1984
414	Bastar	0.031	0.978	57.0	0.077	1989
415	Narayanpur	0.025	0.988	59.1	0.074	1992
416	Dakshin Bastar Dantewada	0.039	0.975	58.7	0.075	1980
417	Bijapur	0.018	0.996	55.3	0.080	1985
Madhya Pradesh						
418	Sheopur	0.025	0.985	59.1	0.074	1996
419	Morena	0.023	0.989	59.0	0.075	1995
420	Bhind	0.026	0.982	59.0	0.074	1991
421	Gwalior	0.032	0.980	59.1	0.074	1995
422	Datia	0.028	0.975	57.6	0.076	1992
423	Shivpuri	0.015	0.993	59.2	0.074	1997
424	Tikamgarh	0.025	0.983	59.4	0.074	1994
425	Chhatarpur	0.023	0.969	56.0	0.078	1994
426	Panna	0.036	0.947	54.8	0.080	1991

(continued)

Table 1.5 (continued)

District code	State/Union Territory/ District	Results of fitting of the logistic model				
		MAPE	R^2	Δt	*r*	t_m
427	Sagar	0.031	0.983	58.4	0.075	1990
428	Damoh	0.023	0.992	59.3	0.074	1989
429	Satna	0.032	0.975	57.5	0.077	1993
430	Rewa	0.025	0.976	58.6	0.075	1994
431	Umaria	0.036	0.963	59.8	0.074	1995
432	Neemuch	0.016	0.998	59.1	0.074	1987
433	Mandsaur	0.019	0.995	58.9	0.075	1988
434	Ratlam	0.036	0.966	58.7	0.075	1992
435	Ujjain	0.022	0.995	59.1	0.074	1990
436	Shajapur	0.028	0.980	59.0	0.075	1992
437	Dewas	0.027	0.990	58.7	0.075	1992
438	Dhar	0.040	0.937	58.6	0.075	1996
439	Indore	0.047	0.950	58.4	0.075	2000
440	Khargone (West Nimar)	0.052	0.923	59.7	0.074	1994
441	Barwani	0.038	0.928	59.6	0.074	1999
442	Rajgarh	0.035	0.956	57.9	0.076	1996
443	Vidisha	0.043	0.919	58.0	0.076	1992
444	Bhopal	0.034	0.993	57.3	0.077	1996
445	Sehore	0.041	0.963	59.6	0.074	1993
446	Raisen	0.037	0.981	59.5	0.074	1991
447	Betul	0.017	0.996	56.9	0.077	1986
448	Harda	0.037	0.952	56.0	0.079	1992
449	Hoshangabad	0.019	0.996	58.3	0.075	1989
450	Katni	0.030	0.977	56.9	0.077	1992
451	Jabalpur	0.028	0.995	59.2	0.074	1986
452	Narsimhapur	0.024	0.991	58.4	0.075	1988
453	Dindori	0.046	0.963	59.1	0.074	1987
454	Mandla	0.029	0.978	57.7	0.076	1989
455	Chhindwara	0.013	0.997	59.4	0.074	1987
456	Seoni	0.028	0.983	57.7	0.076	1989
457	Balaghat	0.017	0.988	58.8	0.075	1984
458	Guna	0.028	0.987	58.8	0.075	1998
459	Ashoknagar	0.041	0.923	57.0	0.077	1993
460	Shahdol	0.026	0.989	59.4	0.074	1988
461	Anuppur	0.008	0.999	53.8	0.082	1987
462	Sidhi	0.031	0.959	59.3	0.074	1996

(continued)

Table 1.5 (continued)

District code	State/Union Territory/ District	Results of fitting of the logistic model				
		MAPE	R^2	Δt	*r*	t_m
463	Singrauli	0.016	0.997	58.7	0.075	1999
464	Jhabua	0.057	0.923	59.2	0.074	1997
465	Alirajpur	0.037	0.969	58.2	0.076	1993
466	Khandwa (East Nimar)	0.041	0.961	57.4	0.077	1991
467	Burhanpur	0.032	0.989	58.6	0.075	1988
Gujarat						
468	Kachchh	0.053	0.946	59.1	0.074	1999
469	Banas Kantha	0.032	0.977	59.8	0.074	1995
470	Patan	0.031	0.977	59.6	0.074	1981
471	Mahesana	0.016	0.993	57.4	0.077	1980
472	Sabar Kantha	0.042	0.978	58.4	0.075	1984
473	Gandhinagar	0.020	0.997	57.6	0.076	1986
474	Ahmadabad	0.044	0.969	57.8	0.076	1992
475	Surendranagar	0.045	0.939	59.5	0.074	1990
476	Rajkot	0.043	0.971	59.5	0.074	1991
477	Jamnagar	0.044	0.980	58.9	0.075	1982
478	Porbandar	0.024	0.973	57.6	0.076	1976
479	Junagadh	0.025	0.991	57.9	0.076	1981
480	Amreli	0.009	0.987	56.1	0.078	1975
481	Bhavnagar	0.021	0.995	59.5	0.074	1988
482	Anand	0.023	0.985	58.8	0.075	1982
483	Kheda	0.032	0.963	52.1	0.084	1981
484	Panch Mahals	0.036	0.961	58.0	0.076	1990
485	Dohad	0.042	0.929	59.0	0.075	1998
486	Vadodara	0.023	0.994	58.9	0.075	1986
487	Narmada	0.033	0.981	59.1	0.074	1984
488	Bharuch	0.032	0.960	58.0	0.076	1988
489	The Dangs	0.067	0.904	59.6	0.074	1993
490	Navsari	0.004	1.000	58.9	0.075	1981
491	Valsad	0.033	0.961	59.4	0.074	1996
492	Surat	0.032	0.980	58.7	0.075	2008
493	Tapi	0.035	0.980	55.7	0.079	1981
Daman and Diu						
494	Diu	0.093	0.680	58.1	0.076	1988
495	Daman	0.080	0.977	59.2	0.074	2028
Dadra and Nagar Haveli						

(continued)

Table 1.5 (continued)

District code	State/Union Territory/ District	Results of fitting of the logistic model				
		MAPE	R^2	Δt	*r*	t_m
496	Dadra & Nagar Haveli	0.070	0.996	59.0	0.074	2015
Maharashtra						
497	Nandurbar	0.043	0.964	57.8	0.076	1995
498	Dhule	0.039	0.961	57.9	0.076	1989
499	Jalgaon	0.020	0.990	58.5	0.075	1988
500	Buldana	0.023	0.979	59.3	0.074	1991
501	Akola	0.024	0.990	58.9	0.075	1986
502	Washim	0.031	0.984	58.2	0.075	1987
503	Amravati	0.019	0.994	58.4	0.075	1985
504	Wardha	0.013	0.997	57.9	0.076	1981
505	Nagpur	0.013	0.998	58.9	0.075	1989
506	Bhandara	0.012	0.997	54.9	0.080	1980
507	Gondiya	0.025	0.930	54.8	0.080	1977
508	Gadchiroli	0.022	0.995	57.1	0.077	1987
509	Chandrapur	0.009	0.999	54.0	0.081	1983
510	Yavatmal	0.022	0.994	57.3	0.077	1986
511	Nanded	0.017	0.995	59.1	0.074	1991
512	Hingoli	0.026	0.977	59.0	0.075	1992
513	Parbhani	0.020	0.993	58.0	0.076	1992
514	Jalna	0.041	0.904	58.3	0.075	1994
515	Aurangabad	0.033	0.959	59.5	0.074	1998
516	Nashik	0.039	0.969	58.1	0.076	1994
517	Thane	0.015	0.998	57.2	0.077	2004
518	Mumbai Suburban	0.028	0.999	51.0	0.086	1982
519	Mumbai	0.010	0.999	25.8	0.171	1958
520	Raigarh	0.025	0.976	58.7	0.075	1994
521	Pune	0.037	0.950	58.2	0.076	1999
522	Ahmadnagar	0.026	0.988	59.5	0.074	1987
523	Bid	0.031	0.967	58.2	0.076	1992
524	Latur	0.023	0.981	58.8	0.075	1992
525	Osmanabad	0.029	0.984	59.5	0.074	1986
526	Solapur	0.027	0.976	58.7	0.075	1988
527	Satara	0.017	0.995	58.5	0.075	1982
528	Ratnagiri	0.019	0.963	41.5	0.106	1978
529	Sindhudurg	0.012	0.971	55.3	0.080	1971
530	Kolhapur	0.016	0.996	58.1	0.076	1984
531	Sangli	0.020	0.994	58.8	0.075	1983

(continued)

Table 1.5 (continued)

District code	State/Union Territory/ District	Results of fitting of the logistic model				
		MAPE	R^2	Δt	*r*	t_m
Andhra Pradesh						
532	Adilabad	0.051	0.885	95.9	0.046	1995
533	Nizamabad	0.020	0.973	75.3	0.058	1983
534	Karimnagar	0.046	0.905	107.0	0.041	1994
535	Medak	0.075	0.890	107.0	0.041	1995
536	Hyderabad	0.061	0.989	68.9	0.064	1984
537	Rangareddy	0.042	0.992	91.9	0.048	2043
538	Mahbubnagar	0.037	0.954	105.0	0.042	2002
539	Nalgonda	0.034	0.906	87.5	0.050	1986
540	Warangal	0.033	0.938	99.9	0.044	1988
541	Khammam	0.030	0.986	85.9	0.051	1985
542	Srikakulam	0.014	0.994	98.9	0.044	1987
543	Vizianagaram	0.019	0.989	94.2	0.047	1980
544	Visakhapatnam	0.024	0.993	79.1	0.056	1990
545	East Godavari	0.027	0.985	91.3	0.048	1980
546	West Godavari	0.030	0.959	87.1	0.051	1977
547	Krishna	0.017	0.990	96.5	0.046	1983
548	Guntur	0.023	0.957	92.5	0.048	1982
549	Prakasam	0.017	0.984	101.0	0.043	1991
550	Sri Potti Sriramulu Nellore	0.011	0.995	82.9	0.053	1987
551	Y. S. R.	0.009	0.992	79.8	0.055	1987
552	Kurnool	0.021	0.979	89.3	0.049	1996
553	Anantapur	0.014	0.989	84.0	0.052	1989
554	Chittoor	0.009	0.998	94.0	0.047	1991
Karnataka						
555	Belgaum	0.022	0.990	57.7	0.076	1987
556	Bagalkot	0.016	0.996	58.4	0.075	1988
557	Bijapur	0.032	0.957	58.0	0.076	1992
558	Bidar	0.017	0.993	59.5	0.074	1989
559	Raichur	0.008	0.998	59.3	0.074	1991
560	Koppal	0.017	0.997	58.3	0.075	1992
561	Gadag	0.017	0.991	59.0	0.075	1982
562	Dharwad	0.024	0.993	58.4	0.075	1985
563	Uttara Kannada	0.016	0.996	57.3	0.077	1975
564	Haveri	0.018	0.988	56.8	0.077	1984

(continued)

Table 1.5 (continued)

District code	State/Union Territory/ District	Results of fitting of the logistic model				
		MAPE	R^2	Δt	*r*	t_m
565	Bellary	0.019	0.995	59.0	0.075	1993
566	Chitradurga	0.020	0.992	58.4	0.075	1983
567	Davanagere	0.004	1.000	57.8	0.076	1981
568	Shimoga	0.027	0.989	59.1	0.074	1973
569	Udupi	0.010	0.972	49.2	0.089	1975
570	Chikmagalur	0.021	0.975	56.3	0.078	1969
571	Tumkur	0.008	0.998	55.6	0.079	1979
572	Bangalore	0.061	0.981	59.0	0.075	2005
573	Mandya	0.005	0.998	52.4	0.084	1972
574	Hassan	0.006	0.998	56.7	0.078	1974
575	Dakshina Kannada	0.019	0.994	57.8	0.076	1982
576	Kodagu	0.027	0.979	57.1	0.077	1967
577	Mysore	0.012	0.998	58.5	0.075	1986
578	Chamarajanagar	0.011	0.998	54.3	0.081	1978
579	Gulbarga	0.027	0.975	59.2	0.074	1993
580	Yadgir	0.021	0.994	59.2	0.074	1997
581	Kolar	0.011	0.998	57.1	0.077	1986
582	Chikkaballapura	0.011	0.998	58.5	0.075	1982
583	Bangalore Rural	0.018	0.994	59.0	0.075	1991
584	Ramanagara	0.009	0.998	53.5	0.082	1977
Goa						
585	North Goa	0.019	0.993	56.0	0.078	1982
586	South Goa	0.027	0.993	56.1	0.078	1981
Lakshadweep						
587	Lakshadweep	0.011	0.999	49.8	0.088	1983
Kerala						
588	Kasaragod	0.007	0.999	57.9	0.076	1980
589	Kannur	0.003	1.000	55.9	0.079	1972
590	Wayanad	0.018	0.998	55.4	0.079	1975
591	Kozhikode	0.009	0.996	56.5	0.078	1976
592	Malappuram	0.011	0.999	57.6	0.076	1986
593	Palakkad	0.006	0.998	55.4	0.079	1980
594	Thrissur	0.013	0.986	55.1	0.080	1973
595	Ernakulam	0.012	0.988	57.5	0.076	1972
596	Idukki	0.019	0.941	52.1	0.084	1963
597	Kottayam	0.009	0.992	54.2	0.081	1969

(continued)

Table 1.5 (continued)

District code	State/Union Territory/ District	Results of fitting of the logistic model				
		MAPE	R^2	Δt	r	t_m
598	Alappuzha	0.005	0.994	56.5	0.078	1965
599	Pathanamthitta	0.009	0.935	49.5	0.089	1960
600	Kollam	0.005	0.998	57.7	0.076	1967
601	Thiruvananthapuram	0.012	0.996	58.8	0.075	1970
Tamil Nadu						
602	Thiruvallur	0.046	0.981	58.9	0.075	2001
603	Chennai	0.019	0.997	58.4	0.075	1976
604	Kancheepuram	0.051	0.965	59.6	0.074	2001
605	Vellore	0.022	0.993	58.3	0.075	1988
606	Tiruvannamalai	0.012	0.987	59.1	0.074	1984
607	Viluppuram	0.021	0.986	58.8	0.075	1988
608	Salem	0.035	0.966	58.8	0.075	1990
609	Namakkal	0.019	0.987	57.4	0.077	1993
610	Erode	0.019	0.989	57.8	0.076	1983
611	The Nilgiris	0.023	0.864	42.1	0.104	1972
612	Dindigul	0.021	0.984	58.1	0.076	1985
613	Karur	0.014	0.957	56.2	0.078	1994
614	Tiruchirappalli	0.021	0.983	57.3	0.077	1982
615	Perambalur	0.018	0.990	59.0	0.075	1988
616	Ariyalur	0.016	0.979	56.6	0.078	1980
617	Cuddalore	0.018	0.981	57.6	0.076	1983
618	Nagapattinam	0.012	0.991	57.0	0.077	1982
619	Thiruvarur	0.009	0.988	58.3	0.075	1979
620	Thanjavur	0.016	0.970	53.2	0.083	1979
621	Pudukkottai	0.021	0.986	56.9	0.077	1981
622	Sivaganga	0.027	0.905	57.0	0.077	1981
623	Madurai	0.026	0.974	57.5	0.076	1984
624	Theni	0.022	0.914	57.4	0.077	1978
625	Virudhunagar	0.012	0.994	58.7	0.075	1984
626	Ramanathapuram	0.028	0.916	58.6	0.075	1977
627	Thoothukkudi	0.022	0.913	53.0	0.083	1980
628	Tirunelveli	0.023	0.980	57.9	0.076	1986
629	Kanniyakumari	0.019	0.931	58.5	0.075	1975
630	Dharmapuri	0.033	0.980	59.1	0.074	1986
631	Krishnagiri	0.030	0.966	59.6	0.074	1992
632	Coimbatore	0.043	0.957	59.2	0.074	1989
633	Tiruppur	0.034	0.961	58.9	0.075	2004

(continued)

Table 1.5 (continued)

District code	State/Union Territory/ District	Results of fitting of the logistic model				
		MAPE	R^2	Δt	r	t_m
Puducherry						
634	Yanam	0.033	0.997	58.7	0.075	2027
635	Puducherry	0.032	0.993	58.7	0.075	1996
636	Mahe	0.015	0.993	58.4	0.075	1986
637	Karaikal	0.026	0.972	58.5	0.075	1991
Andaman and Nicobar Islands						
638	Nicobars	0.036	0.984	27.7	0.159	1976
639	North & Middle Andaman	0.058	0.978	44.2	0.099	1980
640	South Andaman	0.041	0.998	51.9	0.085	1987

Source Author

References

Akinyede O, Soyemi K (2016) Joinpoint regression analysis of pertussis crude incidence rates, Illinois, 1990–2014. Am J Infect Control 44(12):1732–1733

Bhat PNM (1999) Population projections for Delhi: dynamic logistic model versus cohort-component method. Demogr India 28(2):153–165

Burg D, Ausubel JH (2023) Jewish population trajectories between the Jordan River and the Mediterranean Sea. Isr Aff 29(3):653–668

Chatenoud L, Garavello W, Pagan E, Bertuccio P, Gallus S, La Vecchia C, Negri E, Bosetti C (2015) Laryngeal cancer mortality trends in European countries. Int J Cancer 842:833–842

Chaurasia AR (2020) Long-term trend in infant mortality in India: a joinpoint regression analysis for 1971-2018. Indian J Hum Dev 14(3):394–406

Chaurasia AR (2023) Seventy years of mortality transition in India: 1950-2021. Indian J Popul Dev 3(1):1–34

Clegg LX, Hankey BF, Tiwari R, Feuer EJ, Edwards BK (2009) Estimating average annual percent change in trend analysis. Stat Med 20(29):3670–3682

Dorn HF (1950) Pitfalls in population forecasts and projections. J Am Stat Assoc 45(251):311–334

Doucet M, Rochette HD (2016) Prevalence and mortality trends in chronic obstructive pulmonary disease over 2001 to 2011: a public health point of view of the burden. Can Respir J:1–10

Fisher J, Pry R (1971) A simple substitution model of technological change. Technol Forecast Soc Chang 3:75–88

Gillis D, Edwards BPM (2019) The utility of joinpoint regression for estimating population parameters given changes in population structure. Heliyon 5:e02515

Government of India (2014) Census of India 2011. Report of the post enumeration survey. Registrar General and Census Commissioner of India, New Delhi

Hrytsiuk P, Parfeniuk O, Shevchenko I (2023) The logistic dynamics of population growth as a prerequisite for global sustainable development. IOP Conf Ser: Earth Environ Sci 1126:012030

John U, Hanke M (2015) Liver cirrhosis mortality, alcohol consumption and tobacco consumption over a 62-year period in a high alcohol consumption country: a trend analysis. BMC Res Notes 8(1):822

Kim H-J, Chen H-S, Midthune D, Wheeler W, Buckman DW, Green D, Byrne J, Luo J, Feuer EJ (2023) Data-driven choice of a model selection method in joinpoint regression. J Appl Stat 50(9):1992–2013
Kim H-J, Fay MP, Feuer EJ, Midthune DN (2000) Permutation tests for joinpoint regression with applications to cancer rates. Stat Med 19:335–351
Kim H-J, Yu B, Feuer EJ (2009) Selecting the number of change-points in segmented line regression. Stat Sin 19(2):597–609
Kim J, Kim H-J (2016) Consistent model selection in segmented line regression. J Stat Plan Inference 170:106–116
Leach D (1981) Re-evaluation of the logistic curve for human populations. J R Stat Soc Ser (Gen) 144(1):94–103
Lerman PM (1980) Fitting segmented regression models by grid search. J R Stat Society Ser C (Appl Stat) 29(1):77–84
Lewis CD (1982) International and business forecasting methods. A practical guide for exponential smoothing and curve fitting. Butterworth Scientific, London
Lotka AJ (1956) Elements of physical biology. Dover, New York
Marrot LD (2010) Colorectal cancer network (CRCNet) user documentation for surveillance analytic software: joinpoint. Cancer Care Ontario 1:28
Missikpode C, Peek-Asa C, Young T, Swanton A, Leinenkugel K, Torner J (2015) Trends in non-fatal agricultural injuries requiring trauma care. Inj Epidemiol 2(1):30
Mogos MF, Salemi JL, Spooner KK, McFarlin BL, Salihu HM (2016) Differences in mortality between pregnant and nonpregnant women after cardiopulmonary resuscitation. Obstet Gynecol 128(4):880–888
Mondol H, Mallick UK, Biswas Md HA (2018) Mathematical modeling and predicting the current trends of human population growth in Bangladesh. Adv Model Anal 55(2):62–69
National Institute of Health (2020) Joinpoint regression program (version 4.8.0.1). National Institute of Health, National Cancer Institute. Surveillance Research Program, Division of Cancer Control and Population Sciences
Pearl R, Reed LJ (1920) On the rate of growth of the population of United States since 1790 and its mathematical representation. Proc Natl Acad Sci 6(6):275–288
Preston SH, Heuveline P, Guillot M (2001) Demography. Measuring and modeling population processes. Blackwell Publications, Oxford
Puzo Q, Qin P, Mehlum L (2016) Long-term trends of suicide by choice of method in Norway: a joinpoint regression analysis of data from 1969 to 2012. BMC Public Health 16:255
Qiu D, Katanoda K, Marugame T, Sobue T (2009) A joinpoint regression analysis of long-term trends in cancer mortality in Japan (1958–2004). Int J Cancer 124:443–448
Rea F, Pagan E, Compagnoni MM, Cantarutti A, Pigni P, Bagnardi V, Cprrap G (2017) Joinpoint regression analysis with time-on-study as time-scale. Application to three Italian population-based cohort studies. Epidemiol Biostat Public Health 14(3):e12616
Sharif Ullah M, Mostafa G, Jahan N, Khan AH (2019) Analyzing and projection of future Bangladesh population using logistic growth model. Int J Mod Nonlinear Theory Appl 8:53–61
Spiess AN, Neumeyer N (2010) An evaluation of R^2 as an inadequate measure for nonlinear models in pharmacological and biochemical research: a Monte Carlo approach. BMC Pharmacol 10(1):6–17
Verhlust P-F (1838) A note on the law of population growth. *Correspondence Mathematique et physique Publiee par a.* Quetelet (BrussS) 10:113–117
Yung J, Meyer PS, Ausubel JH (1999) Loglet lab for windows (version 1.1) tutorial. Technol Forecast Soc Chang 61(3):273–295
Zhang NR, Siegmund DO (2007) A modified Bayes information criterion with applications to the analysis of comparative genomic hybridization data. Biom 63:22–32

Chapter 2
Population Forecast, 2011–2041

Abstract This chapter forecasts population of 640 districts for the period 2011–2041. The forecast is based on modelling of population growth in each district during the period 1951–2011 presented in the previous chapter. The parameters of population growth model in the district constitute the basis for forecasting population with some adjustments. Summing the population forecasts of 640 districts reveals that the forecasted population of 640 districts in 2041 is a close approximation of the most likely population of India in 2041 projected by the United Nations.

Keywords India · Districts · Population forecast · Population growth modelling

Introduction

Attempts to project or forecast population of districts in India are rare. Since 1958, the Government of India regularly constitutes an expert committee or a technical group after every decennial population census to project the population of the country for the next 25 years. These projections have, however, been limited to projecting the population of the country and its constituent states and Union Territories. There has never been any attempt to project population of districts. The first attempt to project the population of the districts of the country was made by the United Nations Population Fund in 2009. However, this exercise did not cover all districts of the country as they existed in 2009 but was confined to selected states only (UNFPA 2009). Recently, there has been an attempt to project population of all districts of India (Dhar 2022). Both the attempts are based on the ratio method which expresses the population of a smaller area as a proportion of the population of a larger area. Ratio methods are mostly used where there is a perfect hierarchical structure or where the smaller units are mutually exclusive and exhaustive and can be aggregated to equal the larger unit. One requirement of the ratio method is the availability of an independent projection of the population of the larger unit (Smith et al. 2013). The ratio method may also be termed as the top-down approach in which projections are first made for the state or the Union Territory concerned and then the projected population of

A. R. Chaurasia, *Population and Human Development in Districts of India*,
https://doi.org/10.1007/978-981-95-8169-6_2

the state or the Union Territory is apportioned into constituent districts of the state or the Union Territory. This approach pays no attention to the population dynamics specific to the district which may be different from the population dynamics of the state or the Union Territory to which the district belongs. The implicit assumption behind the ratio method is that the population growth pattern that prevails in a district is the same as the population growth pattern that prevails in the state or the Union Territory to which the district belongs. This assumption is, however, difficult to justify. The population growth pattern in the state or the Union Territory is determined by the population growth pattern in the districts that constitute the state or the Union Territory, but the reverse is not true. It is obvious that any approach to project or forecast the population of a district must consider the population dynamics specific to the district. Population projection or forecast of a district based on the population dynamics specific to the district also constitutes what is known as the bottom-up approach of population projection or forecasting. In this approach, the projection or forecast of the population of the smallest area is produced first and then projection or forecast of the population of larger areas is simply the sum of the projected or forecasted population of the smaller areas that constitute the larger area. The major problem with the use of the bottom-up approach, however, is characterising the population growth dynamics of smaller areas. The data required for characterising the population growth dynamics of smaller areas are generally not available.

Projecting or forecasting district population is essentially an exercise in small area population projection or forecasting. Different methods of small area population projection or forecasting have been proposed. A comprehensive review of these methods is given elsewhere (Wilson et al. 2022). These methods can be grouped in the following eight categories: (1) extrapolative and comparative methods; (2) simplified cohort-component methods; (3) methods of model averaging and combining; (4) methods based on the relationship with socioeconomic and spatial variables; (5) housing led population projections; (6) downscaling and disaggregation approaches; (7) small area microsimulation; and (8) machine learning. Data requirement for different methods are different and is a key consideration in the selection of the method. Simple extrapolative models have historically been used for small area population forecasting (Pritchett 1891; Pearl and Reed 1920; Schmitt and Crosetti 1951; White 1954; Isserman 1977; Openshaw and Van Der Knaap 1983; Smith 1987). The advantage of these methods is that the data requirement for their application is minimal and their accuracy is often at par with more complex methods (Smith et al. 2013). On the other hand, comparative methods project or forecast population of small areas based on a relationship with an independent projection or forecast of the population of a larger area that comprises of the small areas. These methods are also known as ratio methods. Both extrapolative methods and comparative or ratio methods, generally project or forecast total population instead of projecting or forecasting age- and sex-specific population, although the two methods can be combined to project or forecast age- and sex-specific population (Siegel and Swanson 2004).

The terms 'projection' and 'forecast' are interchangeably used in the literature. It is, however, important to distinguish between the two. A forecast is the most likely outcome whereas a projection is essentially a hypothetical scenario (Smith et al. 2013). A projection is the outcome of a particular set of assumptions about the future trend. If this set of assumptions is changed, the projection will change. As such, a projection is always a conditional statement, contingent upon the set of assumptions on which it is based. The projection shows what would be the future if a particular set of assumptions holds true. The projection, however, makes no prediction as to whether the set of assumptions will actually hold or not. If this set of assumptions is changed, the projection will change. A projection, therefore, can never be proved wrong by the future trend. If the projected trend is found to differ from the actual trend, then this may be because one or all the assumptions on which the projection is based could not be realised. An advantage of projection, however, is to have different projections based on different set of assumptions about the future. This helps in future scenario analysis. A projection may also be perceived as an exercise in supervised learning as the assumptions behind the projection are set in advance and are subject to modification or change. Setting the assumptions behind a projection in advance, however, introduces some degree of subjectivity.

A population forecast, on the other hand, is the most likely prediction of the future that is contingent upon the past trend. Contrary to the projection, which is non-judgmental, a forecast is explicitly judgmental. Unlike the projection, which is conditional upon assumptions regarding future, a forecast is conditional upon the past. If the past dynamics is characterised or modelled accurately, then, future can be forecasted by just extrapolating the model describing the past trend. A forecast, therefore, shows the most likely future under the assumption that the past pattern will continue to hold in the future also. This means that an accurate characterisation or modelling of the past trend serves the basis for forecasting at a future date. However, if the past trend does not hold in the future, then the forecast will also differ from the actual trend. A forecast may be perceived as an exercise in unsupervised learning as the past trend cannot be modified or changed. Unlike projection, there is no subjectivity in forecast as the model used for extrapolation depends entirely upon the past data which cannot be changed. The challenge in forecasting, however, is to accurately characterise or model the past trend. If there is no consistency in the past trend, then modelling past trend and hence extrapolation for the future may be quite erratic. A population forecast and a population projection may yield similar results if the past population growth dynamics is modelled correctly and if the assumptions on which the population projection is based are realistic. When the focus is on estimating the most likely future population, forecast is preferred over projection. On the other hand, when the focus is on analysing alternative scenarios of future population growth, projection is preferred as different projections can be made from the same set of data under different set of assumptions.

An approach to model population growth may be conceptualised following the demographic transition theory which describes transition of the population from the stage of high stationarity to the stage of low stationarity. The stage of high stationarity is characterised by high birth rate and high death rate so that the difference between the birth rate and the death rate is small and population growth is slow. As transition advances, death rate starts decreasing primarily because of improvements in the living conditions of the people while there is virtually little change in the birth rate or the decrease in the death rate is faster than the decrease in the birth rate so that the difference between the birth rate and the death rate widens and population growth rate starts accelerating. This acceleration is slow in the beginning but gains momentum with time to reach a stage when the growth in population is the most rapid. Subsequently, the birth rate also starts decreasing while the decrease in the death rate slows down because it has reached very low levels so that the gap between the birth rate and the death rate starts decreasing and population growth starts decelerating, although population continues to grow. At the stage of low stationarity, both birth rate and death rate reach very low levels so that population growth slows down to very low level. During the transition, the rate of population growth follows a bell-shape trajectory which implies that growth of population follows an S-shaped trajectory which is characterised by an acceleration in population growth rate up to a certain point and then deceleration beyond that point. The bell-shaped trajectory of population growth rate also implies that population cannot continue to grow indefinitely and it cannot grow beyond a certain limit. Population increases only when the birth rate is higher than the death rate. Population stops increasing when the birth rate is equal to the death rate. On the other hand, population decreases, instead increases, when the birth rate becomes lower than the death rate.

In addition to the growth in population attributed to the change in the birth rate and the death rate, population growth is also influenced by net migration—the net of in-migrants and out-migrants. If the number of in-migrants is greater than the number of out-migrants, the net migration contributes to the increase in population above the increase determined by the difference between the birth rate and the death rate. If the number of in-migrants is less than the number of out-migrants, net migration contributes to the decrease in the population. When the number of net migrants is small compared to the increase in population due to the difference between the birth rate and the death rate, migration hardly affects the S-shaped pattern of population growth. However, when the number of net migrants constitute a substantial proportion of the increase or decrease in population, the S-shaped pattern of population growth may be different. In such a situation, the pattern of population growth may be described by two S-shaped trajectories of population growth, one reflecting the pattern of population growth attributed to the change in the birth rate and the death rate and the other reflecting the pattern of population growth attributed to net migration.

This chapter is based on the modelling of district population growth during the period 1951–2011. The findings of this modelling exercise are presented in the previous chapter which show that the simple logistic growth model provides very good fit to population growth pattern during the period 1951–2011 in all but only a few districts of the country. This chapter uses population growth model for each district to forecast population growth in the district for the period 2011–2041. This forecasting exercise is important as there has been no population census in India since 2011 and the population census is the only source of information about the size of the district population in India. The Government of India carries out a population projection exercise after every population census but this exercise is limited to projecting population of the country and its states/Union Territories only. There has never been any official attempt to project or forecast district population despite emphasis upon decentralised district-based development planning and programming.

The Method

The parameters of the population growth model for different districts of the country presented in the previous chapter have been used for forecasting district population under the assumption that the population growth pattern in the district will continue, at least, in the immediate future. The most critical aspect of fitting the simple logistic growth model is to estimate the parameter U or the upper asymptote of the model which has substantial impact on the forecast based on the model (Smith et al. 2013). There are different methods proposed for the parametrisation of the simple logistic growth model including: (i) accelerating growth function method, (ii) tangent at the inflexion point method, (iii) segmented regression method; (iv) modified segmented regression method; (v) non-significant difference method; and (vi) non-significant difference by simulation method (Passos et al. 2012). The method used in the previous chapter to fit the simple logistic growth model is the tangent at the inflexion point method which gives the size of the population at the point of interaction of the tangent at the inflexion point with the upper asymptote of the model. If s is the slope of the trajectory of population growth when the population growth is maximal, then population at the interaction of the tangent at inflexion point with the upper asymptote, K, is obtained as

$$K = \frac{4 \times s}{r} \tag{2.1}$$

K is, however, an underestimate of the upper asymptote U of the model of the simple logistic growth model so that using K as the upper asymptote of the simple logistic growth model leads to underestimation of population forecast. It has, however, been shown that the upper asymptote of the logistic growth model, U, can be estimated from K as follows (Passos et al. 2012):

$$U = K \times (1 + \exp(-2)) \tag{2.2}$$

Once the asymptote of the logistic growth model is estimated using Eq. (2.2), forecasting the population of the district is straightforward. The validity of the district population forecast obtained in this manner may be judged by summing the forecasted population of all districts of the country to obtain the population forecast of the country and then comparing the population forecast for the country so obtained with the most likely projection for the population of the country obtained through the application of the industry standard cohort component method of population projection. Such a projection exercise in carried out at an interval of two years by the United Nations for all of its member-countries including India.

District Population Forecast, 2011–2041

In each district, the forecasting exercise has been carried out for male and female population separately as the model describing population growth in the district has been found to be different for the male and the female population of the district. The forecasted male and female population of the district is then added to obtain the forecasted total population of the district. The forecasted male, female and total population for each of the 640 districts of the country as they existed at the 2011 population census are given in the appendix Tables 2.4, 2.5 and 2.6 respectively for the years 2016, 2021, 2026, 2031, 2036 and 2041, although forecasting was done annually. Based on the estimates of the forecasted population presented in the appendix tables, average annual growth rate of male, female and total population for the five-year periods 2011–2016; 2016–2021; 2021–2026; 2026–2031; 2031–2036; and 2036–2041 have been calculated. The district population forecasts presented in this chapter give the first-hand idea of the likely growth of population the 640 districts of the country, as they existed at the 2011 population census, during the 30 years between 2011 and 2041 under the assumption that the pattern of district population growth recorded during the period 1951–2011 will be sustained in the 30 years between 2011 and 2041.

Figure 2.1 presents the inter-district variation in the forecasted average annual growth rate of male and female population in the 640 districts of the country for the periods 2011–2016, 2016–2021, 2021–2026, 2026–2031, 2031–2036 and 2036–2041 while Fig. 2.2 reflects the inter-district variation in average annual population growth rate in the total population. The distribution of districts according to the expected average annual population growth rate in male population is summarised in Table 2.1 while that in female and total populations is summarised in Tables 2.2 and 2.3 respectively. The forecasting exercise suggests that the average annual growth rate in the total population will be less than 0.5 per cent per year during the period 2036–2041 in 627 of the 640 districts of the country. There will be only 5 districts where the average annual growth rate in the total population will be equal to or more than 1 per cent per year during the period 2036–2041 whereas in 2 districts, the average annual growth rate of the total population is expected to be more than 2 per cent per year during the period 2036–2041. The forecasting exercise also suggests that the average annual growth rate of the male population will be less than 0.5 per cent per year in 628 districts during the period 2036–2041 but in 620 districts in the female population. There will be 5 districts where the average annual growth rate of the male population is forecasted to be equal to or more than 1 per cent per year during the period 2036–2041 whereas there will be only 1 district where the average annual growth rate of the male population will be more than 2 per cent per year,

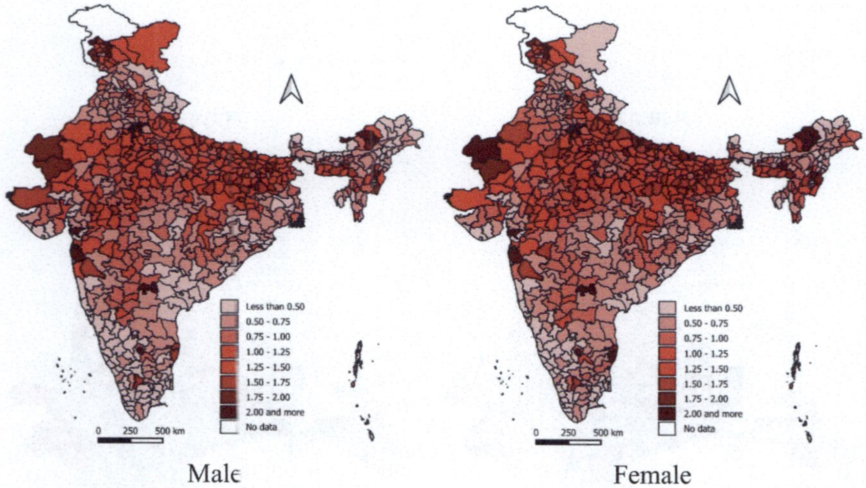

Fig. 2.1 Inter-district variation in the average annual growth rate (per cent) in male and female population in India: 2011–2016; 2016–2021; 2021–2026; 2026–2031; 2031–2036; and 2036–2041. *Source* Author

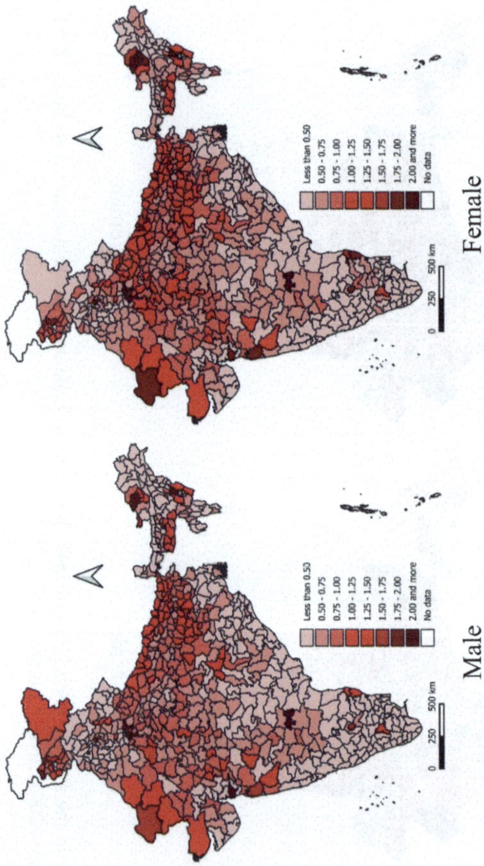

Fig. 2.1 (continued)

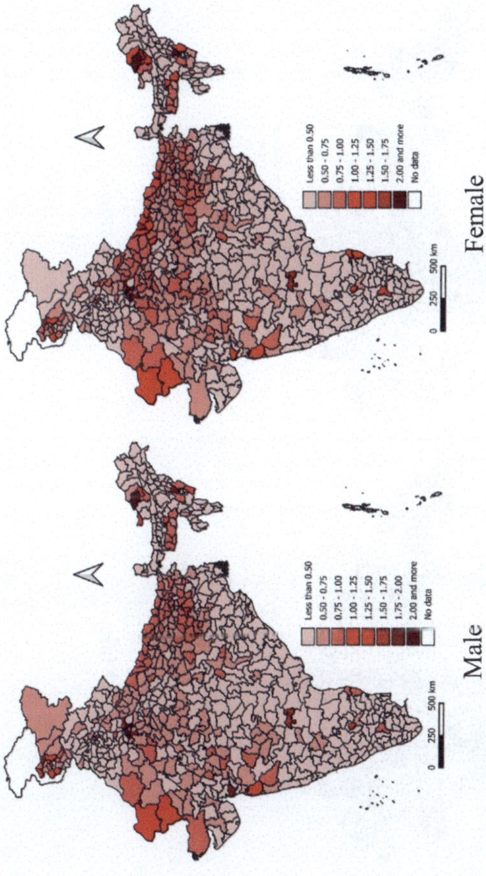

Fig. 2.1 (continued)

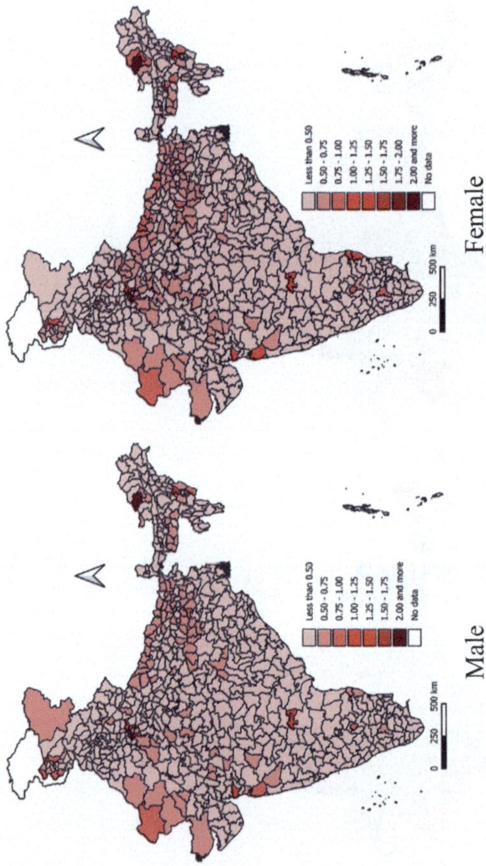

Fig. 2.1 (continued)

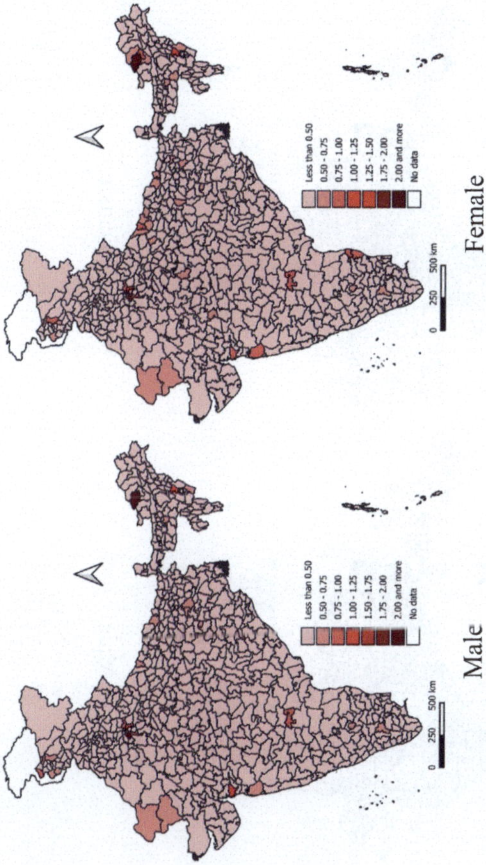

Fig. 2.1 (continued)

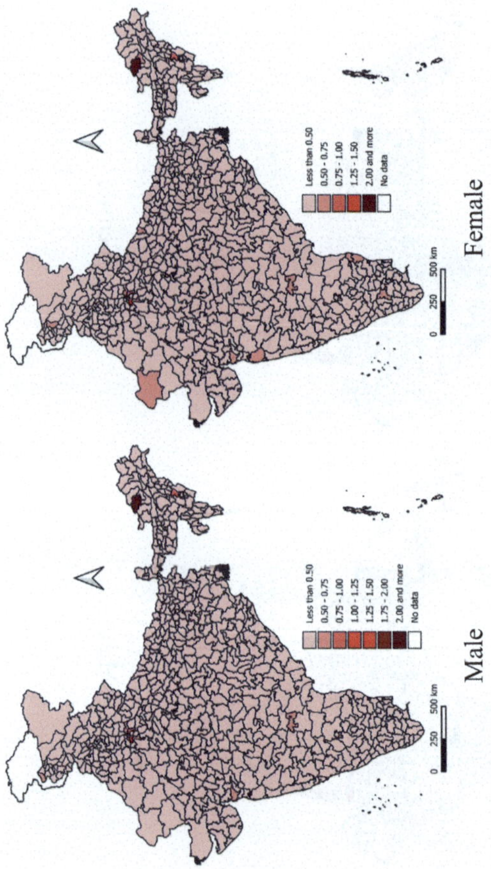

Fig. 2.1 (continued)

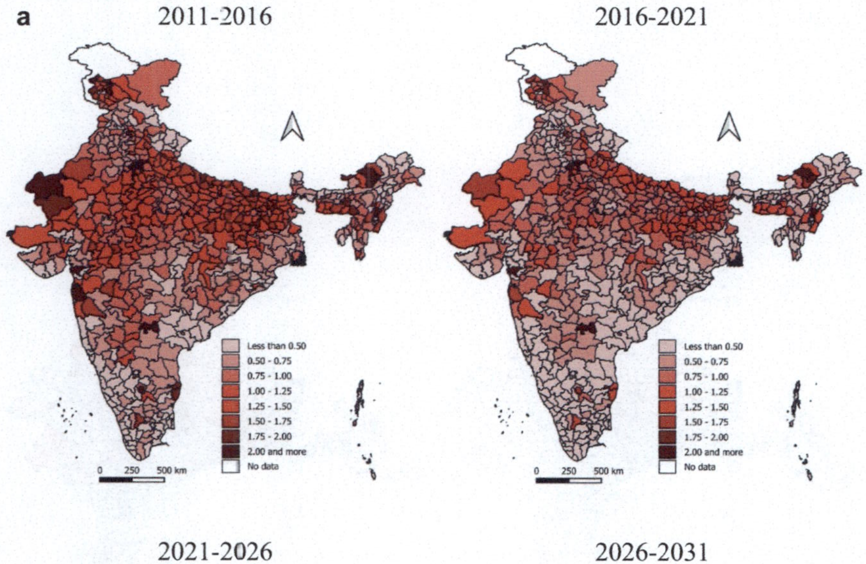

Fig. 2.2 Inter-district variation in average annual growth rate in total population (per cent) in India 2011–2016; 2016–2021; 2021–2026; 2026–2031; 2031–2036; and 2036–2041. *Source* Author

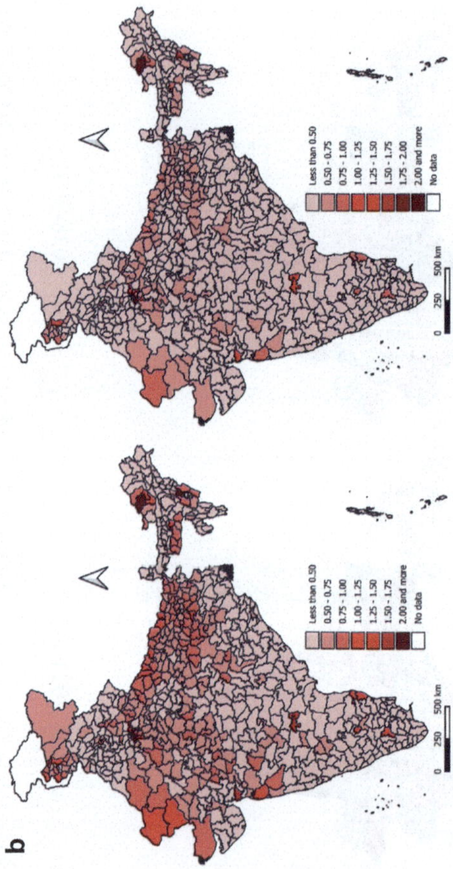

Fig. 2.2 (continued)

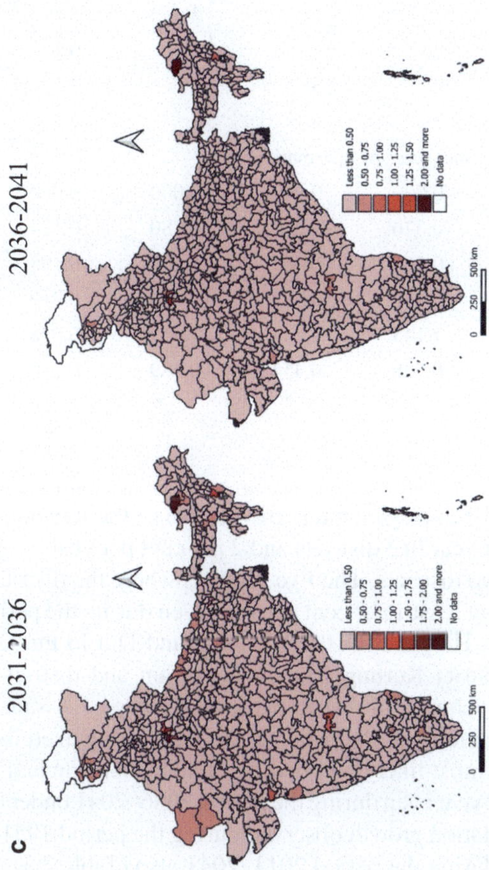

Fig. 2.2 (continued)

Table 2.1 Inter-district variation in the forecast of average annual growth rate in male population

Average annual growth rate (Per cent)	Period					
	2011–2016	2016–2021	2021–2026	2026–2031	2031–2036	2036–2041
	Frequency distribution					
< 0.5	144	269	394	519	612	628
0.5–0.75	162	159	150	99	16	5
0.75–1.0	111	108	70	10	4	2
1.0–1.25	94	66	11	3	3	1
1.25–1.5	70	20	6	2	0	2
1.5–2.0	41	9	3	2	3	1
> = 2.0	18	9	6	5	2	1
	Summary measures of distribution					
Minimum	0.000	0.000	0.000	0.000	0.000	0.000
1st Quartile	0.540	0.376	0.262	0.180	0.124	0.083
Median	0.790	0.567	0.401	0.281	0.195	0.135
3rd Quartile	1.175	0.872	0.632	0.451	0.319	0.222
Maximum	6.768	7.145	7.247	7.069	6.618	5.919
Index of variation	0.608	0.528	0.454	0.389	0.330	0.276

Source Author

On the other hand, the average annual growth rate of the female population will be at least 1 per cent per year in 3 districts and 2 per cent per year or more during the period 2036–2041 in two districts. The five districts where the district population is expected to increase by at least 1 per cent per year even during the period 2036–2041 are: district Gurgaon in Haryana, districts Daman and Diu in the Union Territory of Daman and Diu, district Kurung Kumey in Assam and district Yanam in the Union Territory of Puducherry. In district Kurung Kumey of Assam and in district Yanam of the Union Territory of Puducherry, the forecasting exercise suggests that the population will the most likely be growing at an average annual growth rate of more than 2 per cent per year even during the period 2036–2041 under the assumption that the pattern of population growth observed during the period 1951–2011 in these districts will continue during the period 2011–2041 also (Table 2.3).

The forecasted population of the country may be estimated as the sum of the forecasted population of 640 districts. This exercise suggests that the forecasted population of India will be around 1473 million in 2021; 1544 million in 2031 and 1582 million in 2041. According to the medium or the most likely variant of population projections carried out by the United Nations in the year 2022, the population of the country is projected to 1408 million in 2021, 1526 million in 2031 and 1619 million in 2041 (United Nations 2022). On the other hand, projections prepared by the Government of India after 2011 population census suggest that the population

Table 2.2 Inter-district variation in forecast of average annual growth rate of female population

Average annual growth rate (Per cent)	Period					
	2011–2016	2016–2021	2021–2026	2026–2031	2031–2036	2036–2041
	Frequency distribution					
< 0.5	111	225	363	486	595	620
0.5–0.75	153	167	146	118	27	14
0.75–1.0	118	112	91	18	12	3
1.0–1.25	88	83	19	10	1	0
1.25–1.5	88	25	9	2	2	1
1.5–2.0	59	18	6	3	1	0
> = 2.0	23	10	6	3	2	2
	Summary measures of distribution					
Minimum	0.000	0.000	0.000	0.000	0.000	0.000
1st Quartile	0.592	0.425	0.295	0.206	0.141	0.098
Median	0.872	0.628	0.446	0.315	0.220	0.153
3rd Quartile	1.282	0.946	0.683	0.488	0.346	0.243
Maximum	7.147	7.466	7.454	7.113	6.461	5.557
Index of variation	0.621	0.546	0.474	0.408	0.344	0.283

Source Author

Table 2.3 Summary of inter-district variation in average annual population growth rate in total population

Average annual growth rate (Per cent)	Period					
	2011–2016	2016–2021	2021–2026	2026–2031	2031–2036	2036–2041
	Frequency distribution					
< 0.5	129	250	376	498	599	627
0.5–0.75	150	155	150	112	30	6
0.75–1.0	116	112	80	18	4	2
1.0–1.25	95	77	17	4	2	2
1.25–1.5	80	25	7	1	0	1
1.5–2.0	49	11	3	3	3	0
> = 2.0	21	10	7	4	2	2
	Summary measures of distribution					
Minimum	0.000	0.000	0.000	0.000	0.000	0.000
1st Quartile	0.566	0.400	0.277	0.193	0.134	0.092
Median	0.825	0.592	0.421	0.296	0.206	0.143
3rd Quartile	1.226	0.906	0.659	0.468	0.332	0.233
Maximum	6.962	7.311	7.355	7.092	6.536	5.732
Index of variation	0.612	0.536	0.464	0.398	0.338	0.280

Source Author

of the country is likely to increase to 1479 million by the year 2031. The closeness of the forecasted or the projected population of the country based on two different approaches provides credence to the district population forecasts presented in this chapter.

The present modelling exercise provides forecast of male and female population in every district of the country as male and female population counts have been provided by the Registrar General and Census Commissioner of India. However, the exercise does not provided forecast of district population by age which is a standard requirement for any demographic analysis. There are two possibilities of forecasting district male and female population by age using the forecast of male and female population of the district presented here. The first possibility is to apply the raking method or the method of interactive proportional fitting or the contingency table method (Arriaga et al. 1994) along with the male and female population forecast presented in this monograph. The male and female population forecast for the district may be used as control and the raking method may be applied to obtain male and female population forecast by age. The age-specific male and female population forecast may then be added to obtain age-specific population forecast for the total population of the district.

The second possibility is based on the observation that the age composition of the population also changes with the demographic transition. It is well known that as the demographic transition advances, there is a shift in the age composition of the population. At the early stages of demographic transition, the age-sex population pyramid is triangular with a broad base and a thin top. As the transition advances, the shape of the population pyramid changes which means that the proportionate distribution of the population by age changes. At very advance stage of demographic transition, the triangular shape of the population pyramid is reversed. This suggests that empirical patterns of population age composition at different level of population growth rate can be established, and these empirical patterns can be used to forecast the age composition of male and female population corresponding to the forecast of future male and female population.

Appendix

See Tables 2.4, 2.5 and 2.6.

Table 2.4 Forecast of male population (000) in districts of India, 2011–2041

District code	Name	Male population							
		Census	Forecast						
		2011	2011	2016	2021	2026	2031	2036	2041
Jammu & Kashmir									
001	Kupwara	474.190	528.843	596.737	657.138	707.808	748.263	779.307	802.410
002	Badgam	398.041	442.583	476.233	503.478	524.642	540.559	552.242	560.664
003	Leh (Ladakh)	78.971	88.895	95.235	100.167	103.849	106.512	108.395	109.706
004	Kargil	77.785	87.830	93.150	96.985	99.637	101.417	102.589	103.351
005	Punch	251.899	281.217	307.892	330.529	348.781	362.899	373.473	381.204
006	Rajouri	345.351	383.203	419.483	449.846	474.016	492.500	506.208	516.144
007	Kathua	326.109	362.493	381.056	395.201	405.647	413.182	418.528	422.274
008	Baramula	534.733	601.435	641.853	673.914	698.405	716.580	729.781	739.220
009	Bandipore	207.680	231.973	251.325	267.188	279.635	289.074	296.046	301.096
010	Srinagar	651.124	738.081	783.567	819.201	846.137	865.955	880.247	890.406
011	Ganderbal	158.720	176.150	195.933	213.134	227.290	238.422	246.868	253.103
012	Pulwama	293.064	323.571	347.253	366.235	380.807	391.629	399.468	405.045
013	Shupiyan	136.480	151.508	162.545	171.463	178.378	183.570	187.375	190.115
014	Anantnag	559.767	621.095	690.888	751.376	800.936	839.706	868.952	890.415
015	Kulgam	217.620	249.719	258.277	263.966	267.651	269.999	271.479	272.406
016	Doda	213.641	236.968	254.497	268.596	279.476	287.605	293.534	297.783
017	Ramban	149.132	164.947	178.088	188.770	197.080	203.326	207.900	211.186
018	Kishtwar	120.165	134.126	141.938	148.067	152.712	156.144	158.629	160.405
019	Udhampur	296.784	335.303	357.050	374.007	386.749	396.061	402.728	407.432

(continued)

Table 2.4 (continued)

District code	Name	Male population							
		Census	Forecast						
		2011	2011	2016	2021	2026	2031	2036	2041
020	Reasi	166.461	183.873	194.756	203.154	209.412	213.956	217.193	219.468
021	Jammu	813.821	926.157	967.806	997.949	1019.142	1033.742	1043.658	1050.328
022	Samba	169.124	189.569	198.146	204.604	209.340	212.747	215.163	216.860
Himachal Pradesh									
023	Chamba	261 320	293.974	303.875	311.218	316.540	320.334	323.007	324.874
024	Kangra	750.591	835.098	859.978	878.207	891.276	900.501	906.941	911.402
025	Lahul & Spiti	16.588	20.026	20.031	20.033	20.034	20.035	20.035	20.036
026	Kullu	225.452	255.158	268.108	278.077	285.512	290.929	294.806	297.548
027	Mandi	498.065	553.364	566.854	576.576	583.470	588.305	591.669	593.996
028	Hamirpur	217.070	242.671	249.331	254.244	257.793	260.318	262.095	263.335
029	Una	263.692	292.994	307.177	317.991	325.954	331.668	335.693	338.490
030	Bilaspur	192.764	214.444	220.478	224.859	227.980	230.172	231.698	232.752
031	Solan	308.754	351.079	370.268	384.365	394.342	401.218	405.870	408.979
032	Sirmaur	276.289	308.869	321.317	330.718	337.636	342.628	346.181	348.685
033	Shimla	425.039	475.198	490.137	501.252	509.347	515.153	519.271	522.169
034	Kinnaur	46.249	51.564	52.566	53.280	53.781	54.130	54.371	54.537
Punjab									
035	Gurdaspur	1212.617	1362.645	1403.180	1432.862	1454.143	1469.170	1479.668	1486.947

(continued)

Table 2.4 (continued)

District code	Name	Male population							
		Census	Forecast						
		2011	2011	2016	2021	2026	2031	2036	2041
036	Kapurthala	426.311	482.613	494.857	503.769	510.145	514.651	517.807	520.005
037	Jalandhar	1145.211	1287.219	1329.795	1361.613	1384.843	1401.514	1413.333	1421.638
038	Hoshiarpur	809.057	914.403	934.837	949.740	960.422	967.983	973.287	976.986
039	Shahid	313.291	356.920	360.113	362.052	363.218	363.915	364.330	364.577
040	Fatehgarh Sahib	320.795	359.039	368.989	376.260	381.475	385.167	387.756	389.559
041	Ludhiana	1867.816	2122.927	2216.241	2285.015	2334.355	2369.072	2393.168	2409.733
042	Moga	525.920	589.870	606.660	619.028	627.957	634.312	638.787	641.917
043	Firozpur	1071.637	1196.991	1250.077	1290.512	1320.444	1342.137	1357.618	1368.546
044	Muktsar	475.622	526.725	545.318	559.265	569.491	576.864	582.116	585.826
045	Faridkot	326.671	366.447	378.958	388.192	394.858	399.593	402.918	405.234
046	Bathinda	743.197	821.054	852.527	875.999	893.087	905.311	913.946	919.992
047	Mansa	408.732	454.859	467.576	476.788	483.333	487.921	491.106	493.302
048	Patiala	1002.522	1104.426	1145.101	1175.542	1197.818	1213.853	1225.258	1233.302
049	Amritsar	1318.408	1480.312	1556.506	1615.441	1659.455	1691.473	1714.322	1730.405
050	Tarn Taran	589.369	646.974	673.228	693.310	708.196	718.977	726.654	732.054
051	Rupnagar	357.485	405.593	417.695	426.635	433.101	437.706	440.949	443.216
052	Sahibzada Ajit Singh Nagar	529.253	590.159	643.153	686.481	720.299	745.751	764.386	777.756
053	Sangrur	878.029	983.482	1015.397	1039.061	1056.221	1068.468	1077.108	1083.154

(continued)

Table 2.4 (continued)

District code	Name	Male population							
		Census	Forecast						
		2011	2011	2016	2021	2026	2031	2036	2041
054	Barnala	317.522	353.684	364.689	372.589	378.135	381.967	384.588	386.365
Chandigarh									
055	Chandigarh	580.663	660.573	692.398	715.573	732.045	743.549	751.488	756.921
Uttarakhand									
056	Uttarkashi	168.597	190.228	197.130	202.258	205.981	208.640	210.516	211.828
057	Chamoli	193.991	220.426	224.243	226.787	228.460	229.550	230.255	230.710
058	Rudraprayag	114.589	129.421	131.533	132.901	133.772	134.322	134.668	134.884
059	Tehri Garhwal	297.986	340.584	343.827	345.690	346.749	347.348	347.686	347.876
060	Dehradun	892.199	979.552	1048.656	1103.604	1145.610	1176.764	1199.356	1215.473
061	Garhwal	326.829	375.159	375.691	375.965	376.106	376.179	376.216	376.236
062	Pithoragarh	239.306	270.769	275.137	278.237	280.410	281.918	282.958	283.672
063	Bageshwar	124.326	139.669	141.306	142.421	143.174	143.678	144.014	144.238
064	Almora	291.081	334.597	334.920	335.081	335.161	335.200	335.220	335.230
065	Champawat	131.125	143.730	147.873	150.877	153.017	154.523	155.573	156.301
066	Nainital	493.666	550.763	586.401	614.134	634.936	650.111	660.958	668.599
067	Udham Singh Nagar	858.783	942.284	1005.283	1054.295	1091.176	1118.238	1137.731	1151.586
068	Hardwar	1005.295	1110.801	1188.264	1249.347	1295.641	1329.681	1354.160	1371.483
Haryana									

(continued)

Table 2.4 (continued)

District code	Name	Male population							
		Census	Forecast						
		2011	2011	2016	2021	2026	2031	2036	2041
069	Panchkula	299.679	342.549	361.643	374.499	382.840	388.122	391.416	393.451
070	Ambala	598.703	679.378	710.014	733.234	750.314	762.602	771.303	777.394
071	Yamunanagar	646.718	726.581	761.696	788.270	807.826	821.922	831.931	838.964
072	Kurukshetra	510.976	565.278	585.323	600.222	611.062	618.828	624.330	628.197
073	Kaithal	571.003	640.028	658.898	672.685	682.591	689.624	694.575	698.040
074	Karnal	797.712	887.931	924.148	951.268	971.099	985.349	995.460	1002.571
075	Panipat	646.857	732.611	790.340	836.587	872.139	898.615	917.867	931.627
076	Sonipat	781.299	878.315	909.590	932.284	948.363	959.564	967.274	972.539
077	Jind	713.006	799.630	824.476	842.792	856.045	865.506	872.195	876.893
078	Fatehabad	495.360	547.404	565.079	577.788	586.757	593.004	597.316	600.272
079	Sirsa	682.582	757.868	784.401	803.936	818.042	828.085	835.164	840.119
080	Hisar	931.562	1048.259	1088.935	1119.361	1141.610	1157.610	1168.980	1176.990
081	Bhiwani	866.672	969.798	1006.496	1033.778	1053.599	1067.761	1077.758	1084.757
082	Rohtak	568.479	635.295	656.055	671.470	682.676	690.697	696.375	700.362
083	Jhajjar	514.667	585.014	604.566	619.039	629.528	637.016	642.304	646.009
084	Mahendragarh	486.665	546.277	568.197	584.630	596.647	605.276	611.392	615.686
085	Rewari	474.335	531.375	560.588	583.327	600.458	613.048	622.134	628.604
086	Gurgaon	816.690	909.337	1119.135	1341.720	1562.496	1767.328	1945.921	2093.395

(continued)

Table 2.4 (continued)

District code	Name	Male population							
		Census	Forecast						
		2011	2011	2016	2021	2026	2031	2036	2041
087	Mewat	571 162	641.194	720.615	790.182	847.798	893.342	928.035	953.723
088	Faridabad	966 110	1099.774	1206.114	1290.477	1354.401	1401.172	1434.523	1457.871
089	Palwal	554.497	618.501	661.437	695.079	720.422	738.953	752.209	761.545
National Capital Territory of Delhi									
090	North West	1960.922	2232.809	2379.638	2478.916	2543.480	2584.410	2609.941	2625.704
091	North	475.002	517.006	547.666	573.290	593.017	607.267	617.093	623.653
092	North East	1188.425	1347.616	1414.147	1453.976	1477.058	1490.183	1497.567	1501.695
093	East	907.500	1036.358	1079.902	1108.769	1127.439	1139.321	1146.805	1151.490
094	New Delhi	77.942	77.942	77.942	77.942	77.942	77.942	77.942	77.942
095	Central	307.821	307.821	307.821	307.821	307.821	307.821	307.821	307.821
096	West	1356.240	1539.699	1613.204	1662.582	1694.796	1715.413	1728.447	1736.622
097	South West	1246.046	1418.948	1540.762	1631.094	1694.945	1738.567	1767.679	1786.805
098	South	1467.428	1676.908	1766.042	1827.139	1867.698	1894.051	1910.939	1921.663
Rajasthan									
099	Ganganagar	1043.340	1188.596	1232.623	1265.109	1288.589	1305.305	1317.080	1325.311
100	Hanumangarh	931.184	1038.411	1077.713	1106.821	1127.953	1143.072	1153.777	1161.301
101	Bikaner	1240.801	1402.764	1510.249	1595.881	1661.530	1710.391	1745.962	1771.442
102	Churu	1051.446	1172.980	1233.349	1279.574	1313.954	1338.977	1356.903	1369.599

(continued)

Table 2.4 (continued)

District code	Name	Male population							
		Census	Forecast						
		2011	2011	2016	2021	2026	2031	2036	2041
103	Jhunjhunun	1095.896	1238.236	1281.611	1313.238	1335.798	1351.640	1362.641	1370.223
104	Alwar	1939.026	2170.105	2314.171	2428.135	2515.013	2579.390	2626.099	2659.473
105	Bharatpur	1355.726	1510.986	1589.721	1650.009	1694.788	1727.300	1750.519	1766.906
106	Dhaulpur	653.647	730.447	777.215	814.118	842.199	862.984	878.053	888.818
107	Karauli	783.639	879.397	935.293	979.389	1012.951	1037.807	1055.845	1068.742
108	Sawai Madhopur	704.031	785.641	826.793	858.323	881.748	898.757	910.901	919.469
109	Dausa	857.787	961.567	1031.631	1087.779	1131.006	1163.269	1186.796	1203.664
110	Jaipur	3468.507	3898.193	4191.048	4423.620	4601.289	4733.034	4828.584	4896.775
111	Sikar	1374.990	1547.574	1627.095	1687.893	1733.021	1765.785	1789.194	1805.729
112	Nagaur	1696.325	1902.693	2011.190	2094.884	2157.394	2202.964	2235.603	2258.684
113	Jodhpur	1923.928	2146.161	2297.345	2415.150	2503.482	2567.822	2613.704	2645.931
114	Jaisalmer	361.708	409.414	452.347	488.402	517.211	539.328	555.794	567.772
115	Barmer	1369.022	1526.061	1667.710	1784.967	1877.605	1948.129	2000.318	2038.134
116	Jalor	936.634	1035.167	1105.503	1161.543	1204.537	1236.578	1259.942	1276.710
117	Sirohi	534.231	593.696	633.086	664.431	688.429	706.265	719.233	728.509
118	Pali	1025.422	1149.009	1186.254	1213.763	1233.654	1247.815	1257.787	1264.755
119	Ajmer	1324.085	1479.564	1557.232	1617.584	1662.957	1696.229	1720.184	1737.204
120	Tonk	728.136	814.237	851.990	880.771	902.092	917.556	928.598	936.397

(continued)

Table 2.4 (continued)

District code	Name	Male population							
		Census	Forecast						
		2011	2011	2016	2021	2026	2031	2036	2041
121	Bundi	577.160	646.704	672.964	692.605	706.942	717.223	724.502	729.609
122	Bhilwara	1220.736	1361.176	1433.666	1489.632	1531.399	1561.786	1583.487	1598.780
123	Rajsamand	581.339	642.506	671.052	692.797	708.870	720.489	728.754	734.565
124	Dungarpur	696.532	772.161	820.786	858.475	886.645	907.134	921.744	932.014
125	Banswara	907.754	1009.279	1075.557	1127.504	1166.807	1195.759	1216.668	1231.553
126	Chittaurgarh	783.171	871.055	903.974	928.713	946.867	959.959	969.281	975.860
127	Kota	1021.161	1138.229	1203.122	1252.840	1289.799	1316.661	1335.868	1349.439
128	Baran	633.945	707.546	742.839	769.464	788.939	802.865	812.660	819.473
129	Jhalawar	725.143	805.414	842.693	871.201	892.397	907.829	918.894	926.740
130	Udaipur	1566.801	1739.520	1845.581	1928.802	1991.906	2038.524	2072.305	2096.442
131	Pratapgarh	437.744	486.041	512.021	531.968	546.829	557.655	565.413	570.908
Uttar Pradesh									
132	Saharanpur	1834.106	2049.147	2152.337	2231.522	2290.477	2333.389	2364.114	2385.853
133	Muzaffarnagar	2193.434	2453.300	2562.874	2644.888	2704.529	2746.997	2776.784	2797.458
134	Bijnor	1921.215	2159.190	2264.827	2345.256	2404.727	2447.756	2478.401	2499.984
135	Moradabad	2503.186	2800.154	2986.938	3134.606	3246.987	3330.058	3390.152	3432.948
136	Rampur	1223.889	1367.733	1441.111	1497.763	1540.180	1571.213	1593.537	1609.401
137	Jyotiba Phule Nagar	963.449	1078.004	1146.165	1200.068	1241.219	1271.797	1294.065	1310.044

(continued)

Table 2.4 (continued)

District code	Name	Male population							
		Census	Forecast						
		2011	2011	2016	2021	2026	2031	2036	2041
138	Meerut	1825.743	2049.470	2138.364	2204.876	2253.276	2287.788	2312.043	2328.915
139	Baghpat	700.070	782.395	803.572	819.018	830.060	837.842	843.272	847.033
140	Ghaziabad	2488.834	2784.075	3147.912	3470.403	3740.492	3956.158	4121.903	4245.573
141	Gautam Buddha Nagar	890.214	994.823	1159.108	1315.399	1454.901	1572.507	1666.974	1739.950
142	Bulandshahr	1845.260	2052.992	2126.565	2180.888	2220.004	2247.664	2266.974	2280.335
143	Aligarh	1951.996	2179.254	2307.590	2408.402	2484.816	2541.183	2581.935	2610.971
144	Mahamaya Nagar	836.127	929.568	965.496	992.276	1011.704	1025.524	1035.216	1041.947
145	Mathura	1367.125	1520.649	1610.553	1681.092	1734.450	1773.708	1802.005	1822.101
146	Agra	2364.953	2635.730	2802.154	2934.217	3035.192	3110.226	3164.811	3203.907
147	Firozabad	1332.046	1493.635	1582.536	1652.240	1705.016	1743.929	1772.061	1792.113
148	Mainpuri	993.377	1105.798	1152.296	1187.559	1213.580	1232.398	1245.808	1255.265
149	Budaun	1967.759	2205.206	2328.437	2424.407	2496.472	2549.121	2586.818	2613.421
150	Bareilly	2357.665	2629.995	2788.554	2912.208	3005.148	3073.085	3121.740	3156.078
151	Pilibhit	1072.002	1204.783	1267.626	1316.135	1352.460	1379.050	1398.190	1411.803
152	Shahjahanpur	1606.403	1793.387	1906.462	1996.393	2065.179	2116.238	2153.301	2179.770
153	Kheri	2123.187	2379.450	2560.219	2707.221	2821.791	2908.164	2971.660	3017.480
154	Sitapur	2375.264	2646.848	2819.862	2958.760	3065.992	3146.304	3205.097	3247.423
155	Hardoi	2191.442	2437.495	2564.276	2662.853	2737.011	2791.424	2830.622	2858.488

(continued)

Table 2.4 (continued)

District code	Name	Male population							
		Census	Forecast						
		2011	2011	2016	2021	2026	2031	2036	2041
156	Unnao	1630.087	1823.231	1897.471	1953.046	1993.524	2022.420	2042.754	2056.919
157	Lucknow	2394.476	2677.652	2879.935	3042.462	3167.636	3260.936	3328.795	3377.275
158	Rae Bareli	1752.542	1960.536	2073.706	2162.628	2230.000	2279.647	2315.492	2340.989
159	Farrukhabad	1006.240	1121.616	1175.313	1216.056	1246.014	1267.535	1282.738	1293.352
160	Kannauj	881.776	978.994	1021.847	1054.255	1078.094	1095.274	1107.474	1116.046
161	Etawah	845.856	934.854	969.293	994.909	1013.486	1026.711	1036.004	1042.473
162	Auraiya	740.040	819.313	847.556	868.234	883.019	893.412	900.631	905.603
163	Kanpur Dehat	963.255	1073.809	1112.065	1140.357	1160.774	1175.250	1185.385	1192.419
164	Kanpur Nagar	2459.306	2780.816	2884.242	2961.341	3017.481	3057.664	3086.074	3105.986
165	Jalaun	906.092	1009.401	1048.424	1077.870	1099.549	1115.222	1126.405	1134.309
166	Jhansi	1057.436	1187.435	1232.935	1266.877	1291.609	1309.327	1321.864	1330.658
167	Lalitpur	641.011	716.110	764.998	803.733	833.258	855.106	870.920	882.185
168	Hamirpur	593.537	665.113	684.841	699.325	709.753	717.153	722.353	725.981
169	Mahoba	466.358	520.812	539.954	554.004	564.068	571.151	576.075	579.469
170	Banda	965.876	1067.031	1111.814	1145.504	1170.204	1187.974	1200.584	1209.446
171	Chitrakoot	527.721	586.802	623.496	651.896	673.050	688.361	699.215	706.796
172	Fatehpur	1384.722	1551.064	1609.033	1651.762	1682.454	1704.093	1719.150	1729.531
173	Pratapgarh	1606.085	1797.936	1894.406	1969.473	2025.937	2067.333	2097.116	2118.253

(continued)

Table 2.4 (continued)

District code	Name	Male population							
		Census	Forecast						
		2011	2011	2016	2021	2026	2031	2036	2041
174	Kaushambi	838.485	936.710	998.731	1047.798	1085.111	1112.640	1132.502	1146.602
175	Allahabad	3131.807	3517.001	3714.617	3866.516	3979.465	4061.397	4119.767	4160.819
176	Bara Banki	1707.073	1904.938	2023.731	2118.444	2191.070	2245.110	2284.429	2312.572
177	Faizabad	1259.628	1412.165	1483.990	1539.273	1580.426	1610.304	1631.604	1646.594
178	Ambedkar Nagar	1212.410	1354.978	1424.755	1478.279	1518.006	1546.777	1567.244	1581.621
179	Sultanpur	1914.586	2141.665	2256.697	2345.299	2411.217	2459.005	2493.002	2516.864
180	Bahraich	1843.884	2055.909	2223.389	2361.101	2469.328	2551.400	2611.966	2655.771
181	Shrawasti	593.897	658.706	715.714	763.483	801.557	830.711	852.358	868.065
182	Balrampur	1114.721	1240.611	1330.586	1403.814	1460.862	1503.807	1535.311	1557.985
183	Gonda	1787.146	1997.672	2142.825	2260.756	2352.442	2421.314	2471.722	2507.921
184	Siddharthnagar	1295.095	1441.496	1555.343	1649.977	1724.974	1782.193	1824.595	1855.343
185	Basti	1255.272	1397.612	1466.075	1518.665	1557.754	1586.103	1606.299	1620.504
186	Sant Kabir Nagar	869.656	971.693	1031.561	1078.373	1113.498	1139.045	1157.209	1169.915
187	Mahrajganj	1381.754	1544.835	1653.144	1740.168	1807.330	1857.568	1894.277	1920.642
188	Gorakhpur	2277.777	2545.747	2676.284	2777.224	2852.797	2908.019	2947.659	2975.751
189	Kushinagar	1818.055	2037.297	2186.866	2307.732	2401.360	2471.539	2522.856	2559.705
190	Deoria	1537.436	1738.500	1822.815	1887.680	1936.069	1971.345	1996.631	2014.538
191	Azamgarh	2285.004	2570.278	2717.156	2832.456	2919.872	2984.414	3031.142	3064.495

(continued)

Table 2.4 (continued)

District code	Name	Male population							
		Census	Forecast						
		2011	2011	2016	2021	2026	2031	2036	2041
192	Mau	1114 709	1251.833	1332.436	1396.478	1445.467	1481.866	1508.329	1527.266
193	Ballia	1672.902	1858.434	1954.146	2028.164	2083.546	2123.969	2152.942	2173.439
194	Jaunpur	2220.465	2499.379	2611.211	2695.953	2758.339	2803.298	2835.203	2857.598
195	Ghazipur	1855.075	2065.103	2177.898	2262.792	2324.390	2367.910	2398.082	2418.725
196	Chandauli	1017.905	1142.758	1211.264	1264.896	1305.436	1335.276	1356.811	1372.134
197	Varanasi	1921.357	2156.720	2265.136	2348.375	2410.398	2455.587	2487.976	2510.920
198	Sant Ravidas Nagar	807.099	911.478	944.493	966.897	981.720	991.366	997.574	1001.541
199	Mirzapur	1312.302	1472.200	1559.138	1626.361	1676.565	1713.093	1739.169	1757.531
200	Sonbhadra	971.344	1093.156	1167.213	1225.217	1269.061	1301.320	1324.586	1341.127
201	Etah	947.339	1060.484	1107.971	1144.180	1170.987	1190.404	1204.247	1214.004
202	Kanshiram Nagar	764.165	853.138	890.679	919.304	940.493	955.836	966.769	974.470
Bihar									
203	Pashchim Champaran	2061.110	2256.984	2428.184	2567.363	2675.759	2757.391	2817.325	2860.515
204	Purba Champaran	2681.209	2981.535	3241.908	3459.084	3631.855	3764.198	3862.664	3934.350
205	Sheohar	346.673	389.640	428.283	461.329	488.126	508.936	524.559	535.993
206	Sitamarhi	1803.252	2011.429	2188.009	2334.046	2449.057	2536.193	2600.293	2646.432
207	Madhubani	2329.313	2579.647	2768.482	2921.533	3040.453	3129.857	3195.421	3242.629
208	Supaul	1155.233	1266.987	1342.965	1402.213	1446.934	1479.864	1503.674	1520.663

(continued)

Table 2.4 (continued)

District code	Name	Male population							
		Census	Forecast						
		2011	2011	2016	2021	2026	2031	2036	2041
209	Araria	1463.333	1623.613	1742.296	1835.827	1906.653	1958.683	1996.058	2022.477
210	Kishanganj	866.970	955.354	1022.001	1075.037	1115.681	1145.939	1167.983	1183.790
211	Purnia	1699.370	1877.757	2010.920	2115.331	2193.989	2251.479	2292.575	2321.486
212	Katihar	1600.430	1776.702	1912.880	2022.825	2107.980	2171.837	2218.570	2252.167
213	Madhepura	1047.559	1161.476	1256.487	1333.792	1393.839	1438.820	1471.610	1495.039
214	Saharsa	997.174	1116.539	1206.581	1279.980	1337.221	1380.339	1411.978	1434.750
215	Darbhanga	2059.949	2316.086	2471.702	2595.883	2691.299	2762.501	2814.484	2851.833
216	Muzaffarpur	2527.497	2807.085	3039.363	3230.715	3381.326	3495.670	3580.125	3641.245
217	Gopalganj	1267.666	1422.720	1500.634	1560.830	1605.772	1638.473	1661.825	1678.280
218	Siwan	1675.090	1860.285	1992.439	2099.282	2182.159	2244.404	2290.031	2322.882
219	Saran	2022.821	2246.908	2390.426	2503.561	2589.400	2652.658	2698.282	2730.679
220	Vaishali	1844.535	2056.371	2234.091	2380.542	2495.687	2582.920	2647.165	2693.503
221	Samastipur	2230.003	2476.624	2656.953	2800.666	2910.343	2991.309	3049.626	3090.887
222	Begusarai	1567.660	1742.093	1872.991	1978.771	2060.769	2122.302	2167.365	2199.778
223	Khagaria	883.786	976.572	1058.627	1126.349	1179.742	1220.341	1250.369	1272.127
224	Bhagalpur	1615.663	1784.763	1898.763	1988.721	2057.081	2107.555	2144.035	2169.998
225	Banka	1067.140	1181.911	1264.203	1330.475	1381.761	1420.241	1448.446	1468.768
226	Munger	729.041	803.433	840.619	869.043	890.120	905.399	916.295	923.974

(continued)

Table 2.4 (continued)

District code	Name	Male population							
		Census	Forecast						
		2011	2011	2016	2021	2026	2031	2036	2041
227	Lakhisarai	526.345	580.496	616.446	644.568	665.734	681.208	692.281	700.084
228	Sheikhpura	329.743	366.990	388.073	404.615	417.136	426.357	433.011	437.743
229	Nalanda	1497.060	1660.829	1740.565	1801.866	1847.597	1880.953	1904.886	1921.854
230	Patna	3078.512	3436.273	3661.044	3839.663	3976.420	4078.170	4152.273	4205.406
231	Bhojpur	1430.380	1596.506	1696.109	1774.862	1834.806	1879.124	1911.190	1934.030
232	Buxar	887.977	987.933	1046.592	1092.424	1126.933	1152.198	1170.320	1183.126
233	Kaimur (Bhabua)	847.006	942.320	1016.345	1077.021	1124.647	1160.774	1187.474	1206.831
234	Rohtas	1543.546	1721.172	1814.271	1886.282	1940.203	1979.603	2007.883	2027.921
235	Aurangabad	1318.584	1473.405	1591.824	1689.159	1765.711	1823.852	1866.851	1898.030
236	Gaya	2266.566	2520.206	2707.014	2855.720	2969.075	3052.660	3112.796	3155.300
237	Nawada	1144.668	1286.978	1386.401	1467.319	1530.411	1577.981	1612.946	1638.169
238	Jamui	916.064	1021.126	1099.949	1163.964	1213.772	1251.249	1278.743	1298.541
239	Jehanabad	585.582	653.816	700.819	739.110	769.019	791.625	808.288	820.346
240	Arwal	363.497	405.460	429.996	449.492	464.424	475.542	483.647	489.465
Sikkim									
241	North District	24.730	27.959	28.404	28.662	28.811	28.895	28.943	28.970
242	West District	70.238	79.128	80.803	81.755	82.285	82.576	82.734	82.821
243	South District	76.670	86.483	88.887	90.328	91.172	91.658	91.936	92.094

(continued)

Table 2.4 (continued)

District code	Name	Male population							
		Census	Forecast						
		2011	2011	2016	2021	2026	2031	2036	2041
244	East District	151.432	170.735	176.254	179.585	181.541	182.672	183.319	183.687
Arunachal Pradesh									
245	Tawang	29.151	32.755	36.317	39.176	41.323	42.857	43.914	44.624
246	West Kameng	46.155	52.694	54.020	54.736	55.113	55.309	55.409	55.461
247	East Kameng	38.775	41.683	43.729	45.185	46.180	46.842	47.274	47.554
248	Papum Pare	89.182	100.802	111.053	118.719	124.152	127.856	130.315	131.919
249	Upper Subansiri	41.758	44.488	47.479	49.699	51.271	52.347	53.066	53.540
250	West Siang	58.168	65.852	66.878	67.501	67.875	68.096	68.227	68.304
251	East Siang	50.116	56.193	57.178	57.767	58.115	58.318	58.435	58.504
252	Upper Siang	18.699	21.036	21.322	21.506	21.624	21.698	21.745	21.774
253	Changlang	76.948	86.335	89.428	91.522	92.910	93.816	94.403	94.781
254	Tirap	57.604	64.872	66.341	67.327	67.977	68.400	68.673	68.849
255	Lower Subansiri	41.843	45.657	50.622	54.554	57.454	59.480	60.844	61.739
256	Kurung Kumey	45.318	48.320	67.777	96.882	139.192	198.207	275.950	370.997
257	Dibang Valley	4.414	4.937	4.938	4.938	4.938	4.938	4.938	4.938
258	Lower Dibang Valley	28.053	32.195	32.745	33.061	33.240	33.341	33.398	33.429
259	Lohit	76.221	86.662	89.778	91.852	93.204	94.073	94.626	94.976
260	Anjaw	11.507	12.523	12.667	12.762	12.823	12.862	12.887	12.903

(continued)

Table 2.4 (continued)

District code	Name	Male population							
		Census	Forecast						
		2011	2011	2016	2021	2026	2031	2036	2041
Nagaland									
261	Mon	131.753	157.453	159.009	159.615	159.849	159.939	159.973	159.987
262	Mokokchung	101.092	126.481	127.704	128.313	128.612	128.759	128.831	128.866
263	Zunheboto	71.217	86.663	87.905	88.518	88.816	88.959	89.029	89.062
264	Wokha	84.505	99.686	101.546	102.352	102.696	102.841	102.902	102.928
265	Dimapur	197.394	226.615	241.148	251.269	258.072	262.536	265.419	267.263
266	Phek	83.743	95.994	98.866	100.600	101.623	102.218	102.562	102.759
267	Tuensang	101.933	118.977	122.814	125.013	126.239	126.912	127.279	127.478
268	Longleng	26.502	50.917	50.928	50.930	50.930	50.931	50.931	50.931
269	Kiphire	37.830	53.318	53.332	53.334	53.334	53.334	53.334	53.334
270	Kohima	138.966	157.804	167.049	173.648	178.198	181.263	183.293	184.623
271	Peren	49.714	57.323	58.500	59.082	59.364	59.501	59.566	59.597
Manipur									
272	Senapati	247.323	270.749	316.152	358.570	395.793	426.703	451.216	469.956
273	Tamenglong	72.371	81.297	87.305	92.068	95.696	98.376	100.313	101.689
274	Churachandpur	138.820	155.252	162.506	167.969	171.982	174.877	176.938	178.391
275	Bishnupur	118.782	132.268	135.939	138.593	140.481	141.808	142.734	143.375
276	Thoubal	210.845	234.936	243.715	250.163	254.793	258.063	260.348	261.930

(continued)

Table 2.4 (continued)

District code	Name	Male population							
		Census	Forecast						
		2011	2011	2016	2021	2026	2031	2036	2041
277	Imphal West	255.054	283.021	291.625	297.905	302.404	305.584	307.811	309.359
278	Imphal East	226.094	252.773	260.790	266.651	270.860	273.845	275.942	277.406
279	Ukhrul	94.718	105.355	113.233	119.566	124.462	128.133	130.824	132.763
280	Chandel	74.579	84.997	93.485	100.499	106.025	110.217	113.307	115.537
Mizoram									
281	Mamit	44.828	45.995	46.381	46.597	46.717	46.784	46.820	46.841
282	Kolasib	42.918	47.956	50.466	52.072	53.057	53.644	53.988	54.188
283	Aizawl	199.270	223.416	230.511	234.565	236.823	238.063	238.738	239.104
284	Champhai	63.388	71.672	74.085	75.544	76.398	76.890	77.170	77.328
285	Serchhip	32.851	36.492	38.013	38.963	39.530	39.858	40.045	40.150
286	Lunglei	82.891	92.140	94.542	95.967	96.791	97.260	97.525	97.675
287	Lawngtlai	60.599	67.380	71.377	73.942	75.514	76.451	77.000	77.318
Tripura									
288	Saiha	28.594	32.059	33.140	33.797	34.186	34.414	34.546	34.622
289	West Tripura	879.428	986.586	1016.732	1038.739	1054.529	1065.719	1073.579	1079.065
290	South Tripura	447.544	499.911	510.762	518.267	523.393	526.863	529.199	530.765
291	Dhalai	194.544	214.359	220.448	224.697	227.618	229.605	230.946	231.848
292	North Tripura	352.860	390.173	405.466	416.899	425.247	431.240	435.489	438.474

(continued)

Table 2.4 (continued)

District code	Name	Male population							
		Census	Forecast						
		2011	2011	2016	2021	2026	2031	2036	2041
Meghalaya									
293	West Garo Hills	324.159	361.532	387.498	408.327	424.399	436.433	445.243	451.585
294	East Garo Hills	161.223	180.885	194.291	204.958	213.121	219.183	223.585	226.730
295	South Garo Hills	73.170	80.624	86.677	91.626	95.503	98.440	100.608	102.178
296	West Khasi Hills	193 715	216.300	233.514	247.358	258.043	266.034	271.868	276.054
297	Ribhoi	132.531	148.560	162.998	174.767	183.930	190.813	195.845	199.452
298	East Khasi Hills	410.749	454.608	478.628	497.192	511.110	521.309	528.659	533.892
299	Jaintia Hills	196.285	219.471	238.486	253.901	265.867	274.847	281.417	286.135
Assam									
300	Kokrajhar	452.905	516.042	523.613	528.587	531.823	533.916	535.263	536.128
301	Dhubri	997.348	1099.125	1156.300	1200.398	1233.398	1257.538	1274.905	1287.251
302	Goalpara	513.292	561.585	584.334	601.244	613.516	622.267	628.430	632.732
303	Barpeta	867.004	946.109	980.954	1007.039	1026.115	1039.830	1049.569	1056.425
304	Morigaon	486.651	532.026	554.133	570.767	582.973	591.767	598.018	602.420
305	Nagaon	1439.112	1568.890	1623.592	1663.240	1691.340	1710.942	1724.464	1733.720
306	Sonitpur	983.904	1090.720	1122.679	1145.975	1162.641	1174.406	1182.632	1188.346
307	Lakhimpur	529.674	579.596	595.132	606.378	614.389	620.033	623.976	626.716
308	Dhemaji	351.249	387.455	400.115	409.259	415.754	420.314	423.489	425.687

(continued)

Table 2.4 (continued)

District code	Name	Male population							
		Census	Forecast						
		2011	2011	2016	2021	2026	2031	2036	2041
309	Tinsukia	680.231	749.419	768.193	781.595	791.002	797.526	802.013	805.082
310	Dibrugarh	676.434	748.401	762.475	772.577	779.730	784.747	788.243	790.666
311	Sivasagar	589.216	660.892	675.685	686.431	694.111	699.536	703.336	705.984
312	Jorhat	556.805	619.418	630.576	638.477	643.993	647.803	650.417	652.202
313	Golaghat	543.161	608.434	623.873	635.132	643.211	648.943	652.977	655.799
314	Karbi Anglong	490.167	540.411	556.826	568.779	577.344	583.413	587.678	590.659
315	Dima Hasao	110.802	125.730	130.127	133.336	135.634	137.256	138.391	139.178
316	Cachar	886.284	977.635	1018.404	1049.391	1072.317	1088.944	1100.829	1109.236
317	Karimganj	625.864	689.222	722.517	748.254	767.543	781.669	791.838	799.070
318	Hailakandi	337.890	372.462	389.248	401.964	411.317	418.049	422.820	426.162
319	Bongaigaon	375.818	414.573	430.963	443.247	452.245	458.727	463.340	466.595
320	Chirang	244.860	273.826	276.392	278.051	279.117	279.800	280.235	280.513
321	Kamrup	778.461	856.335	883.214	903.155	917.640	928.000	935.329	940.472
322	Kamrup Metropolitan	647.585	738.348	770.903	794.500	811.206	822.837	830.841	836.304
323	Nalbari	396.006	435.794	443.735	449.341	453.250	455.953	457.811	459.083
324	Baksa	481.330	535.991	544.912	551.112	555.378	558.295	560.279	561.625
325	Darrang	475.273	522.957	546.258	563.763	576.553	585.709	592.168	596.678
326	Udalguri	421.617	471.335	478.291	483.190	486.611	488.988	490.633	491.768

(continued)

Table 2.4 (continued)

District code	Name	Male population							
		Census	Forecast						
		2011	2011	2016	2021	2026	2031	2036	2041
West Bengal									
327	Darjiling	937.259	1052.347	1091.630	1120.934	1142.308	1157.644	1168.519	1176.166
328	Jalpaiguri	1983.064	2216.259	2291.771	2347.902	2388.786	2418.126	2438.959	2453.638
329	Koch Bihar	1451 542	1598.181	1635.571	1662.615	1681.907	1695.533	1705.091	1711.761
330	Uttar Dinajpur	1551.066	1726.037	1825.063	1901.250	1958.061	1999.443	2029.075	2050.034
331	Dakshin Dinajpur	857.199	964.577	996.948	1020.787	1037.989	1050.220	1058.824	1064.833
332	Maldah	2051.541	2283.540	2404.764	2497.911	2567.339	2617.923	2654.169	2679.832
333	Murshidabad	3627.564	4029.723	4231.500	4385.474	4499.568	4582.284	4641.306	4682.945
334	Birbhum	1790.920	1979.598	2046.833	2096.599	2132.624	2158.283	2176.350	2188.968
335	Barddhaman	3966.889	4421.470	4520.012	4590.413	4639.987	4674.540	4698.452	4714.920
336	Nadia	2653.768	2983.671	3080.935	3152.361	3203.835	3240.431	3266.199	3284.219
337	North Twenty	5119.389	5786.341	5979.861	6122.700	6226.133	6299.999	6352.228	6388.900
338	Hugli	2814.653	3154.364	3222.602	3271.910	3307.064	3331.886	3349.295	3361.446
339	Bankura	1838.095	2032.619	2083.471	2120.542	2147.095	2165.876	2179.042	2188.213
340	Puruliya	1496.596	1657.351	1714.352	1757.226	1788.692	1811.373	1827.509	1838.883
341	Haora	2500.819	2806.619	2889.655	2951.181	2995.870	3027.864	3050.532	3066.475
342	Kolkata	2356.766	2764.057	2765.338	2766.044	2766.434	2766.649	2766.768	2766.833
343	South Twenty Four Parganas	4173.778	4657.247	4857.087	5008.205	5119.569	5200.087	5257.502	5298.041

(continued)

Table 2.4 (continued)

District code	Name	Male population							
		Census	Forecast						
		2011	2011	2016	2021	2026	2031	2036	2041
344	Paschim Medinipur	3007.885	3335.360	3426.918	3493.650	3541.423	3575.187	3598.833	3615.287
345	Purba Medinipur	2629.834	2914.802	3004.669	3069.319	3114.867	3146.484	3168.207	3183.027
Jharkhand									
346	Garhwa	683.575	760.129	815.874	860.181	894.012	919.061	937.187	950.089
347	Chatra	533.935	593.696	640.476	678.152	707.236	728.960	744.791	756.120
348	Kodarma	367.222	406.104	446.274	480.214	507.459	528.444	544.097	555.498
349	Giridih	1258.098	1393.297	1512.931	1612.113	1690.650	1750.607	1795.115	1827.474
350	Deoghar	775.022	864.055	932.578	988.833	1033.026	1066.556	1091.330	1109.279
351	Godda	677.927	750.924	799.612	838.802	869.130	891.896	908.597	920.643
352	Sahibganj	589.391	651.379	691.747	723.837	748.418	766.716	780.050	789.617
353	Pakur	452.661	504.002	540.391	570.005	593.097	610.509	623.309	632.544
354	Dhanbad	1405.956	1585.028	1633.649	1669.074	1694.379	1712.200	1724.626	1733.229
355	Bokaro	1072.807	1209.028	1255.085	1289.067	1313.609	1331.060	1343.332	1351.896
356	Lohardaga	232.629	259.410	281.884	300.869	316.137	327.936	336.777	343.250
357	Purbi Singhbhum	1176.902	1307.171	1351.822	1384.574	1408.060	1424.631	1436.188	1444.184
358	Palamu	1006.302	1116.777	1196.406	1260.110	1309.090	1345.612	1372.227	1391.299
359	Latehar	369.666	409.727	439.998	464.374	483.175	497.198	507.398	514.684
360	Hazaribagh	890.881	983.195	1046.115	1095.306	1132.411	1159.651	1179.254	1193.159

(continued)

Table 2.4 (continued)

District code	Name	Male population							
		Census	Forecast						
		2011	2011	2016	2021	2026	2031	2036	2041
361	Ramgarh	494.230	557.051	572.058	582.542	589.746	594.640	597.939	600.152
362	Dumka	668 514	738.031	773.065	800.079	820.205	834.819	845.233	852.556
363	Jamtara	404 830	451.146	479.470	502.244	519.839	533.020	542.666	549.605
364	Ranchi	1494.937	1658.579	1753.743	1827.442	1882.625	1922.912	1951.783	1972.199
365	Khunti	266.335	291.709	306.665	318.589	327.736	334.548	339.512	343.069
366	Gumla	514.390	563.173	597.263	624.846	646.256	662.344	674.140	682.635
367	Simdega	300.309	327.409	339.349	348.501	355.311	360.268	363.818	366.332
368	Pashchimi Singhbhum	749.385	815.643	852.553	881.212	902.702	918.399	929.645	937.593
369	Saraikela-Kharsawan	544.411	596.491	631.353	658.659	679.266	694.384	705.249	712.940
Odisha									
370	Bargarh	749.161	836.525	854.957	868.111	877.345	883.753	888.164	891.183
371	Jharsuguda	296.690	332.739	343.314	351.008	356.478	360.302	362.945	364.757
372	Sambalpur	526.877	587.125	601.710	612.287	619.830	625.146	628.862	631.442
373	Debagarh	158.230	174.596	179.302	182.701	185.113	186.802	187.976	188.785
374	Sundargarh	1061.147	1167.554	1191.828	1208.848	1220.622	1228.693	1234.191	1237.918
375	Kendujhar	906.487	999.032	1029.752	1052.524	1069.068	1080.911	1089.299	1095.197
376	Mayurbhanj	1256.213	1404.759	1454.915	1492.750	1520.549	1540.582	1554.817	1564.832
377	Baleshwar	1185.787	1322.279	1370.066	1405.577	1431.380	1449.826	1462.857	1471.989

(continued)

Table 2.4 (continued)

District code	Name	Male population							
		Census	Forecast						
		2011	2011	2016	2021	2026	2031	2036	2041
378	Bhadrak	760.260	850.865	879.934	901.388	916.882	927.896	935.639	941.039
379	Kendrapara	717.814	795.018	811.821	823.950	832.580	838.657	842.904	845.858
380	Jagatsinghapur	577.865	651.325	664.464	673.692	680.079	684.455	687.431	689.447
381	Cuttack	1352.760	1501.691	1535.932	1560.058	1576.778	1588.232	1596.016	1601.278
382	Jajapur	926.034	1035.082	1067.765	1091.840	1109.205	1121.544	1130.217	1136.267
383	Dhenkanal	612.593	678.720	695.008	706.855	715.335	721.335	725.546	728.485
384	Anugul	655.718	736.033	759.887	777.522	790.292	799.404	805.835	810.341
385	Nayagarh	502.636	554.996	567.178	576.004	582.301	586.744	589.854	592.019
386	Khordha	1167.137	1311.362	1378.928	1430.206	1467.989	1495.227	1514.556	1528.119
387	Puri	865.380	963.251	991.805	1012.792	1027.905	1038.627	1046.155	1051.400
388	Ganjam	1779.218	1987.832	2052.303	2100.127	2134.822	2159.587	2177.060	2189.288
389	Gajapati	282.882	313.887	321.589	327.206	331.223	334.056	336.033	337.405
390	Kandhamal	359.945	403.750	417.927	428.460	436.090	441.514	445.320	447.965
391	Baudh	221.625	245.551	256.501	264.941	271.249	275.858	279.169	281.518
392	Subarnapur	311.312	347.606	358.845	367.228	373.343	377.732	380.844	383.034
393	Balangir	830.097	891.820	921.645	944.085	960.550	972.409	980.837	986.770
394	Nuapada	301.962	332.866	341.819	348.358	353.044	356.357	358.677	360.290
395	Kalahandi	787.101	869.853	908.569	938.327	960.511	976.676	988.259	996.461

(continued)

Table 2.4 (continued)

District code	Name	Male population							
		Census	Forecast						
		2011	2011	2016	2021	2026	2031	2036	2041
396	Rayagada	471.960	520.695	536.540	548.187	556.567	562.503	566.663	569.555
397	Nabarangapur	604.812	672.179	700.284	721.424	736.918	748.061	755.964	761.515
398	Koraput	678.809	756.094	783.887	804.936	820.470	831.717	839.746	845.423
399	Malkangiri	303.624	329.563	339.499	346.764	351.992	355.709	358.331	360.170
Chhattisgarh									
400	Koriya	334.737	376.837	387.854	395.831	401.504	405.487	408.258	410.174
401	Surguja	1193.129	1331.807	1402.725	1457.534	1498.555	1528.524	1550.035	1565.280
402	Jashpur	424.747	466.865	479.069	488.067	494.579	499.229	502.518	504.827
403	Raigarh	750.278	829.284	864.799	891.770	911.660	926.016	936.215	943.381
404	Korba	612.915	688.668	722.091	747.367	765.969	779.387	788.926	795.638
405	Janjgir-Champa	815.717	908.171	978.014	1035.351	1080.116	1113.715	1138.196	1155.650
406	Bilaspur	1351.574	1476.393	1578.010	1659.602	1722.536	1769.590	1803.959	1828.636
407	Kabeerdham	412.058	446.057	477.595	502.723	521.954	536.222	546.566	553.940
408	Rajnandgaon	762.855	839.079	873.563	899.135	917.564	930.573	939.624	945.856
409	Durg	1682.101	1850.406	1911.498	1955.690	1987.004	2008.870	2023.982	2034.352
410	Raipur	2048.186	2229.637	2375.423	2489.894	2576.509	2640.229	2686.142	2718.732
411	Mahasamund	511.967	558.413	575.021	587.225	596.002	602.218	606.572	609.599
412	Dhamtari	397.897	445.681	462.862	475.764	485.209	491.994	496.802	500.178

(continued)

Table 2.4 (continued)

District code	Name	Male population							
		Census	Forecast						
		2011	2011	2016	2021	2026	2031	2036	2041
413	Uttar Bastar Kanker	373.338	412.322	424.612	433.553	439.938	444.438	447.579	449.758
414	Bastar	698.487	774.250	803.967	826.229	842.498	854.175	862.447	868.253
415	Narayanpur	70.104	78.135	82.129	85.201	87.494	89.167	90.369	91.221
416	Dakshin Bastar Dantewada	264.142	289.636	295.529	299.770	302.782	304.899	306.378	307.406
417	Bijapur	128.663	145.793	150.009	152.941	154.940	156.284	157.179	157.771
Madhya Pradesh									
418	Sheopur	361.784	405.767	432.651	453.945	470.220	482.322	491.138	497.466
419	Morena	1068.417	1191.738	1258.534	1309.929	1348.208	1376.031	1395.900	1409.908
420	Bhind	926.843	1027.310	1068.705	1099.646	1122.177	1138.271	1149.612	1157.527
421	Gwalior	1090.327	1211.435	1282.344	1337.523	1379.067	1409.574	1431.567	1447.214
422	Datia	420.157	467.490	487.804	503.023	514.114	522.035	527.611	531.497
423	Shivpuri	919.795	1032.150	1097.852	1149.431	1188.462	1217.184	1237.888	1252.592
424	Tikamgarh	760.355	851.448	898.716	935.338	962.818	982.949	997.438	1007.733
425	Chhatarpur	936.121	1048.107	1106.783	1151.737	1185.017	1209.036	1226.053	1237.952
426	Panna	533.480	595.307	623.129	644.023	659.275	670.179	677.859	683.212
427	Sagar	1256.257	1398.834	1456.707	1499.736	1530.918	1553.095	1568.660	1579.481
428	Damoh	661.873	738.190	768.548	791.421	808.235	820.374	829.022	835.125
429	Satna	1157.495	1293.944	1362.582	1415.064	1453.964	1482.136	1502.199	1516.316

(continued)

Table 2.4 (continued)

District code	Name	Male population							
		Census	Forecast						
		2011	2011	2016	2021	2026	2031	2036	2041
430	Rewa	1225.100	1369.249	1451.759	1516.552	1565.707	1602.031	1628.354	1647.161
431	Umaria	330.674	367.169	389.583	407.158	420.475	430.306	437.425	442.508
432	Neemuch	422.653	473.463	490.694	503.536	512.897	519.613	524.374	527.722
433	Mandsaur	682.851	766.894	795.951	817.636	833.450	844.787	852.817	858.455
434	Ratlam	738.241	818.221	856.614	885.808	907.388	923.008	934.143	941.995
435	Ujjain	1016.289	1136.375	1183.575	1219.233	1245.514	1264.531	1278.111	1287.717
436	Shajapur	780.520	873.495	914.197	945.256	968.296	985.032	997.004	1005.475
437	Dewas	805.359	899.554	943.417	976.353	1000.392	1017.574	1029.673	1038.104
438	Dhar	1112.725	1240.069	1321.525	1385.655	1434.308	1470.191	1496.108	1514.546
439	Indore	1699.527	1891.064	2055.590	2190.403	2296.104	2376.150	2435.189	2477.889
440	Khargone (West Nimar)	953.121	1053.443	1107.584	1149.021	1179.807	1202.179	1218.175	1229.482
441	Barwani	699.340	775.976	833.662	880.202	916.247	943.292	963.103	977.362
442	Rajgarh	790.212	879.463	932.374	973.885	1005.296	1028.421	1045.104	1056.964
443	Vidisha	769.568	853.795	892.879	922.161	943.475	958.665	969.329	976.737
444	Bhopal	1236.130	1387.076	1485.446	1561.264	1617.652	1658.485	1687.487	1707.804
445	Sehore	683.743	761.229	803.307	835.911	860.411	878.401	891.387	900.647
446	Raisen	700.358	783.191	821.498	850.814	872.633	888.538	899.957	908.067
447	Betul	799.236	894.990	924.630	946.461	962.218	973.423	981.309	986.818

(continued)

Table 2.4 (continued)

District code	Name	Male population							
		Census	Forecast						
		2011	2011	2016	2021	2026	2031	2036	2041
448	Harda	294.838	329.318	346.829	360.386	370.562	378.025	383.405	387.237
449	Hoshangabad	648.725	727.822	754.970	774.957	789.326	799.480	806.569	811.476
450	Katni	662.013	733.358	768.879	795.852	815.719	830.027	840.164	847.264
451	Jabalpur	1277.278	1423.529	1469.836	1504.103	1528.973	1546.772	1559.380	1568.249
452	Narsimhapur	568.810	637.014	661.041	678.999	692.107	701.509	708.169	712.845
453	Dindori	351.913	384.572	397.764	407.537	414.623	419.682	423.253	425.754
454	Mandla	525.272	580.602	601.488	617.060	628.392	636.495	642.216	646.220
455	Chhindwara	1064.468	1199.073	1242.128	1274.292	1297.790	1314.680	1326.677	1335.129
456	Seoni	695.879	768.727	797.046	818.046	833.248	844.061	851.656	856.945
457	Balaghat	842.178	933.899	956.364	972.758	984.512	992.833	998.671	1002.740
458	Guna	649.362	725.479	776.495	816.911	847.736	870.573	887.130	898.947
459	Ashoknagar	443.837	492.865	518.346	538.009	552.729	563.502	571.255	576.768
460	Shahdol	540.021	601.030	623.157	639.669	651.716	660.363	666.495	670.808
461	Anuppur	379.114	428.247	441.828	451.355	457.897	462.324	465.290	467.264
462	Sidhi	575.912	644.115	686.654	720.297	745.945	764.957	778.759	788.626
463	Singrauli	613.637	690.982	751.986	801.709	840.524	869.815	891.359	906.908
464	Jhabua	515.023	568.899	608.872	640.657	664.989	683.078	696.235	705.654
465	Alirajpur	362.542	407.336	431.055	449.586	463.591	473.913	481.379	486.707

(continued)

Table 2.4 (continued)

District code	Name	Male population							
		Census	Forecast						
		2011	2011	2016	2021	2026	2031	2036	2041
466	Khandwa (East Nimar)	674.329	743.393	775.008	798.635	815.832	828.108	836.751	842.777
467	Burhanpur	388.504	429.676	446.548	459.124	468.284	474.847	479.492	482.753
Gujarat									
468	Kachchh	1096.737	1203.506	1303.614	1385.957	1450.681	1499.763	1535.981	1562.170
469	Banas Kantha	1610.379	1791.236	1904.050	1992.656	2059.877	2109.545	2145.532	2171.238
470	Patan	694.397	762.626	779.459	791.478	799.926	805.799	809.851	812.632
471	Mahesana	1056.520	1181.175	1208.459	1228.205	1242.274	1252.186	1259.115	1263.932
472	Sabar Kantha	1244.231	1370.328	1415.248	1448.519	1472.660	1489.917	1502.120	1510.686
473	Gandhinagar	723.364	813.005	840.794	861.225	875.935	886.364	893.680	898.772
474	Ahmadabad	3788.051	4205.097	4428.009	4598.498	4725.013	4816.808	4882.329	4928.552
475	Surendranagar	909.917	1010.397	1051.188	1081.913	1104.477	1120.743	1132.310	1140.458
476	Rajkot	1974.445	2186.576	2293.227	2374.769	2435.374	2479.477	2511.082	2533.481
477	Jamnagar	1114.192	1223.305	1256.068	1279.868	1296.866	1308.858	1317.247	1323.079
478	Porbandar	300.209	331.762	337.241	341.128	343.855	345.752	347.064	347.969
479	Junagadh	1404.356	1553.487	1591.226	1618.598	1638.151	1651.968	1661.655	1668.410
480	Amreli	771.049	855.979	868.040	876.485	882.340	886.372	889.134	891.021
481	Bhavnagar	1490.201	1660.875	1727.168	1776.712	1812.847	1838.738	1857.052	1869.891
482	Anand	1087.224	1203.134	1229.973	1249.560	1263.634	1273.635	1280.685	1285.627

(continued)

Table 2.4 (continued)

District code	Name	Male population							
		Census	Forecast						
		2011	2011	2016	2021	2026	2031	2036	2041
483	Kheda	1185.727	1312.775	1342.055	1363.366	1378.646	1389.485	1397.115	1402.457
484	Panch Mahals	1226.961	1358.178	1410.436	1448.809	1476.270	1495.561	1508.938	1518.130
485	Dohad	1068.651	1190.506	1285.189	1361.887	1421.439	1466.173	1498.949	1522.527
486	Vadodara	2153.736	2402.385	2481.196	2539.201	2581.025	2610.736	2631.621	2646.192
487	Narmada	301.086	332.558	341.284	347.602	352.096	355.252	357.449	358.969
488	Bharuch	805.707	898.114	929.583	953.082	970.209	982.471	991.140	997.212
489	The Dangs	113.821	125.871	132.598	137.790	141.677	144.520	146.567	148.021
490	Navsari	678.165	764.125	781.550	794.180	803.197	809.565	814.029	817.141
491	Valsad	887.222	996.526	1063.797	1117.449	1158.661	1189.412	1211.866	1228.002
492	Surat	3402.224	3844.802	4443.772	4992.044	5464.027	5849.334	6150.485	6377.949
493	Tapi	402.188	442.098	451.226	457.778	462.417	465.669	467.934	469.503
Daman & Diu									
494	Diu	25.642	28.282	29.420	30.267	30.884	31.324	31.635	31.852
495	Daman	124.659	140.850	180.183	221.047	259.641	292.948	319.526	339.439
Dadra & Nagar Haveli									
496	Dadra & Nagar Haveli	193.760	218.932	269.318	322.308	374.414	422.373	463.904	498.009
Maharashtra									
497	Nandurbar	833.170	916.808	973.174	1017.562	1051.266	1076.155	1094.159	1106.988

(continued)

Table 2.4 (continued)

District code	Name	Male population							
		Census	Forecast						
		2011	2011	2016	2021	2026	2031	2036	2041
498	Dhule	1054.031	1157.766	1202.760	1236.604	1261.443	1279.347	1292.084	1301.062
499	Jalgaon	2197.365	2444.505	2530.234	2593.630	2639.404	2671.887	2694.656	2710.478
500	Buldana	1337.560	1489.048	1551.410	1598.293	1632.561	1657.098	1674.408	1686.493
501	Akola	932.334	1042.997	1074.162	1096.894	1113.153	1124.621	1132.629	1138.183
502	Washim	620.302	685.857	713.941	735.061	750.521	761.615	769.463	774.959
503	Amravati	1480.768	1660.038	1704.833	1737.168	1760.065	1776.060	1787.129	1794.737
504	Wardha	668.385	756.674	771.729	782.503	790.101	795.404	799.079	801.613
505	Nagpur	2384.975	2687.811	2796.964	2878.936	2939.045	2982.357	3013.172	3034.899
506	Bhandara	605.520	685.139	697.475	706.193	712.269	716.464	719.341	721.306
507	Gondiya	661.554	725.882	736.209	743.491	748.570	752.084	754.504	756.163
508	Gadchiroli	541.328	610.016	631.005	646.512	657.725	665.705	671.322	675.245
509	Chandrapur	1123.834	1281.075	1309.153	1328.520	1341.673	1350.511	1356.408	1360.324
510	Yavatmal	1419.965	1588.242	1639.852	1678.124	1705.912	1725.778	1739.826	1749.682
511	Nanded	1730.075	1947.438	2044.703	2119.054	2174.258	2214.366	2243.048	2263.328
512	Hingoli	606.294	673.616	706.588	731.954	750.892	764.717	774.644	781.689
513	Parbhani	942.870	1046.331	1098.860	1139.251	1169.370	1191.321	1207.052	1218.191
514	Jalna	1011.473	1120.749	1181.045	1227.913	1263.180	1289.074	1307.746	1321.036
515	Aurangabad	1924.469	2151.397	2316.985	2449.653	2551.748	2627.906	2683.404	2723.164

(continued)

Table 2.4 (continued)

District code	Name	Male population							
		Census	Forecast						
		2011	2011	2016	2021	2026	2031	2036	2041
516	Nashik	3157.186	3529.030	3753.214	3928.886	4061.660	4159.299	4229.662	4279.631
517	Thane	5865.078	6650.034	7390.760	7989.772	8448.120	8784.214	9023.044	9188.995
518	Mumbai Suburban	5031.323	5762.245	5906.939	6004.253	6068.775	6111.152	6138.810	6156.790
519	Mumbai	1684.608	2064.263	2064.263	2064.263	2064.263	2064.263	2064.263	2064.263
520	Raigarh	1344.345	1498.973	1592.878	1668.205	1726.354	1769.926	1801.852	1824.864
521	Pune	4924.105	5499.478	5977.276	6368.008	6673.345	6903.589	7072.577	7194.163
522	Ahmadnagar	2342.825	2619.527	2708.866	2774.418	2821.443	2854.633	2877.790	2893.819
523	Bid	1349.106	1493.909	1567.218	1623.225	1664.708	1694.735	1716.109	1731.143
524	Latur	1273.140	1427.373	1503.590	1562.403	1606.375	1638.483	1661.524	1677.853
525	Osmanabad	861.535	961.119	991.972	1014.846	1031.434	1043.269	1051.616	1057.456
526	Solapur	2227.852	2499.842	2591.438	2660.051	2710.192	2746.177	2771.667	2789.557
527	Satara	1510.842	1703.310	1743.952	1772.734	1792.753	1806.503	1815.865	1822.202
528	Ratnagiri	761.121	888.123	896.035	900.886	903.827	905.596	906.655	907.288
529	Sindhudurg	417.332	476.198	481.754	485.666	488.375	490.229	491.487	492.337
530	Kolhapur	1980.658	2224.514	2288.588	2335.582	2369.409	2393.432	2410.330	2422.135
531	Sangli	1435.728	1614.627	1657.489	1688.863	1711.418	1727.426	1738.682	1746.546
Andhra Pradesh									
532	Adilabad	1369.597	1551.785	1601.969	1639.062	1665.938	1685.130	1698.694	1708.210

(continued)

Table 2.4 (continued)

District code	Name	Male population							
		Census	Forecast						
		2011	2011	2016	2021	2026	2031	2036	2041
533	Nizamabad	1250.641	1410.935	1442.318	1464.670	1480.355	1491.249	1498.760	1503.913
534	Karimnagar	1880.800	2132.802	2182.645	2217.110	2240.513	2256.208	2266.648	2273.553
535	Medak	1523.030	1715.912	1781.695	1830.857	1866.654	1892.227	1910.250	1922.830
536	Hyderabad	2018.575	2332.269	2369.250	2391.343	2404.370	2411.992	2416.431	2419.010
537	Rangareddy	2701.008	3038.783	3489.621	3897.899	4245.887	4527.492	4745.959	4909.961
538	Mahbubnagar	2050.386	2296.762	2382.668	2447.061	2494.162	2528.002	2552.002	2568.868
539	Nalgonda	1759.772	1997.717	2037.342	2063.836	2081.262	2092.601	2099.927	2104.639
540	Warangal	1759.281	1994.158	2043.136	2078.389	2103.319	2120.731	2132.786	2141.082
541	Khammam	1390.988	1579.395	1615.109	1640.347	1657.921	1670.032	1678.320	1683.963
542	Srikakulam	1341.738	1513.405	1541.646	1561.571	1575.393	1584.870	1591.316	1595.676
543	Vizianagaram	1161.477	1319.706	1334.505	1344.169	1350.402	1354.392	1356.932	1358.545
544	Visakhapatnam	2138.910	2416.624	2489.810	2542.028	2578.498	2603.590	2620.678	2632.233
545	East Godavari	2569.688	2915.256	2951.020	2973.921	2988.416	2997.524	3003.220	3006.772
546	West Godavari	1964.918	2244.735	2268.713	2283.792	2293.179	2298.986	2302.565	2304.765
547	Krishna	2267.375	2561.419	2616.717	2656.038	2683.556	2702.601	2715.680	2724.615
548	Guntur	2440.521	2735.168	2785.235	2820.877	2845.909	2863.325	2875.363	2883.644
549	Prakasam	1714.764	1922.392	1970.277	2004.960	2029.605	2046.881	2058.875	2067.148
550	Sri Potti Sriramulu Nellore	1492.974	1665.028	1708.029	1739.471	1762.030	1777.995	1789.186	1796.977

(continued)

Table 2.4 (continued)

District code	Name	Male population							
		Census	Forecast						
		2011	2011	2016	2021	2026	2031	2036	2041
551	Y. S. R	1451.777	1625.783	1668.802	1700.280	1722.868	1738.852	1750.050	1757.840
552	Kurnool	2039.227	2283.792	2365.642	2426.887	2471.672	2503.873	2526.745	2542.852
553	Anantapur	2064.495	2319.333	2389.633	2441.680	2479.426	2506.392	2525.451	2538.820
554	Chittoor	2090.204	2339.620	2400.912	2445.371	2477.002	2499.199	2514.624	2525.271
Karnataka									
555	Belgaum	2423.063	2701.157	2787.635	2851.544	2897.761	2930.664	2953.827	2970.004
556	Bagalkot	950.111	1061.671	1102.701	1133.767	1156.703	1173.324	1185.207	1193.621
557	Bijapur	1111.022	1229.698	1286.849	1330.199	1362.056	1384.925	1401.068	1412.328
558	Bidar	870.665	980.617	1020.434	1050.600	1072.876	1089.016	1100.551	1108.713
559	Raichur	964.511	1088.455	1136.723	1172.863	1199.199	1218.013	1231.263	1240.502
560	Koppal	699.926	786.923	827.332	858.615	882.098	899.322	911.742	920.587
561	Gadag	537.147	600.535	613.711	623.251	630.047	634.832	638.175	640.497
562	Dharwad	937.206	1040.055	1069.613	1091.119	1106.485	1117.320	1124.892	1130.148
563	Uttara Kannada	726.256	816.939	828.258	836.219	841.767	845.607	848.254	850.072
564	Haveri	819.128	915.061	936.919	952.410	963.193	970.605	975.657	979.080
565	Bellary	1236.954	1378.939	1449.167	1502.864	1542.691	1571.567	1592.161	1606.676
566	Chitradurga	840.843	945.710	967.494	982.994	993.840	1001.342	1006.489	1010.000
567	Davanagere	986.400	1115.051	1138.601	1155.079	1166.448	1174.216	1179.488	1183.050

(continued)

Table 2.4 (continued)

District code	Name	Male population							
		Census	Forecast						
		2011	2011	2016	2021	2026	2031	2036	2041
568	Shimoga	877.415	980.735	993.047	1001.616	1007.533	1011.596	1014.377	1016.275
569	Udupi	562.131	624.534	632.268	637.658	641.378	643.927	645.666	646.849
570	Chikmagalur	566 622	650.015	655.553	659.343	661.921	663.668	664.849	665.646
571	Tumkur	1350.594	1542.095	1564.816	1580.551	1591.322	1598.637	1603.578	1606.904
572	Bangalore	5022.661	5554.775	6208.467	6766.237	7217.410	7566.850	7828.505	8019.511
573	Mandya	905.085	1031.318	1039.110	1044.141	1047.369	1049.432	1050.748	1051.586
574	Hassan	883.667	1006.720	1019.092	1027.681	1033.591	1037.632	1040.384	1042.253
575	Dakshina Kannada	1034.714	1149.898	1175.262	1193.683	1206.866	1216.201	1222.762	1227.349
576	Kodagu	274.508	310.816	312.893	314.327	315.311	315.985	316.445	316.758
577	Mysore	1511.600	1692.388	1744.646	1783.277	1811.276	1831.280	1845.427	1855.358
578	Chamarajanagar	512.231	580.181	587.058	591.637	594.654	596.629	597.916	598.752
579	Gulbarga	1301.755	1449.943	1514.955	1563.952	1599.811	1625.495	1643.608	1656.243
580	Yadgir	590.329	656.141	694.202	723.936	746.317	762.698	774.442	782.738
581	Kolar	776.396	869.677	895.252	914.258	928.087	937.995	945.015	949.951
582	Chikkaballapura	636.437	715.514	731.892	743.695	752.065	757.932	762.010	764.830
583	Bangalore Rural	509.172	566.424	589.726	607.361	620.346	629.713	636.373	641.058
584	Ramanagara	548.008	619.324	625.680	629.876	632.619	634.403	635.557	636.302
Goa									

(continued)

Table 2.4 (continued)

District code	Name	Male population							
		Census	Forecast						
		2011	2011	2016	2021	2026	2031	2036	2041
585	North Goa	416.677	470.020	479.848	486.816	491.683	495.050	497.361	498.941
586	South Goa	322.463	361.358	368.577	373.686	377.256	379.728	381.429	382.594
Lakshadweep									
587	Lakshadweep	33.123	37.766	38.636	39.228	39.623	39.884	40.056	40.168
Kerala									
588	Kasaragod	628.613	711.493	725.378	735.325	742.365	747.305	750.750	753.142
589	Kannur	1181.446	1341.363	1350.090	1355.619	1359.103	1361.293	1362.666	1363.526
590	Wayanad	401.684	459.673	466.904	471.984	475.522	477.970	479.657	480.817
591	Kozhikode	1470.942	1660.155	1680.779	1695.083	1704.919	1711.643	1716.220	1719.328
592	Malappuram	1960.328	2207.658	2277.812	2329.396	2366.599	2393.058	2411.689	2424.717
593	Palakkad	1359.478	1527.468	1553.664	1572.235	1585.234	1594.251	1600.467	1604.733
594	Thrissur	1480.763	1666.188	1683.405	1695.368	1703.615	1709.271	1713.135	1715.769
595	Ernakulam	1619.557	1816.334	1836.047	1849.883	1859.520	1866.196	1870.804	1873.977
596	Idukki	552.808	640.735	643.697	645.635	646.899	647.723	648.259	648.608
597	Kottayam	968.289	1097.848	1104.286	1108.751	1111.832	1113.950	1115.404	1116.399
598	Alappuzha	1013.142	1153.877	1158.312	1161.244	1163.176	1164.446	1165.279	1165.826
599	Pathanamthitta	561.716	655.249	655.999	656.455	656.733	656.901	657.003	657.065
600	Kollam	1246.968	1428.046	1435.950	1441.208	1444.692	1446.995	1448.513	1449.514

(continued)

Table 2.4 (continued)

District code	Name	Male population							
		Census	Forecast						
		2011	2011	2016	2021	2026	2031	2036	2041
601	Thiruvananthapuram	1581.678	1801.867	1817.527	1828.337	1835.754	1840.822	1844.275	1846.623
Tamil Nadu									
602	Thiruvallur	1876 062	2063.936	2231.066	2367.619	2474.435	2555.180	2614.653	2657.625
603	Chennai	2335.844	2638.976	2674.952	2699.370	2715.788	2726.757	2734.055	2738.896
604	Kancheepuram	2012.958	2186.979	2359.268	2499.219	2607.977	2689.623	2749.337	2792.184
605	Vellore	1961.688	2179.450	2241.287	2287.007	2320.104	2343.699	2360.337	2371.979
606	Tiruvannamalai	1235.889	1360.617	1387.018	1405.856	1419.069	1428.226	1434.520	1438.821
607	Viluppuram	1740.319	1913.283	1966.930	2006.685	2035.485	2056.005	2070.455	2080.547
608	Salem	1781.571	1966.724	2027.177	2071.813	2104.019	2126.871	2142.896	2154.039
609	Namakkal	869.280	962.593	993.770	1016.873	1033.539	1045.330	1053.557	1059.242
610	Erode	1129.868	1254.625	1280.175	1298.594	1311.664	1320.833	1327.215	1331.633
611	The Nilgiris	360.143	422.713	423.760	424.342	424.665	424.844	424.943	424.997
612	Dindigul	1080.938	1190.746	1215.168	1233.005	1245.809	1254.889	1261.270	1265.728
613	Karur	528.184	575.169	591.742	604.040	612.905	619.163	623.515	626.511
614	Tiruchirappalli	1352.284	1489.216	1515.737	1534.328	1547.174	1555.961	1561.932	1565.970
615	Perambalur	282.157	311.863	320.928	327.725	332.706	336.295	338.851	340.655
616	Ariyalur	374.703	416.853	423.181	427.735	430.969	433.244	434.834	435.939
617	Cuddalore	1311.697	1441.193	1469.311	1489.384	1503.494	1513.305	1520.075	1524.722

(continued)

Table 2.4 (continued)

District code	Name	Male population							
		Census	Forecast						
		2011	2011	2016	2021	2026	2031	2036	2041
618	Nagapattinam	798.127	890.969	904.111	913.419	919.923	924.426	927.522	929.643
619	Thiruvarur	626.693	695.742	704.032	709.827	713.834	716.585	718.462	719.740
620	Thanjavur	1182.416	1314.604	1331.563	1343.411	1351.597	1357.208	1361.034	1363.634
621	Pudukkottai	803.188	897.165	915.543	928.749	938.094	944.637	949.184	952.326
622	Sivaganga	668.672	719.997	733.603	743.482	750.542	755.532	759.031	761.470
623	Madurai	1526.475	1664.736	1700.773	1726.557	1744.707	1757.339	1766.060	1772.047
624	Theni	625.683	687.403	697.567	704.794	709.873	713.413	715.865	717.558
625	Virudhunagar	967.709	1074.278	1100.049	1118.977	1132.633	1142.357	1149.217	1154.025
626	Ramanathapuram	682.658	733.713	744.016	751.118	755.960	759.237	761.444	762.926
627	Thoothukkudi	865.021	942.500	955.022	963.873	970.052	974.328	977.269	979.285
628	Tirunelveli	1520.912	1671.108	1716.016	1749.527	1773.975	1791.518	1803.957	1812.704
629	Kanniyakumari	926.345	1017.537	1028.999	1037.076	1042.718	1046.636	1049.346	1051.215
630	Dharmapuri	774.303	854.593	881.020	900.522	914.624	924.669	931.751	936.705
631	Krishnagiri	960.232	1065.658	1118.008	1158.467	1188.788	1210.993	1226.980	1238.349
632	Coimbatore	1729.297	1890.004	1946.374	1988.212	2018.620	2040.383	2055.790	2066.613
633	Tiruppur	1246.159	1386.016	1501.031	1598.059	1675.968	1736.083	1781.056	1813.931
Puducherry									
634	Yanam	27.301	30.597	38.691	47.680	57.032	66.123	74.395	81.483

(continued)

Table 2.4 (continued)

District code	Name	Male population							
		Census	Forecast						
		2011	2011	2016	2021	2026	2031	2036	2041
635	Puducherry	468.258	519.498	551.323	576.219	595.045	608.917	618.945	626.096
636	Mahe	19.143	21.420	21.998	22.424	22.730	22.948	23.100	23.207
637	Karaikal	97.809	108.703	112.788	115.870	118.138	119.776	120.944	121.768
Andaman & Nicobar Islands									
638	Nicobars	20.727	24.968	25.025	25.053	25.067	25.074	25.078	25.079
639	North & Middle Andaman	54.861	64.418	65.446	66.101	66.513	66.772	66.933	67.034
640	South Andaman	127.283	145.559	151.345	155.478	158.370	160.363	161.724	162.646

Source Author

Table 2.5 Forecast of female population (000) in districts of India, 2011–2041

District code	Name	Female population							
		Census	Forecast						
		2011	2011	2016	2021	2026	2031	2031	2041
Jammu & Kashmir									
001	Kupwara	396.164	447.617	489.876	523.640	549.232	567.865	581.039	590.159
002	Badgam	355.704	400.693	429.020	451.596	468.918	481.826	491.235	497.985
003	Leh (Ladakh)	54.516	62.966	64.409	65.303	65.844	66.168	66.361	66.475
004	Kargil	63.017	71.341	75.192	77.925	79.788	81.023	81.826	82.342
005	Punch	224.936	252.087	274.870	294.029	309.361	321.151	329.943	336.349
006	Rajouri	297.064	329.285	357.070	379.979	398.016	411.709	421.819	429.131
007	Kathua	290.326	322.497	338.069	349.992	358.862	365.318	369.944	373.221
008	Baramula	473.306	536.109	570.916	598.229	618.894	634.103	645.070	652.862
009	Bandipore	184.552	207.070	224.712	239.087	250.292	258.727	264.912	269.360
010	Srinagar	585.705	653.374	700.056	737.278	765.769	786.904	802.219	813.131
011	Ganderbal	138.726	154.769	169.876	182.602	192.813	200.693	206.591	210.908
012	Pulwama	267.376	297.498	317.419	333.129	345.050	353.834	360.167	364.663
013	Shupiyan	129.735	145.668	157.961	168.094	176.084	182.167	186.674	189.948
014	Anantnag	518.925	576.793	649.380	714.207	768.836	812.660	846.449	871.714
015	Kulgam	206.863	238.062	246.416	251.784	255.142	257.206	258.462	259.222
016	Doda	196.295	218.291	235.014	248.619	259.238	267.260	273.171	277.448
017	Ramban	134.581	149.102	162.434	173.433	182.072	188.600	193.390	196.830
018	Kishtwar	110.531	123.273	130.987	137.069	141.691	145.105	147.577	149.339
019	Udhampur	258.201	287.820	305.545	319.373	329.784	337.414	342.896	346.779

(continued)

Table 2.5 (continued)

District code	Name	Female population							
		Census	Forecast						
		2011	2011	2016	2021	2026	2031	2031	2041
020	Reasi	148.206	163.776	174.864	183.608	190.238	195.117	198.629	201.117
021	Jammu	716.137	809.444	846.641	874.485	894.764	909.242	919.430	926.527
022	Samba	149.774	166.592	172.768	177.333	180.633	182.981	184.633	185.786
Himachal Pradesh									
023	Chamba	257.760	288.508	299.378	307.448	313.293	317.451	320.370	322.402
024	Kangra	759.484	846.453	869.868	887.108	899.567	908.450	914.723	919.122
025	Lahul & Spiti	14.975	16.834	16.871	16.895	16.911	16.921	16.927	16.931
026	Kullu	212.451	240.045	253.407	263.759	271.506	277.152	281.190	284.037
027	Mandi	501.712	561.208	574.940	584.689	591.497	596.198	599.418	601.611
028	Hamirpur	237.698	263.682	268.813	272.525	275.174	277.046	278.360	279.277
029	Una	257.481	285.991	296.645	304.669	310.559	314.801	317.816	319.937
030	Bilaspur	189.192	210.763	216.147	220.054	222.843	224.809	226.185	227.141
031	Solan	271.566	303.393	319.314	331.630	340.866	347.633	352.506	355.973
032	Sirmaur	253.566	283.581	296.017	305.405	312.302	317.268	320.793	323.270
033	Shimla	388.971	434.010	449.255	460.593	468.830	474.711	478.860	481.762
034	Kinnaur	37.872	42.826	43.501	43.967	44.284	44.498	44.641	44.737
Punjab									
035	Gurdaspur	1085.706	1216.251	1248.361	1271.508	1287.888	1299.328	1307.244	1312.686

(continued)

Table 2.5 (continued)

District code	Name	Female population							
		Census	Forecast						
		2011	2011	2016	2021	2026	2031	2031	2041
036	Kapurthala	388.857	435.056	445.594	453.071	458.284	461.876	464.330	465.997
037	Jalandhar	1048.379	1166.576	1208.161	1239.534	1262.637	1279.349	1291.284	1299.727
038	Hoshiarpur	777.568	875.812	898.965	916.043	928.397	937.211	943.435	947.800
039	Shahid	299.019	337.600	344.094	348.611	351.701	353.789	355.190	356.124
040	Fatehgarh Sahib	279.368	311.170	320.035	326.533	331.211	334.535	336.874	338.511
041	Ludhiana	1630.923	1821.382	1906.591	1968.925	2013.186	2043.955	2065.029	2079.318
042	Moga	469.826	523.969	538.277	548.606	555.916	561.018	564.544	566.965
043	Firozpur	957.437	1064.842	1106.700	1137.602	1159.831	1175.524	1186.457	1194.004
044	Muktsar	426.274	472.459	489.538	502.192	511.350	517.865	522.443	525.632
045	Faridkot	290.837	324.855	335.630	343.554	349.260	353.308	356.148	358.125
046	Bathinda	645.328	715.335	741.486	760.866	774.907	784.916	791.967	796.893
047	Mansa	361.019	402.130	414.404	423.467	430.030	434.716	438.028	440.352
048	Patiala	893.164	979.853	1015.948	1042.859	1062.488	1076.579	1086.578	1093.617
049	Amritsar	1172.248	1309.185	1378.960	1433.671	1475.146	1505.789	1528.002	1543.882
050	Tarn Taran	530.258	582.695	611.084	633.351	650.211	662.639	671.619	678.015
051	Rupnagar	327.142	367.747	379.332	387.882	394.059	398.453	401.544	403.702
052	Sahibzada Ajit Singh Nagar	465.375	518.519	572.395	617.948	654.582	682.873	704.047	719.523
053	Sangrur	777.140	865.027	894.504	916.357	932.199	943.495	951.457	957.023

(continued)

Table 2.5 (continued)

District code	Name	Female population							
		Census	Forecast						
		2011	2011	2016	2021	2026	2031	2031	2041
054	Barnala	278.005	309.916	320.665	328.666	334.491	338.664	341.619	343.695
Chandigarh									
055	Chandigarh	474.787	535.434	564.863	586.597	602.217	613.224	620.873	626.137
Uttarakhand									
056	Uttarkashi	161.489	182.313	189.916	195.389	199.213	201.830	203.595	204.774
057	Chamoli	197.614	224.275	229.677	233.551	236.278	238.175	239.482	240.377
058	Rudraprayag	127.696	144.882	148.673	151.475	153.504	154.955	155.981	156.701
059	Tehri Garhwal	320.945	365.054	372.830	378.414	382.357	385.108	387.011	388.320
060	Dehradun	804.495	892.449	968.978	1031.420	1080.191	1117.002	1144.072	1163.602
061	Garhwal	360.442	413.222	415.443	416.923	417.904	418.552	418.979	419.260
062	Pithoragarh	244.133	275.291	279.789	283.012	285.291	286.888	287.999	288.769
063	Bageshwar	135.572	153.250	155.730	157.485	158.712	159.562	160.148	160.550
064	Almora	331.425	382.587	384.237	385.266	385.904	386.298	386.542	386.692
065	Champawat	128.523	143.728	148.583	152.137	154.687	156.489	157.751	158.627
066	Nainital	460.939	514.502	553.763	585.198	609.401	627.479	640.680	650.161
067	Udham Singh Nagar	790.119	873.671	941.280	994.334	1034.445	1063.923	1085.140	1100.182
068	Hardwar	885.127	984.462	1070.262	1140.853	1196.358	1238.465	1269.548	1292.034
Haryana									

(continued)

Table 2.5 (continued)

District code	Name	Female population							
		Census	Forecast						
		2011	2011	2016	2021	2026	2031	2031	2041
069	Panchkula	261.614	297.340	318.823	334.249	344.849	351.915	356.531	359.505
070	Ambala	529.647	595.646	618.091	634.742	646.795	655.364	661.379	665.564
071	Yamunanagar	567.487	636.036	667.235	690.842	708.211	720.727	729.614	735.856
072	Kurukshetra	453.679	498.595	516.649	529.963	539.565	546.381	551.165	554.496
073	Kaithal	503.301	557.891	576.122	589.517	599.178	606.054	610.901	614.295
074	Karnal	707.612	784.485	820.425	847.639	867.707	882.219	892.565	899.868
075	Panipat	558.580	627.831	681.278	724.528	758.003	783.030	801.261	814.292
076	Sonipat	668.702	744.375	773.257	794.879	810.669	821.990	830.003	835.621
077	Jind	621.146	690.702	713.256	729.775	741.631	750.018	755.891	759.973
078	Fatehabad	446.651	491.399	508.289	520.457	529.053	535.039	539.169	541.999
079	Sirsa	612.607	679.328	704.183	722.485	735.693	745.086	751.698	756.317
080	Hisar	812.369	907.922	946.211	975.071	996.293	1011.616	1022.535	1030.241
081	Bhiwani	767.773	857.653	890.677	915.404	933.498	946.519	955.776	962.300
082	Rohtak	492.725	546.443	563.768	576.301	585.166	591.337	595.585	598.487
083	Jhajjar	443.738	498.118	513.493	524.825	533.010	538.838	542.945	545.818
084	Mahendragarh	435.423	489.617	507.173	520.147	529.516	536.168	540.834	544.081
085	Rewari	425.997	475.126	497.271	513.870	525.940	534.527	540.540	544.705
086	Gurgaon	697.742	771.856	946.493	1130.631	1312.004	1479.056	1623.670	1742.295

(continued)

Table 2.5 (continued)

District code	Name	Female population							
		Census	Forecast						
		2011	2011	2016	2021	2026	2031	2031	2041
087	Mewat	518.101	579.070	655.738	724.213	781.957	828.341	864.166	891.005
088	Faridabad	843.623	956.703	1068.133	1159.386	1230.232	1282.988	1321.066	1347.935
089	Palwal	488.211	543.730	587.781	623.530	651.325	672.221	687.540	698.562
National Capital Territory of Delhi									
090	North West	1695.617	1915.145	2057.305	2153.907	2216.702	2256.353	2280.931	2295.993
091	North	412.976	456.363	489.660	517.377	538.736	554.238	565.006	572.260
092	North East	1053.199	1190.153	1260.537	1303.835	1329.478	1344.324	1352.806	1357.616
093	East	801.846	910.395	958.058	991.157	1013.489	1028.264	1037.913	1044.161
094	New Delhi	64.062	64.062	64.062	64.062	64.062	64.062	64.062	64.062
095	Central	274.499	274.499	274.499	274.499	274.499	274.499	274.499	274.499
096	West	1187.003	1339.508	1419.460	1476.172	1515.067	1541.130	1558.324	1569.551
097	South West	1046.912	1183.919	1306.915	1401.251	1469.655	1517.264	1549.461	1570.814
098	South	1264.501	1428.792	1527.300	1598.013	1646.819	1679.596	1701.208	1715.285
Rajasthan									
099	Ganganagar	925.828	1052.765	1095.236	1126.807	1149.761	1166.182	1177.793	1185.937
100	Hanumangarh	843.508	938.046	974.420	1001.324	1020.823	1034.748	1044.586	1051.486
101	Bikaner	1123.136	1268.931	1371.137	1453.272	1516.676	1564.120	1598.800	1623.718
102	Churu	988.101	1102.254	1157.195	1198.978	1229.860	1252.209	1268.135	1279.360

(continued)

Table 2.5 (continued)

District code	Name	Female population							
		Census	Forecast						
		2011	2011	2016	2021	2026	2031	2031	2041
103	Jhunjhunun	1041.149	1177.556	1222.793	1256.352	1280.674	1298.003	1310.202	1318.716
104	Alwar	1735.153	1940.220	2076.188	2185.010	2268.839	2331.531	2377.385	2410.383
105	Bharatpur	1192.736	1326.552	1407.952	1471.717	1519.972	1555.545	1581.266	1599.603
106	Dhaulpur	552.869	621.412	668.975	706.981	736.060	757.577	773.109	784.121
107	Karauli	674.609	757.689	809.855	851.547	883.621	907.579	925.084	937.668
108	Sawai Madhopur	631.520	706.025	747.442	779.732	804.073	821.961	834.861	844.038
109	Dausa	776.622	870.616	938.888	994.095	1036.858	1068.892	1092.293	1109.075
110	Jaipur	3157.671	3542.774	3826.251	4052.362	4225.372	4353.588	4446.373	4512.370
111	Sikar	1302.343	1468.664	1543.343	1600.476	1642.919	1673.767	1695.832	1711.436
112	Nagaur	1611.418	1805.539	1905.880	1982.297	2038.647	2079.216	2107.924	2127.992
113	Jodhpur	1763.237	1962.159	2113.110	2233.660	2326.152	2394.960	2444.982	2480.741
114	Jaisalmer	308.211	347.731	389.701	425.988	455.692	478.935	496.490	509.398
115	Barmer	1234.729	1372.287	1500.169	1605.873	1689.281	1752.717	1799.625	1833.594
116	Jalor	892.096	989.952	1056.047	1108.390	1148.340	1177.979	1199.510	1214.913
117	Sirohi	502.115	560.093	597.024	626.610	649.425	666.507	679.017	688.030
118	Pali	1012.151	1132.814	1172.340	1201.052	1221.414	1235.610	1245.389	1252.070
119	Ajmer	1258.967	1400.815	1483.314	1547.983	1596.845	1632.749	1658.592	1676.920
120	Tonk	693.190	774.375	814.889	846.257	869.797	887.054	899.487	908.334

(continued)

Table 2.5 (continued)

District code	Name	Female population							
		Census	Forecast						
		2011	2011	2016	2021	2026	2031	2031	2041
121	Bundi	533.746	599.404	628.672	650.969	667.473	679.432	687.964	693.984
122	Bhilwara	1187.787	1323.154	1401.156	1462.300	1508.546	1542.587	1567.146	1584.609
123	Rajsamand	575.258	637.308	664.230	684.660	699.742	710.649	718.421	723.901
124	Dungarpur	692.020	769.167	815.350	851.268	878.269	898.053	912.279	922.369
125	Banswara	889.731	981.287	1041.140	1086.738	1120.291	1144.356	1161.299	1173.074
126	Chittaurgarh	761.167	847.145	882.331	908.858	928.341	942.380	952.356	959.376
127	Kota	929.853	1038.984	1106.971	1160.150	1200.365	1230.007	1251.445	1266.739
128	Baran	588.810	657.375	696.571	727.090	750.027	766.810	778.852	787.371
129	Jhalawar	685.986	759.008	796.696	825.181	846.039	860.958	871.453	878.748
130	Udaipur	1501.519	1668.531	1765.935	1841.445	1898.056	1939.444	1969.147	1990.184
131	Pratapgarh	430.104	478.286	507.068	529.478	546.354	558.747	567.680	574.034
Uttar Pradesh									
132	Saharanpur	1632.276	1822.668	1933.607	2020.736	2086.817	2135.622	2170.963	2196.191
133	Muzaffarnagar	1950.078	2184.571	2304.098	2396.832	2466.483	2517.534	2554.284	2580.398
134	Bijnor	1761.498	1983.421	2104.293	2199.040	2270.764	2323.638	2361.861	2389.101
135	Moradabad	2268.820	2534.721	2742.581	2911.726	3043.359	3142.288	3214.704	3266.698
136	Rampur	1111.530	1240.219	1324.012	1390.477	1441.261	1478.964	1506.362	1525.963
137	Jyotiba Phule Nagar	876.772	984.471	1060.937	1122.968	1171.222	1207.549	1234.229	1253.469

(continued)

Table 2.5 (continued)

District code	Name	Female population							
		Census	Forecast						
		2011	2011	2016	2021	2026	2031	2031	2041
138	Meerut	1617.946	1814.175	1904.910	1974.072	2025.208	2062.171	2088.455	2106.928
139	Baghpat	602.978	675.710	697.957	714.584	726.728	735.449	741.635	745.986
140	Ghaziabad	2192.811	2461.719	2824.983	3154.719	3436.009	3663.541	3839.800	3971.821
141	Gautam Buddha Nagar	757.901	847.140	1000.913	1150.883	1287.480	1404.363	1499.150	1572.729
142	Bulandshahr	1653.911	1837.863	1921.449	1985.125	2032.175	2066.161	2090.312	2107.274
143	Aligarh	1721.893	1918.121	2054.116	2164.270	2249.907	2314.385	2361.770	2395.978
144	Mahamaya Nagar	728.581	811.037	850.936	881.290	903.611	919.618	930.892	938.732
145	Mathura	1180.059	1313.333	1413.926	1496.290	1560.801	1609.591	1645.526	1671.478
146	Agra	2053.844	2284.003	2454.441	2593.160	2701.423	2783.180	2843.400	2886.945
147	Firozabad	1166.110	1302.909	1395.651	1469.962	1527.125	1569.733	1600.753	1622.950
148	Mainpuri	875.152	974.149	1026.461	1066.772	1096.767	1118.512	1133.979	1144.832
149	Budaun	1714.137	1916.957	2063.874	2184.715	2279.660	2351.611	2404.656	2442.977
150	Bareilly	2090.694	2345.494	2529.702	2680.945	2799.865	2890.230	2957.130	3005.709
151	Pilibhit	959.005	1078.535	1144.801	1196.440	1235.272	1263.699	1284.101	1298.537
152	Shahjahanpur	1400.135	1562.605	1687.492	1789.879	1869.964	1930.330	1974.578	2006.349
153	Kheri	1898.056	2132.304	2319.171	2473.250	2594.336	2685.949	2753.293	2801.754
154	Sitapur	2108.728	2345.318	2525.744	2672.266	2785.964	2871.115	2933.208	2977.611
155	Hardoi	1901.403	2113.276	2258.823	2376.718	2468.277	2537.084	2587.527	2623.838

(continued)

Table 2.5 (continued)

District code	Name	Female population							
		Census	Forecast						
		2011	2011	2016	2021	2026	2031	2031	2041
156	Unnao	1478.280	1650.130	1727.777	1786.542	1829.618	1860.459	1882.170	1897.272
157	Lucknow	2195.362	2466.723	2693.131	2881.233	3030.212	3143.797	3227.910	3288.863
158	Rae Bareli	1653 017	1853.000	1965.740	2054.982	2122.963	2173.246	2209.638	2235.560
159	Farrukhabad	878.964	979.659	1042.439	1092.466	1130.818	1159.358	1180.128	1195.002
160	Kannauj	774.840	864.538	910.560	946.087	972.635	991.995	1005.863	1015.671
161	Etawah	735.954	812.848	849.531	877.332	897.774	912.471	922.870	930.144
162	Auraiya	639.505	710.199	739.948	762.152	778.267	789.728	797.761	803.335
163	Kanpur Dehat	832.929	925.976	963.404	991.328	1011.587	1025.989	1036.079	1043.077
164	Kanpur Nagar	2121.462	2407.589	2505.715	2578.122	2630.190	2666.944	2692.548	2710.221
165	Jalaun	783.832	874.770	916.793	949.263	973.591	991.401	1004.220	1013.333
166	Jhansi	941.167	1053.959	1100.050	1134.669	1159.969	1178.091	1190.884	1199.825
167	Lalitpur	580.581	649.701	706.367	753.378	790.599	818.994	840.049	855.333
168	Hamirpur	510.748	571.393	590.278	604.124	614.037	621.013	625.864	629.209
169	Mahoba	409.600	457.509	480.048	497.490	510.573	520.158	527.059	531.967
170	Banda	833.534	922.743	969.312	1005.182	1031.955	1051.471	1065.455	1075.351
171	Chitrakoot	464.009	516.420	552.802	582.150	604.894	621.978	634.513	643.551
172	Fatehpur	1248.011	1397.602	1459.120	1505.971	1540.624	1565.705	1583.574	1596.162
173	Pratapgarh	1603.056	1796.473	1898.508	1978.753	2039.609	2084.498	2116.938	2140.034

(continued)

Table 2.5 (continued)

District code	Name	Female population							
		Census	Forecast						
		2011	2011	2016	2021	2026	2031	2031	2041
174	Kaushambi	761.111	845.160	906.785	956.030	993.659	1021.438	1041.429	1055.553
175	Allahabad	2822.584	3163.868	3376.916	3546.627	3676.736	3773.586	3844.108	3894.638
176	Bara Banki	1553.626	1734.767	1869.766	1979.268	2063.892	2126.876	2172.455	2204.771
177	Faizabad	1211.368	1353.157	1436.098	1499.525	1545.992	1578.974	1601.864	1617.501
178	Ambedkar Nagar	1185.478	1331.686	1414.137	1479.329	1528.960	1565.667	1592.239	1611.175
179	Sultanpur	1882.531	2107.720	2250.052	2364.949	2453.961	2520.749	2569.667	2604.868
180	Bahraich	1643.847	1822.327	2002.338	2156.705	2282.364	2380.388	2454.344	2508.748
181	Shrawasti	523.464	581.654	653.458	719.324	776.104	822.486	858.738	886.105
182	Balrampur	1033.944	1146.732	1260.822	1359.744	1440.889	1504.479	1552.549	1587.907
183	Gonda	1646.773	1838.797	2005.399	2145.916	2258.278	2344.356	2408.157	2454.300
184	Siddharthnagar	1264.202	1405.139	1550.394	1678.017	1783.902	1867.651	1931.424	1978.594
185	Basti	1209.192	1349.427	1438.480	1509.869	1564.759	1605.626	1635.327	1656.538
186	Sant Kabir Nagar	845.527	948.658	1012.665	1062.532	1099.639	1126.315	1145.022	1157.914
187	Mahrajganj	1302.949	1458.443	1577.848	1676.653	1754.764	1814.316	1858.474	1890.548
188	Gorakhpur	2163.118	2425.952	2565.024	2673.906	2756.104	2816.467	2859.903	2890.703
189	Kushinagar	1746.489	1961.174	2110.578	2231.679	2325.614	2396.024	2447.456	2484.321
190	Deoria	1563.510	1760.074	1850.048	1919.093	1970.333	2007.423	2033.787	2052.288
191	Azamgarh	2328.909	2619.159	2772.477	2893.241	2985.020	3052.894	3102.082	3137.209

(continued)

Table 2.5 (continued)

District code	Name	Female population							
		Census	Forecast						
		2011	2011	2016	2021	2026	2031	2031	2041
192	Mau	1091.259	1223.531	1300.383	1361.275	1407.751	1442.220	1467.245	1485.133
193	Ballia	1566.872	1749.850	1839.218	1908.928	1961.500	2000.147	2028.028	2047.869
194	Jaunpur	2273.739	2558.707	2688.203	2787.856	2862.084	2916.044	2954.577	2981.747
195	Ghazipur	1765.193	1980.169	2096.283	2187.718	2257.162	2308.468	2345.606	2372.092
196	Chandauli	934.851	1051.055	1118.757	1172.511	1213.618	1244.160	1266.370	1282.269
197	Varanasi	1754.984	1977.300	2094.183	2186.012	2255.714	2307.246	2344.611	2371.322
198	Sant Ravidas Nagar	771.114	870.127	921.066	959.920	988.539	1009.081	1023.553	1033.616
199	Mirzapur	1184.568	1330.234	1421.488	1494.601	1550.929	1593.027	1623.783	1645.881
200	Sonbhadra	891.215	1004.701	1088.583	1156.794	1209.985	1250.123	1279.666	1301.016
201	Etah	827.141	925.928	980.889	1024.466	1057.746	1082.440	1100.375	1113.200
202	Kanshiram Nagar	672.554	750.359	796.674	833.661	862.070	883.247	898.682	909.749
Bihar									
203	Pashchim Champaran	1873.932	2077.761	2255.625	2403.221	2519.993	2608.941	2674.756	2722.417
204	Purba Champaran	2418.162	2693.412	2946.687	3159.304	3328.470	3457.413	3552.519	3620.980
205	Sheohar	309.573	345.321	385.435	421.514	452.056	476.613	495.552	509.696
206	Sitamarhi	1620.322	1808.369	1992.314	2150.211	2278.223	2377.287	2451.223	2504.927
207	Madhubani	2158.066	2406.192	2586.478	2734.144	2849.774	2937.170	3001.475	3047.862
208	Supaul	1073.793	1176.081	1250.860	1309.502	1353.914	1386.669	1410.357	1427.245

(continued)

Table 2.5 (continued)

District code	Name	Female population							
		Census	Forecast						
		2011	2011	2016	2021	2026	2031	2031	2041
209	Araria	1348.236	1477.892	1586.299	1672.672	1738.860	1788.083	1823.879	1849.490
210	Kishanganj	823.430	907.098	974.108	1027.470	1068.316	1098.646	1120.662	1136.382
211	Purnia	1565.249	1735.718	1867.086	1972.498	2053.755	2114.474	2158.800	2190.611
212	Katihar	1470.599	1628.333	1747.772	1842.739	1915.216	1968.816	2007.542	2035.052
213	Madhepura	954.203	1056.261	1144.823	1217.333	1273.962	1316.578	1347.762	1370.116
214	Saharsa	903.487	1004.664	1086.752	1153.937	1206.466	1246.087	1275.168	1296.091
215	Darbhanga	1877.436	2110.870	2257.095	2375.331	2467.085	2536.042	2586.630	2623.087
216	Muzaffarpur	2273.565	2505.254	2694.443	2848.332	2968.018	3057.896	3123.629	3170.779
217	Gopalganj	1294.346	1444.713	1544.383	1625.216	1688.128	1735.539	1770.410	1795.600
218	Siwan	1655.374	1843.171	1954.654	2042.300	2108.543	2157.134	2191.999	2216.621
219	Saran	1929.041	2143.865	2286.751	2401.440	2489.676	2555.374	2603.106	2637.171
220	Vaishali	1650.486	1821.345	1958.153	2069.997	2157.554	2223.801	2272.643	2307.970
221	Samastipur	2031.563	2256.671	2413.051	2538.050	2633.818	2704.833	2756.228	2792.774
222	Begusarai	1402.881	1564.632	1686.488	1785.727	1863.001	1921.095	1963.628	1994.167
223	Khagaria	783.100	864.386	943.116	1009.410	1062.560	1103.520	1134.135	1156.498
224	Bhagalpur	1422.103	1576.630	1681.689	1764.710	1827.709	1874.058	1907.388	1930.963
225	Banka	967.623	1074.963	1152.987	1216.304	1265.546	1302.590	1329.766	1349.339
226	Munger	638.724	706.224	743.053	771.728	793.294	809.094	820.449	828.498

(continued)

Table 2.5 (continued)

District code	Name	Female population							
		Census	Forecast						
		2011	2011	2016	2021	2026	2031	2031	2041
227	Lakhisarai	474.567	520.139	554.470	582.066	603.357	619.271	630.885	639.215
228	Sheikhpura	306.599	340.461	362.636	380.516	394.359	404.742	412.347	417.821
229	Nalanda	1380.593	1521.627	1597.881	1656.367	1699.767	1731.188	1753.533	1769.223
230	Patna	2759.953	3067.178	3300.142	3489.248	3636.228	3746.654	3827.521	3885.639
231	Bhojpur	1298.027	1438.189	1523.523	1590.687	1641.558	1678.981	1705.924	1725.021
232	Buxar	818.375	906.932	971.545	1024.236	1065.408	1096.517	1119.433	1135.999
233	Kaimur (Bhabua)	779.378	865.701	944.272	1010.556	1063.803	1104.919	1135.709	1158.242
234	Rohtas	1416.372	1577.906	1673.778	1749.207	1806.417	1848.616	1879.104	1900.806
235	Aurangabad	1221.389	1361.796	1474.714	1568.366	1642.526	1699.128	1741.127	1771.642
236	Gaya	2124.852	2362.566	2549.294	2702.202	2821.973	2912.562	2979.283	3027.469
237	Nawada	1074.478	1198.599	1285.458	1355.767	1410.322	1451.282	1481.279	1502.851
238	Jamui	844.341	940.664	1020.705	1087.592	1140.925	1181.882	1212.437	1234.741
239	Jehanabad	539.731	603.907	645.689	679.467	705.653	725.303	739.690	750.034
240	Arwal	337.346	375.573	397.599	415.063	428.405	438.313	445.517	450.673
Sikkim									
241	North District	18.979	21.140	21.403	21.556	21.643	21.693	21.722	21.738
242	West District	66.197	74.721	76.773	77.999	78.712	79.120	79.352	79.483
243	South District	70.180	79.913	82.276	83.650	84.427	84.860	85.099	85.230

(continued)

Table 2.5 (continued)

District code	Name	Female population							
		Census	Forecast						
		2011	2011	2016	2021	2026	2031	2031	2041
244	East District	132.151	148.332	153.768	157.119	159.124	160.303	160.990	161.386
Arunachal Pradesh									
245	Tawang	20.826	23.395	24.994	26.195	27.053	27.647	28.046	28.312
246	West Kameng	37.792	42.384	44.104	45.288	46.081	46.603	46.942	47.160
247	East Kameng	39.915	43.458	47.364	50.465	52.776	54.417	55.544	56.300
248	Papum Pare	87.391	98.727	113.087	124.748	133.582	139.930	144.319	147.274
249	Upper Subansiri	41.690	45.326	50.521	54.787	58.050	60.413	62.057	63.168
250	West Siang	54.106	61.185	62.730	63.770	64.457	64.906	65.196	65.383
251	East Siang	49.098	55.079	56.994	58.219	58.982	59.448	59.730	59.900
252	Upper Siang	16.621	18.205	18.469	18.638	18.746	18.815	18.858	18.884
253	Changlang	71.278	80.200	83.904	86.450	88.152	89.269	89.993	90.459
254	Tirap	54.371	60.833	62.781	64.101	64.973	65.539	65.902	66.133
255	Lower Subansiri	41.187	45.327	52.073	58.024	62.855	66.521	69.161	70.992
256	Kurung Kumey	46.758	50.384	72.026	104.618	151.872	216.735	299.390	395.289
257	Dibang Valley	3.590	3.786	3.805	3.817	3.824	3.829	3.832	3.833
258	Lower Dibang Valley	26.027	29.726	30.538	30.989	31.235	31.367	31.438	31.476
259	Lohit	69.505	78.782	82.932	85.802	87.728	88.995	89.817	90.346
260	Anjaw	9.660	10.682	11.153	11.502	11.749	11.917	12.030	12.104

(continued)

Table 2.5 (continued)

District code	Name	Female population							
		Census	Forecast						
		2011	2011	2016	2021	2026	2031	2031	2041
Nagaland									
261	Mon	118.507	140.242	141.498	141.960	142.128	142.189	142.211	142.219
262	Mokokchung	93.530	116.182	116.820	117.085	117.195	117.241	117.259	117.267
263	Zunheboto	69.540	83.558	84.727	85.294	85.565	85.693	85.754	85.783
264	Wokha	81.838	95.950	98.540	99.781	100.360	100.628	100.751	100.807
265	Dimapur	181.417	206.969	223.085	234.220	241.582	246.307	249.283	251.135
266	Phek	79.675	91.099	94.530	96.614	97.844	98.557	98.966	99.200
267	Tuensang	94.663	109.796	113.501	115.565	116.681	117.275	117.588	117.753
268	Longleng	23.982	45.508	45.520	45.522	45.522	45.522	45.522	45.522
269	Kiphire	36.174	49.152	49.168	49.170	49.171	49.171	49.171	49.171
270	Kohima	129.022	146.086	155.209	161.491	165.651	168.333	170.033	171.099
271	Peren	45.505	53.354	54.386	54.883	55.118	55.229	55.281	55.305
Manipur									
272	Senapati	231.825	257.175	302.230	345.006	383.090	415.106	440.745	460.492
273	Tamenglong	68.280	76.272	82.582	87.742	91.777	94.824	97.064	98.679
274	Churachandpur	135.323	150.581	158.801	165.053	169.668	173.000	175.368	177.031
275	Bishnupur	118.617	131.982	135.966	138.885	140.986	142.480	143.532	144.268
276	Thoubal	211.323	235.548	245.504	252.966	258.420	262.334	265.106	267.050

(continued)

Table 2.5 (continued)

District code	Name	Female population							
		Census	Forecast						
		2011	2011	2016	2021	2026	2031	2031	2041
277	Imphal West	262.938	290.040	300.462	308.137	313.662	317.575	320.316	322.219
278	Imphal East	230.019	254.907	264.162	270.966	275.867	279.343	281.783	283.482
279	Ukhrul	89.280	99.479	109.183	117.398	124.013	129.130	132.966	135.775
280	Chandel	69.603	79.899	86.099	90.562	93.620	95.644	96.954	97.790
Mizoram									
281	Mamit	41.536	42.763	43.366	43.712	43.909	44.019	44.081	44.116
282	Kolasib	41.037	45.874	49.077	51.186	52.494	53.276	53.733	53.997
283	Aizawl	201.039	225.022	236.807	244.254	248.783	251.474	253.051	253.967
284	Champhai	62.357	69.873	72.807	74.655	75.773	76.435	76.821	77.044
285	Serchhip	32.086	35.997	37.544	38.491	39.041	39.351	39.523	39.618
286	Lunglei	78.537	87.036	89.598	91.127	92.015	92.522	92.810	92.972
287	Lawngtlai	57.295	63.869	68.813	72.096	74.151	75.390	76.120	76.544
Tripura									
288	Saiha	27.980	31.474	32.879	33.757	34.289	34.603	34.788	34.895
289	West Tripura	846.311	945.373	973.560	993.871	1008.254	1018.316	1025.295	1030.107
290	South Tripura	428.457	477.950	489.332	497.409	503.068	506.996	509.706	511.567
291	Dhalai	183.686	201.796	207.677	211.788	214.616	216.542	217.844	218.719
292	North Tripura	341.087	375.169	391.059	403.025	411.817	418.163	422.682	425.871

(continued)

Table 2.5 (continued)

District code	Name	Female population							
		Census	Forecast						
		2011	2011	2016	2021	2026	2031	2031	2041
Meghalaya									
293	West Garo Hills	319.132	355.211	382.210	404.023	420.953	433.688	443.045	449.800
294	East Garo Hills	156.694	175.347	188.896	199.752	208.106	214.339	218.881	222.135
295	South Garo Hills	69.164	76.282	81.496	85.662	88.856	91.229	92.952	94.182
296	West Khasi Hills	189.746	212.128	230.781	246.080	258.095	267.211	273.948	278.832
297	Ribhoi	126.309	141.589	156.493	168.936	178.849	186.455	192.123	196.257
298	East Khasi Hills	415.173	458.113	488.399	512.157	530.103	543.275	552.741	559.440
299	Jaintia Hills	198.839	223.024	244.776	262.846	277.169	288.106	296.221	302.115
Assam									
300	Kokrajhar	434.237	493.651	501.918	507.441	511.092	513.488	515.054	516.074
301	Dhubri	951.410	1047.288	1101.442	1143.050	1174.089	1196.737	1212.997	1224.536
302	Goalpara	494.891	541.702	565.586	583.538	596.693	606.153	612.865	617.583
303	Barpeta	826.618	901.542	936.662	962.781	981.733	995.242	1004.747	1011.376
304	Morigaon	470.772	513.977	537.941	556.161	569.641	579.417	586.402	591.342
305	Nagaon	1384.656	1516.395	1585.534	1637.690	1675.997	1703.583	1723.168	1736.932
306	Sonitpur	940.206	1041.610	1079.523	1107.593	1127.921	1142.409	1152.617	1159.752
307	Lakhimpur	512.463	563.576	582.325	596.171	606.209	613.387	618.472	622.048
308	Dhemaji	334.884	368.971	381.647	390.657	396.944	401.277	404.236	406.244

(continued)

Table 2.5 (continued)

District code	Name	Female population							
		Census	Forecast						
		2011	2011	2016	2021	2026	2031	2031	2041
309	Tinsukia	647.698	713.175	739.138	758.373	772.309	782.244	789.244	794.136
310	Dibrugarh	649.901	716.336	735.225	749.062	759.023	766.103	771.090	774.581
311	Sivasagar	561.834	626.070	643.367	656.011	665.076	671.483	675.967	679.084
312	Jorhat	535.451	594.296	608.990	619.538	626.967	632.129	635.684	638.115
313	Golaghat	523.727	582.719	600.505	613.607	623.084	629.847	634.627	637.984
314	Karbi Anglong	466.146	513.808	532.788	546.730	556.775	563.912	568.933	572.439
315	Dima Hasao	103.300	116.529	122.140	126.372	129.482	131.726	133.321	134.445
316	Cachar	850.333	936.700	979.581	1012.399	1036.804	1054.569	1067.299	1076.321
317	Karimganj	602.822	662.579	694.126	718.242	736.137	749.124	758.398	764.946
318	Hailakandi	321.406	352.186	368.113	380.102	388.868	395.139	399.557	402.636
319	Bongaigaon	362.986	398.471	415.838	428.896	438.473	445.370	450.271	453.722
320	Chirang	237.302	263.292	266.229	268.178	269.464	270.308	270.861	271.222
321	Kamrup	739.081	813.183	841.683	862.964	878.497	889.650	897.561	903.126
322	Kamrup Metropolitan	606.353	683.845	731.438	768.385	796.145	816.494	831.142	841.550
323	Nalbari	375.633	411.090	419.740	425.990	430.448	433.597	435.807	437.350
324	Baksa	468.745	522.271	532.723	540.131	545.324	548.936	551.435	553.158
325	Darrang	453.227	497.400	520.843	538.431	551.249	560.393	566.819	571.286
326	Udalguri	410.051	455.683	463.263	468.554	472.212	474.726	476.446	477.619

(continued)

Table 2.5 (continued)

District code	Name	Female population							
		Census	Forecast						
		2011	2011	2016	2021	2026	2031	2031	2041
West Bengal									
327	Darjiling	909.564	1018.821	1067.445	1104.495	1131.963	1151.917	1166.199	1176.313
328	Jalpaiguri	1889.782	2107.518	2183.078	2238.657	2278.680	2307.061	2326.969	2340.826
329	Koch Bihar	1367.544	1506.421	1541.358	1566.466	1584.267	1596.767	1605.485	1611.537
330	Uttar Dinajpur	1456.068	1621.877	1718.711	1793.921	1850.530	1892.140	1922.195	1943.630
331	Dakshin Dinajpur	819.077	921.689	953.360	976.674	993.485	1005.426	1013.818	1019.672
332	Maldah	1937.304	2144.600	2244.815	2317.989	2369.832	2405.782	2430.343	2446.951
333	Murshidabad	3476.243	3862.878	4074.037	4237.711	4360.624	4450.750	4515.684	4561.878
334	Birbhum	1711.484	1886.497	1951.593	1999.684	2034.397	2059.036	2076.316	2088.333
335	Barddhaman	3750.574	4166.902	4286.167	4373.482	4436.291	4480.903	4512.305	4534.268
336	Nadia	2513.332	2823.748	2915.801	2982.885	3030.843	3064.657	3088.267	3104.642
337	North Twenty	4890.392	5506.001	5716.362	5870.315	5980.514	6058.147	6112.226	6149.604
338	Hugli	2704.492	3017.984	3094.000	3148.353	3186.587	3213.175	3231.518	3244.102
339	Bankura	1758.579	1941.533	1992.610	2030.217	2057.399	2076.784	2090.478	2100.086
340	Puruliya	1433.119	1581.753	1639.184	1682.412	1714.088	1736.845	1752.964	1764.266
341	Haora	2349.210	2613.655	2709.997	2782.095	2834.807	2872.692	2899.588	2918.516
342	Kolkata	2139.928	2413.207	2433.256	2446.632	2455.486	2461.318	2465.146	2467.653
343	South Twenty Four Parganas	3988.183	4434.661	4638.087	4792.634	4906.924	4989.769	5048.954	5090.798

(continued)

Table 2.5 (continued)

District code	Name	Female population							
		Census	Forecast						
		2011	2011	2016	2021	2026	2031	2031	2041
344	Paschim Medinipur	2905.572	3218.184	3308.711	3373.920	3420.003	3452.130	3474.318	3489.542
345	Purba Medinipur	2466.041	2742.669	2830.296	2894.454	2940.457	2972.952	2995.662	3011.416
Jharkhand									
346	Garhwa	639.209	712.885	768.070	812.610	847.083	872.908	891.787	905.343
347	Chatra	508.951	564.562	611.023	649.074	678.885	701.437	718.052	730.054
348	Kodarma	349.037	387.696	420.283	447.136	468.254	484.260	496.055	504.569
349	Giridih	1187.376	1318.537	1405.461	1473.451	1524.599	1561.960	1588.667	1607.464
350	Deoghar	717.051	794.657	860.255	914.145	956.379	988.271	1011.686	1028.524
351	Godda	635.624	698.821	748.629	789.367	821.235	845.306	863.011	875.781
352	Sahibganj	561.176	619.229	660.500	693.505	718.856	737.731	751.461	761.281
353	Pakur	447.761	490.671	529.154	560.857	585.799	604.718	618.676	628.763
354	Dhanbad	1278.531	1442.346	1505.995	1553.360	1587.667	1612.034	1629.100	1640.937
355	Bokaro	989.523	1109.825	1163.295	1203.728	1233.489	1254.962	1270.235	1280.986
356	Lohardaga	229.161	254.609	276.236	294.445	309.046	320.302	328.718	334.869
357	Purbi Singhbhum	1117.017	1239.811	1291.530	1330.230	1358.425	1378.569	1392.761	1402.660
358	Palamu	933.567	1040.144	1115.811	1176.440	1223.060	1257.785	1283.042	1301.096
359	Latehar	357.312	394.886	424.090	447.597	465.722	479.239	489.071	496.095
360	Hazaribagh	843.614	938.461	994.343	1037.498	1069.688	1093.083	1109.767	1121.504

(continued)

Table 2.5 (continued)

District code	Name	Female population							
		Census	Forecast						
		2011	2011	2016	2021	2026	2031	2031	2041
361	Ramgarh	455.213	511.944	533.015	548.787	560.329	568.636	574.544	578.709
362	Dumka	652.928	722.621	763.341	795.835	820.764	839.321	852.825	862.492
363	Jamtara	386.212	431.319	458.536	480.204	496.743	508.967	517.789	524.045
364	Ranchi	1419.316	1570.732	1670.057	1746.738	1803.734	1844.913	1874.060	1894.391
365	Khunti	265.550	289.427	304.675	316.912	326.350	333.411	338.572	342.281
366	Gumla	510.823	555.805	584.054	605.941	622.273	634.120	642.539	648.435
367	Simdega	299.259	327.256	337.709	345.617	351.446	355.664	358.673	360.799
368	Pashchimi Singhbhum	752.953	825.362	869.422	904.371	930.998	950.669	964.873	974.961
369	Saraikela-Kharsawan	520.645	575.978	616.435	649.229	674.668	693.739	707.672	717.660
Odisha									
370	Bargarh	732.094	818.826	837.678	851.380	861.170	868.081	872.918	876.283
371	Jharsuguda	282.815	317.293	329.078	337.889	344.300	348.874	352.093	354.335
372	Sambalpur	514.222	572.440	588.544	600.374	608.902	614.965	619.235	622.222
373	Debagarh	154.290	171.084	176.195	179.970	182.705	184.659	186.041	187.011
374	Sundargarh	1032.290	1139.070	1172.161	1196.484	1214.052	1226.582	1235.439	1241.660
375	Kendujhar	895.246	987.334	1018.537	1041.486	1058.004	1069.710	1077.915	1083.620
376	Mayurbhanj	1263.525	1396.822	1451.426	1493.208	1524.286	1546.918	1563.147	1574.656
377	Baleshwar	1134.742	1260.802	1302.011	1331.773	1352.808	1367.449	1377.530	1384.421

(continued)

Table 2.5 (continued)

District code	Name	Female population							
		Census	Forecast						
		2011	2011	2016	2021	2026	2031	2031	2041
378	Bhadrak	746.077	833.328	862.646	884.389	900.151	911.390	919.311	924.846
379	Kendrapara	722.547	799.703	815.368	826.329	833.887	839.044	842.539	844.896
380	Jagatsinghapur	559.106	626.075	638.357	647.014	653.026	657.156	659.974	661.887
381	Cuttack	1271.710	1412.274	1450.388	1478.131	1497.944	1511.902	1521.642	1528.392
382	Jajapur	901.158	1007.223	1038.623	1061.584	1078.008	1089.572	1097.624	1103.186
383	Dhenkanal	580.218	644.775	660.003	671.127	679.121	684.798	688.797	691.597
384	Anugul	618.103	692.955	716.167	733.467	746.087	755.148	761.581	766.111
385	Nayagarh	460.153	508.455	517.236	523.554	528.041	531.197	533.403	534.938
386	Khordha	1084.536	1205.828	1269.505	1316.857	1350.902	1374.790	1391.266	1402.496
387	Puri	833.350	925.581	952.391	972.184	986.507	996.720	1003.927	1008.976
388	Ganjam	1749.813	1960.196	2020.130	2064.476	2096.549	2119.366	2135.409	2146.594
389	Gajapati	294.935	326.445	335.296	341.804	346.488	349.807	352.133	353.750
390	Kandhamal	373.165	416.624	435.377	449.874	460.738	468.691	474.414	478.482
391	Baudh	219.537	242.848	253.762	262.132	268.352	272.868	276.090	278.361
392	Subarnapur	298.871	334.576	344.502	351.814	357.086	360.828	363.456	365.286
393	Balangir	818.900	884.951	915.458	938.318	954.996	966.928	975.344	981.223
394	Nuapada	308.420	337.761	346.797	353.337	357.978	361.228	363.482	365.034
395	Kalahandi	789.768	870.334	908.554	937.867	959.671	975.526	986.864	994.876

(continued)

Table 2.5 (continued)

District code	Name	Female population						
		Census	Forecast					
		2011	2011	2016	2021	2026	2031	2031	2041
396	Rayagada	495.951	545.443	565.412	580.379	591.321	599.177	604.744	608.652
397	Nabarangapur	616.134	682.339	714.860	739.738	758.233	771.692	781.336	788.168
398	Koraput	700.838	776.502	811.239	837.989	857.966	872.547	883.012	890.432
399	Malkangiri	309.568	336.907	349.738	359.281	366.235	371.228	374.773	377.272
Chhattisgarh									
400	Koriya	324.180	363.657	376.304	385.591	392.268	396.997	400.311	402.615
401	Surguja	1166.757	1299.628	1370.507	1425.517	1466.841	1497.128	1518.928	1534.417
402	Jashpur	426.922	465.771	476.225	483.651	488.837	492.416	494.865	496.532
403	Raigarh	743.706	821.348	857.888	885.932	906.800	921.973	932.822	940.487
404	Korba	593.725	667.583	703.380	730.916	751.478	766.494	777.282	784.941
405	Janjgir-Champa	803.990	895.317	960.960	1015.275	1058.122	1090.665	1114.679	1132.024
406	Bilaspur	1312.055	1429.681	1522.107	1594.981	1650.207	1690.817	1720.021	1740.689
407	Kabeerdham	410.468	444.706	478.243	505.573	526.932	543.085	555.000	563.628
408	Rajnandgaon	774.278	856.371	894.913	924.375	946.236	962.101	973.430	981.428
409	Durg	1661.771	1835.873	1909.345	1964.415	2004.717	2033.698	2054.276	2068.757
410	Raipur	2015.686	2178.666	2323.522	2437.637	2524.149	2587.846	2633.742	2666.300
411	Mahasamund	520.787	565.633	582.166	594.377	603.208	609.498	613.929	617.027
412	Dhamtari	401.884	451.655	470.256	484.442	494.970	502.626	508.112	512.002

(continued)

Table 2.5 (continued)

District code	Name	Female population							
		Census	Forecast						
		2011	2011	2016	2021	2026	2031	2031	2041
413	Uttar Bastar Kanker	375.603	414.907	427.947	437.531	444.442	449.357	452.818	455.238
414	Bastar	714.712	789.551	822.473	847.398	865.783	879.085	888.575	895.278
415	Narayanpur	69.716	78.009	82.006	85.063	87.331	88.976	90.149	90.977
416	Dakshin Bastar Dantewada	269.496	298.413	305.392	310.434	314.019	316.538	318.293	319.510
417	Bijapur	126.567	144.177	148.591	151.781	154.041	155.622	156.716	157.469
Madhya Pradesh									
418	Sheopur	326.077	364.361	388.575	407.482	421.706	432.111	439.566	444.829
419	Morena	897.553	997.517	1060.212	1108.775	1145.026	1171.345	1190.072	1203.205
420	Bhind	776.162	858.232	896.305	924.993	945.991	961.034	971.647	979.054
421	Gwalior	941.709	1046.571	1117.667	1174.197	1217.482	1249.680	1273.119	1289.913
422	Datia	366.597	408.045	429.967	447.064	459.968	469.468	476.335	481.234
423	Shivpuri	806.255	901.767	968.036	1021.528	1062.958	1094.031	1116.779	1133.139
424	Tikamgarh	684.811	766.144	813.767	850.984	879.034	899.605	914.389	924.861
425	Chhatarpur	826.254	928.128	988.756	1036.713	1073.229	1100.235	1119.778	1133.702
426	Panna	483.040	539.236	566.928	588.227	604.118	615.707	624.021	629.913
427	Sagar	1122.201	1248.076	1305.596	1349.499	1382.093	1405.798	1422.780	1434.816
428	Damoh	602.346	669.005	697.003	718.182	733.794	745.084	753.135	758.820
429	Satna	1071.440	1198.319	1264.854	1316.474	1355.250	1383.678	1404.149	1418.701

(continued)

Table 2.5 (continued)

District code	Name	Female population							
		Census	Forecast						
		2011	2011	2016	2021	2026	2031	2031	2041
430	Rewa	1140.006	1276.695	1352.085	1411.219	1456.019	1489.074	1512.989	1530.049
431	Umaria	314.084	347.647	368.748	385.220	397.639	406.758	413.327	417.992
432	Neemuch	403.414	452.488	469.589	482.276	491.472	498.025	502.640	505.862
433	Mandsaur	657.560	739.727	770.485	793.656	810.682	822.965	831.710	837.878
434	Ratlam	716.828	796.962	838.719	870.592	894.156	911.167	923.238	931.700
435	Ujjain	970.575	1084.970	1135.562	1174.025	1202.462	1223.056	1237.747	1248.116
436	Shajapur	732.151	817.294	858.831	890.788	914.620	931.984	944.421	953.220
437	Dewas	758.356	847.360	893.415	928.862	955.330	974.647	988.513	998.347
438	Dhar	1073.068	1192.561	1273.794	1337.996	1386.824	1422.884	1448.942	1467.477
439	Indore	1577.070	1752.183	1912.218	2043.633	2146.684	2224.609	2281.930	2323.246
440	Khargone (West Nimar)	919.925	1018.968	1079.666	1127.038	1162.773	1189.047	1208.000	1221.487
441	Barwani	686.541	764.401	823.531	871.320	908.342	936.093	956.384	970.953
442	Rajgarh	755.602	835.686	892.041	937.088	971.716	997.545	1016.379	1029.888
443	Vidisha	689.307	763.352	804.968	837.300	861.616	879.461	892.324	901.476
444	Bhopal	1134.931	1274.701	1380.129	1464.489	1529.445	1577.994	1613.481	1638.997
445	Sehore	627.589	699.635	740.141	771.309	794.507	811.351	823.361	831.815
446	Raisen	631.239	702.295	738.833	766.889	787.790	803.013	813.919	821.639
447	Betul	776.125	870.758	900.671	922.806	938.842	950.281	958.351	964.000

(continued)

Table 2.5 (continued)

District code	Name	Female population							
		Census	Forecast						
		2011	2011	2016	2021	2026	2031	2031	2041
448	Harda	275.627	307.059	324.573	338.185	348.413	355.906	361.292	365.113
449	Hoshangabad	592.625	664.007	693.051	714.948	731.010	742.555	750.733	756.467
450	Katni	630.029	696.907	732.035	758.905	778.814	793.219	803.465	810.664
451	Jabalpur	1186.011	1316.904	1365.041	1400.760	1426.686	1445.205	1458.282	1467.441
452	Narsimhapur	523.044	584.054	606.800	623.898	636.434	645.458	651.867	656.378
453	Dindori	352.611	385.349	399.527	410.115	417.840	423.382	427.311	430.071
454	Mandla	529.633	583.727	606.903	624.423	637.320	646.630	653.256	657.925
455	Chhindwara	1026.454	1148.281	1190.959	1222.962	1246.405	1263.286	1275.291	1283.754
456	Seoni	683.252	752.829	780.621	801.026	815.630	825.894	833.016	837.913
457	Balaghat	859.520	951.520	974.305	990.577	1001.985	1009.879	1015.292	1018.981
458	Guna	592.157	661.043	715.332	759.652	794.332	820.577	839.943	853.968
459	Ashoknagar	401.234	443.332	468.237	487.313	501.427	511.606	518.812	523.845
460	Shahdol	526.042	583.472	607.971	626.314	639.694	649.268	656.026	660.749
461	Anuppur	370.123	418.229	434.626	446.660	455.279	461.344	465.559	468.463
462	Sidhi	551.121	613.908	656.805	691.184	717.696	737.542	752.070	762.531
463	Singrauli	564.636	635.957	692.695	738.938	774.991	802.137	822.045	836.367
464	Jhabua	510.025	563.766	605.688	639.423	665.523	685.109	699.473	709.830
465	Alirajpur	366.457	411.089	436.229	455.868	470.689	481.585	489.442	495.030

(continued)

Table 2.5 (continued)

District code	Name	Female population							
		Census	Forecast						
		2011	2011	2016	2021	2026	2031	2031	2041
466	Khandwa (East Nimar)	635.732	698.887	730.975	755.263	773.137	786.019	795.165	801.590
467	Burhanpur	369.343	407.747	423.694	435.531	444.116	450.239	454.555	457.570
Gujarat									
468	Kachchh	995.634	1085.066	1159.864	1219.944	1266.250	1300.814	1326.004	1344.041
469	Banas Kantha	1510.127	1675.838	1781.341	1864.004	1926.539	1972.601	2005.868	2029.554
470	Patan	649.337	712.171	727.342	738.315	746.137	751.654	755.517	758.208
471	Mahesana	978.544	1089.145	1111.053	1126.896	1138.196	1146.176	1151.772	1155.677
472	Sabar Kantha	1184.358	1298.814	1338.550	1367.939	1389.265	1404.529	1415.346	1422.957
473	Gandhinagar	667.839	747.180	772.282	790.820	804.225	813.774	820.502	825.207
474	Ahmadabad	3426.174	3791.506	3996.800	4154.433	4271.897	4357.489	4418.844	4462.309
475	Surendranagar	846.351	939.394	978.838	1008.643	1030.572	1046.392	1057.643	1065.564
476	Rajkot	1830.113	2029.216	2125.216	2198.501	2252.907	2292.467	2320.801	2340.873
477	Jamnagar	1045.927	1149.883	1179.578	1200.954	1216.082	1226.661	1233.998	1239.055
478	Porbandar	285.240	315.241	320.506	324.283	326.961	328.845	330.163	331.081
479	Junagadh	1338.726	1482.507	1518.793	1544.994	1563.619	1576.712	1585.844	1592.179
480	Amreli	743.141	833.518	843.384	850.034	854.475	857.423	859.373	860.659
481	Bhavnagar	1390.164	1545.923	1600.756	1640.865	1669.549	1689.734	1703.776	1713.468
482	Anand	1005.521	1107.838	1133.733	1152.350	1165.510	1174.700	1181.065	1185.447

(continued)

Table 2.5 (continued)

District code	Name	Female population							
		Census	Forecast						
		2011	2011	2016	2021	2026	2031	2031	2041
483	Kheda	1114.158	1223.737	1250.835	1270.154	1283.711	1293.120	1299.601	1304.041
484	Panch Mahals	1163.815	1288.338	1344.003	1386.192	1417.319	1439.832	1455.883	1467.208
485	Dohad	1058.435	1175.075	1267.927	1342.783	1400.664	1443.990	1475.639	1498.349
486	Vadodara	2011.890	2233.995	2314.363	2374.160	2417.682	2448.852	2470.919	2486.415
487	Narmada	289.211	317.955	326.796	333.271	337.927	341.231	343.554	345.177
488	Bharuch	745.312	828.428	856.621	877.685	893.053	904.072	911.875	917.351
489	The Dangs	114.470	127.028	134.312	139.949	144.175	147.267	149.491	151.070
490	Navsari	651.507	733.159	750.255	762.697	771.606	777.912	782.338	785.428
491	Valsad	818.456	911.014	965.170	1007.815	1040.254	1064.289	1081.751	1094.257
492	Surat	2679.098	2995.046	3374.040	3704.999	3978.343	4193.898	4357.764	4478.898
493	Tapi	404.834	444.464	453.250	459.279	463.355	466.084	467.898	469.100
Daman & Diu									
494	Diu	26.432	29.591	30.534	31.214	31.694	32.026	32.254	32.410
495	Daman	66.514	73.195	86.340	99.262	111.155	121.455	129.917	136.571
Dadra & Nagar Haveli									
496	Dadra & Nagar Haveli	149.949	166.046	189.756	211.177	229.427	244.213	255.713	264.376
Maharashtra									
497	Nandurbar	815.125	899.101	952.243	993.446	1024.261	1046.691	1062.697	1073.959

(continued)

Table 2.5 (continued)

District code	Name	Female population							
		Census	Forecast						
		2011	2011	2016	2021	2026	2031	2031	2041
498	Dhule	996.831	1095.638	1137.738	1169.115	1191.918	1208.190	1219.649	1227.645
499	Jalgaon	2032.552	2256.936	2328.190	2380.461	2417.970	2444.463	2462.966	2475.788
500	Buldana	1248.698	1394.328	1450.393	1492.764	1523.943	1546.438	1562.437	1573.701
501	Akola	881.572	986.236	1018.952	1043.133	1060.627	1073.087	1081.864	1087.998
502	Washim	576.858	638.177	662.413	680.657	694.050	703.702	710.566	715.401
503	Amravati	1407.577	1573.581	1622.984	1659.406	1685.682	1704.343	1717.449	1726.582
504	Wardha	632.339	711.706	725.046	734.406	740.876	745.304	748.312	750.347
505	Nagpur	2268.595	2555.582	2674.325	2764.674	2831.579	2880.136	2914.862	2939.438
506	Bhandara	594.814	671.756	684.580	693.750	700.212	704.721	707.844	709.997
507	Gondiya	660.953	726.510	736.477	743.409	748.177	751.432	753.642	755.138
508	Gadchiroli	531.614	597.510	619.060	635.149	646.893	655.326	661.311	665.522
509	Chandrapur	1080.473	1226.131	1256.989	1278.918	1294.235	1304.806	1312.042	1316.965
510	Yavatmal	1352.383	1505.414	1553.264	1588.545	1614.011	1632.109	1644.830	1653.703
511	Nanded	1631.217	1833.160	1923.345	1992.105	2043.020	2079.910	2106.220	2124.773
512	Hingoli	571.051	633.860	662.267	683.746	699.522	710.866	718.899	724.524
513	Parbhani	893.216	994.913	1042.750	1079.156	1106.039	1125.450	1139.242	1148.928
514	Jalna	947.573	1052.678	1105.990	1147.106	1177.831	1200.255	1216.340	1227.735
515	Aurangabad	1776.813	1984.426	2141.489	2268.183	2366.224	2439.685	2493.407	2532.002

(continued)

Table 2.5 (continued)

District code	Name	Female population							
		Census	Forecast						
		2011	2011	2016	2021	2026	2031	2031	2041
516	Nashik	2950.001	3290.949	3497.687	3659.052	3780.525	3869.498	3933.369	3978.561
517	Thane	5195.070	5863.220	6626.859	7292.286	7840.825	8272.688	8600.547	8842.642
518	Mumbai Suburban	4325.639	4902.044	5045.390	5142.713	5207.718	5250.663	5278.829	5297.214
519	Mumbai	1400.803	1620.860	1622.550	1623.540	1624.120	1624.460	1624.658	1624.775
520	Raigarh	1289.855	1431.066	1500.177	1553.855	1594.227	1623.861	1645.228	1660.437
521	Pune	4505.303	5000.835	5412.874	5750.026	6014.366	6214.763	6362.844	6470.208
522	Ahmadnagar	2200.334	2466.964	2552.378	2616.196	2662.840	2696.385	2720.230	2737.041
523	Bid	1235.943	1371.549	1431.379	1476.674	1510.011	1534.042	1551.105	1563.092
524	Latur	1181.056	1321.279	1386.678	1437.000	1474.606	1502.099	1521.879	1535.945
525	Osmanabad	796.041	888.348	913.863	932.651	946.211	955.859	962.652	967.401
526	Solapur	2089.904	2343.235	2428.682	2492.450	2538.872	2572.058	2595.476	2611.849
527	Satara	1492.899	1681.652	1716.621	1741.953	1760.033	1772.801	1781.750	1787.989
528	Ratnagiri	853.948	996.387	999.434	1001.207	1002.234	1002.827	1003.169	1003.366
529	Sindhudurg	432.319	502.053	502.382	502.580	502.697	502.768	502.809	502.834
530	Kolhapur	1895.343	2124.381	2183.711	2226.988	2257.980	2279.882	2295.217	2305.884
531	Sangli	1386.415	1558.364	1602.191	1634.457	1657.765	1674.374	1686.093	1694.305
Andhra Pradesh									
532	Adilabad	1371.642	1552.069	1607.308	1648.210	1677.827	1698.929	1713.789	1724.170

(continued)

Table 2.5 (continued)

District code	Name	Female population							
		Census	Forecast						
		2011	2011	2016	2021	2026	2031	2031	2041
533	Nizamabad	1300.694	1462.457	1501.137	1529.308	1549.481	1563.751	1573.759	1580.736
534	Karimnagar	1895.469	2142.681	2201.372	2243.339	2272.757	2293.092	2307.012	2316.479
535	Medak	1510.258	1690.568	1757.883	1807.764	1843.698	1869.063	1886.711	1898.867
536	Hyderabad	1924 748	2203.099	2237.240	2256.969	2268.217	2274.580	2278.164	2280.177
537	Rangareddy	2595 733	2900.586	3334.648	3726.375	4058.462	4325.384	4530.889	4683.926
538	Mahbubnagar	2002.642	2238.321	2321.349	2383.116	2427.924	2459.839	2482.275	2497.903
539	Nalgonda	1729.037	1957.784	2010.106	2047.911	2074.722	2093.484	2106.492	2115.453
540	Warangal	1753.295	1972.349	2023.667	2059.986	2085.197	2102.464	2114.180	2122.082
541	Khammam	1406.382	1582.391	1626.614	1658.334	1680.681	1696.227	1706.946	1714.292
542	Srikakulam	1361.376	1535.045	1568.236	1592.450	1609.779	1622.012	1630.563	1636.500
543	Vizianagaram	1182.997	1335.070	1351.324	1361.894	1368.674	1372.984	1375.709	1377.426
544	Visakhapatnam	2151.679	2417.376	2506.390	2572.166	2619.578	2653.146	2676.611	2692.868
545	East Godavari	2584.608	2921.671	2961.738	2987.795	3004.527	3015.184	3021.936	3026.200
546	West Godavari	1972.048	2235.829	2261.201	2277.403	2287.642	2294.069	2298.087	2300.592
547	Krishna	2250.023	2532.123	2595.060	2640.827	2673.537	2696.625	2712.780	2724.014
548	Guntur	2447.292	2728.624	2786.361	2827.893	2857.316	2877.935	2892.275	2902.195
549	Prakasam	1682.684	1876.472	1927.097	1964.455	1991.468	2010.717	2024.289	2033.788
550	Sri Potti Sriramulu Nellore	1470.583	1644.480	1687.662	1719.044	1741.406	1757.117	1768.045	1775.595

(continued)

Table 2.5 (continued)

District code	Name	Female population							
		Census	Forecast						
		2011	2011	2016	2021	2026	2031	2031	2041
551	Y. S. R	1430.692	1600.810	1648.928	1684.482	1710.192	1728.494	1741.377	1750.376
552	Kurnool	2014.236	2238.062	2326.912	2393.396	2441.855	2476.507	2500.946	2518.016
553	Anantapur	2016.653	2258.643	2335.262	2392.024	2433.105	2462.337	2482.889	2497.215
554	Chittoor	2083.860	2338.239	2417.401	2476.681	2520.035	2551.196	2573.315	2588.878
Karnataka									
555	Belgaum	2356.598	2628.415	2720.352	2788.637	2838.173	2873.499	2898.383	2915.762
556	Bagalkot	939.641	1046.306	1086.544	1116.871	1139.153	1155.217	1166.642	1174.689
557	Bijapur	1066.309	1178.174	1236.834	1282.353	1316.538	1341.585	1359.608	1372.408
558	Bidar	832.635	931.523	967.692	994.764	1014.518	1028.666	1038.663	1045.661
559	Raichur	964.301	1083.809	1136.487	1176.551	1206.146	1227.539	1242.764	1253.477
560	Koppal	689.994	772.986	811.178	840.482	862.305	878.196	889.581	897.642
561	Gadag	527.423	586.931	600.122	609.650	616.415	621.162	624.463	626.746
562	Dharwad	909.817	1007.407	1041.990	1067.657	1086.301	1099.632	1109.059	1115.671
563	Uttara Kannada	710.913	797.715	809.663	818.037	823.846	827.847	830.589	832.462
564	Haveri	778.540	870.547	895.650	914.192	927.622	937.210	943.986	948.740
565	Bellary	1215.641	1350.011	1421.528	1476.388	1517.168	1546.778	1567.913	1582.816
566	Chitradurga	818.613	917.727	943.093	961.751	975.223	984.822	991.595	996.343
567	Davanagere	959.097	1079.762	1106.807	1126.091	1139.619	1149.001	1155.456	1159.873

(continued)

Table 2.5 (continued)

District code	Name	Female population							
		Census	Forecast						
		2011	2011	2016	2021	2026	2031	2031	2041
568	Shimoga	875.338	970.697	983.283	991.695	997.267	1000.936	1003.344	1004.919
569	Udupi	615.230	692.165	697.693	701.247	703.514	704.952	705.862	706.437
570	Chikmagalur	571.339	652.971	660.574	665.894	669.589	672.142	673.899	675.107
571	Tumkur	1328.386	1510.381	1538.644	1558.883	1573.178	1583.176	1590.120	1594.921
572	Bangalore	4598.890	5102.090	5718.508	6247.061	6676.328	7009.850	7260.172	7443.213
573	Mandya	900.684	1025.549	1036.806	1044.410	1049.504	1052.898	1055.150	1056.642
574	Hassan	892.754	1016.160	1030.211	1039.983	1046.711	1051.313	1054.445	1056.570
575	Dakshina Kannada	1054.935	1174.460	1200.231	1218.604	1231.499	1240.450	1246.616	1250.840
576	Kodagu	279.911	315.696	318.959	321.249	322.847	323.956	324.723	325.253
577	Mysore	1489.527	1660.591	1717.403	1758.997	1788.765	1809.726	1824.315	1834.388
578	Chamarajanagar	508.560	572.205	582.435	589.702	594.794	598.327	600.762	602.433
579	Gulbarga	1264.571	1405.405	1473.422	1525.345	1563.757	1591.519	1611.247	1625.098
580	Yadgir	583.942	650.750	692.060	724.289	748.382	765.823	778.159	786.740
581	Kolar	760.005	852.124	877.814	896.885	910.734	920.630	927.621	932.519
582	Chikkaballapura	618.667	690.298	706.736	718.686	727.232	733.271	737.503	740.452
583	Bangalore Rural	481.751	534.250	555.669	571.887	583.836	592.465	598.604	602.928
584	Ramanagara	534.628	603.247	612.755	619.440	624.079	627.269	629.450	630.933
Goa									

(continued)

Table 2.5 (continued)

District code	Name	Female population							
		Census	Forecast						
		2011	2011	2016	2021	2026	2031	2031	2041
585	North Goa	401.331	450.999	461.040	468.158	473.116	476.529	478.857	480.438
586	South Goa	318.074	357.982	367.180	373.895	378.716	382.136	384.542	386.224
Lakshadweep									
587	Lakshadweep	31.350	35.652	36.426	36.946	37.291	37.517	37.663	37.758
Kerala									
588	Kasaragod	678.762	761.421	780.032	793.482	803.054	809.792	814.499	817.769
589	Kannur	1341.557	1510.575	1533.698	1550.116	1561.657	1569.715	1575.314	1579.191
590	Wayanad	415.736	467.100	475.614	481.480	485.478	488.184	490.006	491.228
591	Kozhikode	1615.351	1801.975	1836.631	1861.507	1879.139	1891.524	1900.169	1906.176
592	Malappuram	2152.592	2407.858	2495.135	2559.440	2605.774	2638.626	2661.654	2677.666
593	Palakkad	1450.456	1631.982	1662.829	1684.752	1700.106	1710.751	1718.079	1723.099
594	Thrissur	1640.437	1840.281	1862.944	1878.916	1890.068	1897.804	1903.147	1906.825
595	Ernakulam	1662.831	1861.746	1886.817	1904.703	1917.345	1926.220	1932.423	1936.743
596	Idukki	556.166	642.397	646.598	649.389	651.236	652.455	653.258	653.786
597	Kottayam	1006.262	1134.788	1143.041	1148.693	1152.538	1155.141	1156.898	1158.082
598	Alappuzha	1114.647	1254.863	1262.937	1268.487	1272.284	1274.871	1276.630	1277.823
599	Pathanamthitta	635.696	727.894	730.692	732.578	733.845	734.694	735.263	735.644
600	Kollam	1388.407	1561.737	1577.991	1589.322	1597.165	1602.567	1606.275	1608.814

(continued)

Table 2.5 (continued)

District code	Name	Female population							
		Census	Forecast						
		2011	2011	2016	2021	2026	2031	2031	2041
601	Thiruvananthapuram	1719.749	1948.735	1973.603	1991.200	2003.547	2012.159	2018.140	2022.283
Tamil Nadu									
602	Thiruvallur	1852.042	2042.167	2254.362	2445.467	2610.314	2747.302	2857.656	2944.346
603	Chennai	2310.888	2600.100	2662.571	2709.982	2745.534	2771.957	2791.464	2805.795
604	Kancheepuram	1985.294	2176.622	2417.902	2639.910	2835.008	2999.650	3133.908	3240.360
605	Vellore	1974.643	2195.352	2276.331	2340.397	2389.945	2427.597	2455.827	2476.782
606	Tiruvannamalai	1228.986	1366.917	1408.256	1440.583	1465.316	1483.922	1497.743	1507.914
607	Viluppuram	1718.054	1894.584	1969.310	2030.030	2078.121	2115.442	2143.951	2165.466
608	Salem	1700.485	1855.549	1924.782	1979.942	2022.908	2055.793	2080.624	2099.183
609	Namakkal	857.321	937.838	974.012	1003.234	1026.261	1044.052	1057.592	1067.777
610	Erode	1121.376	1241.052	1279.697	1309.930	1333.107	1350.601	1363.652	1373.301
611	The Nilgiris	375.251	437.734	442.380	445.673	447.993	449.619	450.757	451.550
612	Dindigul	1078.837	1186.815	1220.398	1246.791	1267.122	1282.544	1294.104	1302.695
613	Karur	536.309	583.584	602.265	617.259	629.009	638.046	644.895	650.030
614	Tiruchirappalli	1370.006	1515.722	1559.894	1594.189	1620.324	1639.961	1654.559	1665.323
615	Perambalur	283.066	312.132	323.156	332.002	338.940	344.286	348.347	351.401
616	Ariyalur	380.191	421.102	430.042	436.853	441.959	445.742	448.521	450.548
617	Cuddalore	1294.217	1426.975	1466.859	1497.721	1521.161	1538.713	1551.718	1561.278

(continued)

Table 2.5 (continued)

District code	Name	Female population							
		Census	Forecast						
		2011	2011	2016	2021	2026	2031	2031	2041
618	Nagapattinam	818.323	908.577	928.036	942.995	954.314	962.777	969.048	973.664
619	Thiruvarur	637.584	708.771	721.985	732.069	739.658	745.312	749.492	752.564
620	Thanjavur	1223.474	1364.052	1394.752	1418.488	1436.560	1450.161	1460.307	1467.826
621	Pudukkottai	815.157	910.614	933.433	950.616	963.337	972.636	979.370	984.215
622	Sivaganga	670.429	732.894	747.393	758.362	766.530	772.541	776.925	780.103
623	Madurai	1511.777	1646.735	1696.045	1734.474	1763.887	1786.089	1802.674	1814.966
624	Theni	620.216	671.274	680.793	687.606	692.431	695.821	698.191	699.841
625	Virudhunagar	974.579	1089.089	1122.700	1149.174	1169.626	1185.188	1196.897	1205.629
626	Ramanathapuram	670.787	745.041	761.166	773.536	782.888	789.880	795.066	798.889
627	Thoothukkudi	885.155	978.022	993.782	1005.679	1014.544	1021.085	1025.877	1029.370
628	Tirunelveli	1556.321	1712.128	1759.656	1797.153	1826.172	1848.297	1864.976	1877.441
629	Kanniyakumari	944.029	1036.939	1053.647	1066.105	1075.302	1082.040	1086.950	1090.514
630	Dharmapuri	732.540	805.734	834.928	857.961	875.772	889.333	899.535	907.142
631	Krishnagiri	919.577	1017.415	1075.168	1122.880	1161.151	1191.128	1214.175	1231.641
632	Coimbatore	1728.748	1886.732	1964.385	2026.044	2073.846	2110.222	2137.513	2157.771
633	Tiruppur	1232.893	1347.926	1465.413	1570.026	1659.356	1732.947	1791.803	1837.768
Puducherry									
634	Yanam	28.325	31.886	42.572	56.045	72.244	90.641	110.230	129.709

(continued)

Table 2.5 (continued)

District code	Name	Female population							
		Census	Forecast						
		2011	2011	2016	2021	2026	2031	2031	2041
635	Puducherry	482.031	534.093	576.234	610.458	637.110	657.195	671.959	682.615
636	Mahe	22.673	25.110	25.867	26.430	26.840	27.134	27.343	27.489
637	Karaikal	102.413	113.390	118.717	122.842	125.937	128.205	129.839	131.001
Andaman & Nicobar Islands									
638	Nicobar	16.115	20.432	20.456	20.466	20.470	20.472	20.473	20.474
639	North & Middle Andaman	50.736	58.773	59.895	60.587	61.007	61.262	61.415	61.506
640	South Andaman	110.859	125.794	131.519	135.466	138.119	139.873	141.019	141.762

Source Author

Table 2.6 Forecast of total population (000) in districts of India, 2011–2041

District Code	Name	Total population							
		Census	Forecast						
		2011	2011	2016	2021	2026	2031	2036	2041
Jammu & Kashmir									
001	Kupwara	870.354	976.460	1086.613	1180.778	1257.040	1316.129	1360.345	1392.569
002	Badgam	753.745	843.276	905.253	955.074	993.560	1022.384	1043.476	1058.648
003	Leh (Ladakh)	133.487	151.861	159.644	165.470	169.693	172.680	174.756	176.180
004	Kargil	140.802	159.170	168.342	174.911	179.425	182.440	184.415	185.693
005	Punch	476.835	533.304	582.762	624.558	658.142	684.050	703.416	717.553
006	Rajouri	642.415	712.488	776.553	829.824	872.031	904.209	928.027	945.275
007	Kathua	616.435	684.990	719.124	745.193	764.508	778.500	788.472	795.496
008	Baramula	1008.039	1137.543	1212.770	1272.143	1317.299	1350.683	1374.851	1392.082
009	Bandipore	392.232	439.043	476.037	506.275	529.927	547.800	560.957	570.457
010	Srinagar	1236.829	1391.454	1483.623	1556.478	1611.906	1652.859	1682.467	1703.538
011	Ganderbal	297.446	330.919	365.809	395.735	420.103	439.115	453.459	464.010
012	Pulwama	560.440	621.069	664.673	699.364	725.857	745.462	759.635	769.709
013	Shupiyan	266.215	297.176	320.507	339.558	354.462	365.736	374.049	380.063
014	Anantnag	1078.692	1197.887	1340.268	1465.583	1569.771	1652.366	1715.402	1762.129
015	Kulgam	424.483	487.781	504.693	515.750	522.793	527.205	529.942	531.628
016	Doda	409.936	455.259	489.510	517.216	538.714	554.865	566.705	575.231
017	Ramban	283.713	314.049	340.522	362.203	379.152	391.926	401.290	408.016
018	Kishtwar	230.696	257.399	272.926	285.136	294.403	301.249	306.206	309.743
019	Udhampur	554.985	623.123	662.595	693.380	716.533	733.475	745.624	754.211

(continued)

Table 2.6 (continued)

District Code	Name	Total population							
		Census	Forecast						
		2011	2011	2016	2021	2026	2031	2036	2041
020	Reasi	314.667	347.649	369.620	386.762	399.650	409.073	415.822	420.585
021	Jammu	1529.958	1735.601	1814.447	1872.434	1913.907	1942.983	1963.087	1976.855
022	Samba	318.898	356.161	370.914	381.937	389.973	395.728	399.796	402.646
Himachal Pradesh									
023	Chamba	519.080	582.482	603.253	618.666	629.833	637.785	643.377	647.276
024	Kangra	1510.075	1681.551	1729.846	1765.315	1790.843	1808.951	1821.664	1830.524
025	Lahul & Spiti	31.564	36.860	36.902	36.928	36.945	36.956	36.963	36.967
026	Kullu	437.903	495.204	521.515	541.836	557.018	568.081	575.996	581.586
027	Mandi	999.777	1114.572	1141.794	1161.264	1174.968	1184.504	1191.087	1195.608
028	Hamirpur	454.768	506.353	518.144	526.769	532.967	537.364	540.454	542.613
029	Una	521.173	578.985	603.822	622.660	636.513	646.470	653.509	658.427
030	Bilaspur	381.956	425.207	436.625	444.913	450.823	454.982	457.882	459.893
031	Solan	580.320	654.472	689.583	715.995	735.208	748.850	758.376	764.952
032	Sirmaur	529.855	592.451	617.334	636.123	649.937	659.896	666.974	671.955
033	Shimla	814.010	909.208	939.392	961.845	978.177	989.864	998.131	1003.930
034	Kinnaur	84.121	94.389	96.067	97.247	98.065	98.628	99.012	99.273
Punjab									
035	Gurdaspur	2298.323	2578.896	2651.541	2704.371	2742.031	2768.497	2786.911	2799.633

(continued)

Table 2.6 (continued)

District Code	Name	Total population							
		Census	Forecast						
		2011	2011	2016	2021	2026	2031	2036	2041
036	Kapurthala	815.168	917.669	940.452	956.840	968.429	976.527	982.137	986.002
037	Jalandhar	2193.590	2453.795	2537.957	2601.147	2647.480	2680.864	2704.617	2721.366
038	Hoshiarpur	1586.625	1790.215	1833.802	1865.783	1888.819	1905.194	1916.722	1924.786
039	Shahid	612.310	694.520	704.207	710.664	714.919	717.704	719.520	720.701
040	Fatehgarh Sahib	600.163	670.208	689.024	702.793	712.686	719.702	724.630	728.070
041	Ludhiana	3498.739	3944.308	4122.833	4253.940	4347.541	4413.027	4458.197	4489.051
042	Moga	995.746	1113.839	1144.937	1167.634	1183.874	1195.330	1203.332	1208.881
043	Firozpur	2029.074	2261.832	2356.777	2428.113	2480.274	2517.661	2544.076	2562.550
044	Muktsar	901.896	999.184	1034.856	1061.457	1080.841	1094.729	1104.559	1111.458
045	Faridkot	617.508	691.302	714.587	731.746	744.118	752.900	759.065	763.360
046	Bathinda	1388.525	1536.390	1594.013	1636.864	1667.994	1690.227	1705.913	1716.885
047	Mansa	769.751	856.989	881.980	900.254	913.363	922.636	929.133	933.653
048	Patiala	1895.686	2084.279	2161.049	2218.401	2260.306	2290.432	2311.836	2326.919
049	Amritsar	2490.656	2789.497	2935.466	3049.112	3134.602	3197.263	3242.324	3274.287
050	Tarn Taran	1119.627	1229.669	1284.312	1326.661	1358.408	1381.616	1398.272	1410.068
051	Rupnagar	684.627	773.340	797.027	814.518	827.161	836.159	842.493	846.917
052	Sahibzada Ajit Singh Nagar	994.628	1108.678	1215.548	1304.429	1374.881	1428.624	1468.432	1497.279
053	Sangrur	1655.169	1848.510	1909.901	1955.418	1988.420	2011.963	2028.565	2040.177

(continued)

Table 2.6 (continued)

District Code	Name	Total population							
		Census	Forecast						
		2011	2011	2016	2021	2026	2031	2036	2041
054	Barnala	595.527	663.600	685.354	701.254	712.625	720.631	726.207	730.061
Chandigarh									
055	Chandigarh	1055.450	1196.008	1257.261	1302.170	1334.262	1356.773	1372.361	1383.058
Uttarakhand									
056	Uttarkashi	330.086	372.542	387.046	397.647	405.195	410.470	414.111	416.602
057	Chamoli	391.605	444.701	453.920	460.338	464.739	467.725	469.737	471.087
058	Rudraprayag	242.285	274.303	280.207	284.375	287.277	289.277	290.648	291.585
059	Tehri Garhwal	618.931	705.639	716.657	724.104	729.107	732.457	734.697	736.196
060	Dehradun	1696.694	1872.001	2017.635	2135.025	2225.801	2293.765	2343.428	2379.075
061	Garhwal	687.271	788.381	791.134	792.888	794.011	794.731	795.195	795.495
062	Pithoragarh	483.439	546.060	554.926	561.249	565.701	568.806	570.957	572.441
063	Bageshwar	259.898	292.919	297.036	299.907	301.886	303.240	304.162	304.788
064	Almora	622.506	717.185	719.158	720.347	721.065	721.499	721.762	721.922
065	Champawat	259.648	287.458	296.456	303.014	307.704	311.012	313.324	314.929
066	Nainital	954.605	1065.265	1140.164	1199.333	1244.337	1277.590	1301.638	1318.760
067	Udham Singh Nagar	1648.902	1815.955	1946.563	2048.629	2125.620	2182.161	2222.871	2251.768
068	Hardwar	1890.422	2095.263	2258.526	2390.201	2491.999	2568.147	2623.708	2663.516
Haryana									

(continued)

Table 2.6 (continued)

District Code	Name	Total population							
		Census	Forecast						
		2011	2011	2016	2021	2026	2031	2036	2041
069	Panchkula	561.293	639.890	680.466	708.747	727.688	740.037	747.947	752.956
070	Ambala	1128.350	1275.024	1328.105	1367.976	1397.108	1417.966	1432.682	1442.958
071	Yamunanagar	1214.205	1362.617	1428.931	1479.112	1516.036	1542.649	1561.545	1574.820
072	Kurukshetra	964.655	1063.873	1101.973	1130.185	1150.627	1165.210	1175.495	1182.693
073	Kaithal	1074.304	1197.918	1235.021	1262.202	1281.769	1295.678	1305.476	1312.334
074	Karnal	1505.324	1672.417	1744.573	1798.907	1838.806	1867.568	1888.026	1902.438
075	Panipat	1205.437	1360.442	1471.619	1561.115	1630.143	1681.645	1719.129	1745.918
076	Sonipat	1450.001	1622.691	1682.848	1727.163	1759.032	1781.554	1797.277	1808.159
077	Jind	1334.152	1490.332	1537.733	1572.567	1597.676	1615.524	1628.086	1636.866
078	Fatehabad	942.011	1038.803	1073.368	1098.245	1115.810	1128.044	1136.485	1142.271
079	Sirsa	1295.189	1437.196	1488.584	1526.421	1553.735	1573.171	1586.862	1596.436
080	Hisar	1743.931	1956.181	2035.146	2094.433	2137.904	2169.227	2191.514	2207.231
081	Bhiwani	1634.445	1827.451	1897.172	1949.181	1987.097	2014.280	2033.534	2047.056
082	Rohtak	1061.204	1181.738	1219.823	1247.772	1267.843	1282.034	1291.960	1298.849
083	Jhajjar	958.405	1083.132	1118.060	1143.863	1162.539	1175.855	1185.249	1191.827
084	Mahendragarh	922.088	1035.894	1075.370	1104.777	1126.162	1141.443	1152.226	1159.767
085	Rewari	900.332	1006.501	1057.859	1097.197	1126.398	1147.575	1162.674	1173.309
086	Gurgaon	1514.432	1681.193	2065.628	2472.351	2874.501	3246.384	3569.591	3835.689

(continued)

Table 2.6 (continued)

District Code	Name	Total population							
		Census	Forecast						
		2011	2011	2016	2021	2026	2031	2036	2041
087	Mewat	1089.263	1220.264	1376.353	1514.395	1629.755	1721.683	1792.200	1844.728
088	Faridabad	1809.733	2056.477	2274.246	2449.863	2584.633	2684.160	2755.589	2805.805
089	Palwal	1042.708	1162.231	1249.218	1318.610	1371.747	1411.174	1439.749	1460.107
National Capital Territory of Delhi									
090	North West	3656.539	4147.954	4436.943	4632.823	4760.182	4840.763	4890.872	4921.697
091	North	887.978	973.369	1037.326	1090.667	1131.753	1161.505	1182.099	1195.912
092	North East	2241.624	2537.770	2674.684	2757.811	2806.536	2834.508	2850.373	2859.311
093	East	1709.346	1946.753	2037.959	2099.926	2140.928	2167.585	2184.719	2195.651
094	New Delhi	142.004	142.004	142.004	142.004	142.004	142.004	142.004	142.004
095	Central	582.320	582.320	582.320	582.320	582.320	582.320	582.320	582.320
096	West	2543.243	2879.208	3032.664	3138.754	3209.863	3256.543	3286.771	3306.173
097	South West	2292.958	2602.867	2847.677	3032.344	3164.601	3255.831	3317.140	3357.618
098	South	2731.929	3105.700	3293.342	3425.153	3514.516	3573.647	3612.146	3636.948
Rajasthan									
099	Ganganagar	1969.168	2241.361	2327.859	2391.917	2438.350	2471.487	2494.873	2511.248
100	Hanumangarh	1774.692	1976.457	2052.133	2108.145	2148.776	2177.819	2198.363	2212.787
101	Bikaner	2363.937	2671.695	2881.386	3049.153	3178.207	3274.512	3344.762	3395.160
102	Churu	2039.547	2275.235	2390.544	2478.552	2543.815	2591.186	2625.038	2648.960

(continued)

Table 2.6 (continued)

District Code	Name	Total population							
		Census	Forecast						
		2011	2011	2016	2021	2026	2031	2036	2041
103	Jhunjhunun	2137.045	2415.792	2504.405	2569.591	2616.472	2649.643	2672.843	2688.939
104	Alwar	3674.179	4110.325	4390.358	4613.146	4783.851	4910.920	5003.484	5069.856
105	Bharatpur	2548.462	2837.538	2997.673	3121.726	3214.760	3282.845	3331.785	3366.508
106	Dhaulpur	1206.516	1351.859	1446.189	1521.099	1578.259	1620.561	1651.163	1672.939
107	Karauli	1458.248	1637.086	1745.148	1830.935	1896.572	1945.387	1980.929	2006.410
108	Sawai Madhopur	1335.551	1491.666	1574.235	1638.055	1685.822	1720.717	1745.762	1763.508
109	Dausa	1634.409	1832.183	1970.520	2081.874	2167.864	2232.162	2279.090	2312.739
110	Jaipur	6626.178	7440.967	8017.299	8475.982	8826.661	9086.622	9274.957	9409.144
111	Sikar	2677.333	3016.238	3170.439	3288.369	3375.940	3439.552	3485.027	3517.164
112	Nagaur	3307.743	3708.232	3917.070	4077.181	4196.040	4282.180	4343.527	4386.676
113	Jodhpur	3687.165	4108.321	4410.455	4648.810	4829.634	4962.782	5058.686	5126.672
114	Jaisalmer	669.919	757.145	842.047	914.390	972.902	1018.263	1052.284	1077.170
115	Barmer	2603.751	2898.348	3167.880	3390.840	3566.886	3700.845	3799.942	3871.728
116	Jalor	1828.730	2025.118	2161.550	2269.933	2352.877	2414.557	2459.452	2491.622
117	Sirohi	1036.346	1153.789	1230.111	1291.040	1337.853	1372.773	1398.249	1416.539
118	Pali	2037.573	2281.823	2358.594	2414.815	2455.068	2483.425	2503.176	2516.824
119	Ajmer	2583.052	2880.379	3040.546	3165.567	3259.803	3328.978	3378.776	3414.124
120	Tonk	1421.326	1588.612	1666.880	1727.027	1771.890	1804.609	1828.086	1844.732

(continued)

Table 2.6 (continued)

District Code	Name	Total population							
		Census	Forecast						
		2011	2011	2016	2021	2026	2031	2036	2041
121	Bundi	1110.906	1246.107	1301.637	1343.574	1374.415	1396.655	1412.466	1423.592
122	Bhilwara	2408.523	2684.330	2834.822	2951.932	3039.945	3104.373	3150.633	3183.390
123	Rajsamand	1156.597	1279.814	1335.282	1377.457	1408.612	1431.138	1447.175	1458.466
124	Dungarpur	1388.552	1541.328	1636.137	1709.743	1764.914	1805.188	1834.023	1854.383
125	Banswara	1797.485	1990.566	2116.697	2214.241	2287.098	2340.115	2377.967	2404.628
126	Chittaurgarh	1544.338	1718.200	1786.306	1837.570	1875.208	1902.339	1921.638	1935.236
127	Kota	1951.014	2177.213	2310.093	2412.990	2490.164	2546.668	2587.313	2616.178
128	Baran	1222.755	1364.920	1439.409	1496.554	1538.966	1569.675	1591.513	1606.844
129	Jhalawar	1411.129	1564.421	1639.388	1696.382	1738.436	1768.788	1790.346	1805.488
130	Udaipur	3068.420	3408.051	3611.516	3770.247	3889.962	3977.968	4041.452	4086.626
131	Pratapgarh	867.848	964.327	1019.089	1061.445	1093.183	1116.402	1133.094	1144.942
Uttar Pradesh									
132	Saharanpur	3466.382	3871.815	4085.944	4252.258	4377.295	4469.011	4535.076	4582.044
133	Muzaffarnagar	4143.512	4637.872	4866.972	5041.720	5171.013	5264.530	5331.068	5377.856
134	Bijnor	3682.713	4142.611	4369.120	4544.296	4675.490	4771.394	4840.262	4889.085
135	Moradabad	4772.006	5334.875	5729.520	6046.332	6290.346	6472.346	6604.857	6699.646
136	Rampur	2335.819	2607.952	2765.123	2888.240	2981.441	3050.177	3099.899	3135.364
137	Jyotiba Phule Nagar	1840.221	2062.475	2207.102	2323.036	2412.441	2479.346	2528.294	2563.513

(continued)

Table 2.6 (continued)

District Code	Name	Total population							
		Census	Forecast						
		2011	2011	2016	2021	2026	2031	2036	2041
138	Meerut	3443.689	3863.645	4043.274	4178.948	4278.484	4349.960	4400.498	4435.843
139	Baghpat	1303.048	1458.105	1501.529	1533.602	1556.788	1573.291	1584.907	1593.020
140	Ghaziabad	4681.645	5245.795	5972.894	6625.121	7176.501	7619.699	7961.703	8217.394
141	Gautam Buddha Nagar	1648.115	1841.963	2160.021	2466.282	2742.381	2976.870	3166.124	3312.679
142	Bulandshahr	3499.171	3890.855	4048.014	4166.013	4252.178	4313.826	4357.286	4387.609
143	Aligarh	3673.889	4097.375	4361.707	4572.672	4734.723	4855.568	4943.705	5006.949
144	Mahamaya Nagar	1564.708	1740.605	1816.432	1873.566	1915.315	1945.142	1966.109	1980.679
145	Mathura	2547.184	2833.982	3024.479	3177.382	3295.251	3383.299	3447.531	3493.579
146	Agra	4418.797	4919.733	5256.595	5527.377	5736.614	5893.406	6008.210	6090.852
147	Firozabad	2498.156	2796.545	2978.188	3122.202	3232.141	3313.662	3372.814	3415.063
148	Mainpuri	1868.529	2079.947	2178.757	2254.331	2310.348	2350.910	2379.787	2400.097
149	Budaun	3681.896	4122.164	4392.312	4609.122	4776.132	4900.732	4991.475	5056.398
150	Bareilly	4448.359	4975.489	5318.256	5593.152	5805.013	5963.315	6078.870	6161.787
151	Pilibhit	2031.007	2283.318	2412.427	2512.574	2587.732	2642.749	2682.292	2710.341
152	Shahjahanpur	3006.538	3355.992	3593.955	3786.272	3935.143	4046.568	4127.878	4186.119
153	Kheri	4021.243	4511.755	4879.390	5180.471	5416.127	5594.113	5724.953	5819.234
154	Sitapur	4483.992	4992.166	5345.606	5631.026	5851.956	6017.419	6138.305	6225.034
155	Hardoi	4092.845	4550.772	4823.099	5039.571	5205.288	5328.508	5418.148	5482.326

(continued)

Table 2.6 (continued)

District Code	Name	Total population							
		Census	Forecast						
		2011	2011	2016	2021	2026	2031	2036	2041
156	Unnao	3108.367	3473.361	3625.248	3739.588	3823.142	3882.879	3924.924	3954.191
157	Lucknow	4589.838	5144.374	5573.067	5923.695	6197.848	6404.734	6556.705	6666.138
158	Rae Bareli	3405.559	3813.536	4039.446	4217.611	4352.962	4452.894	4525.129	4576.550
159	Farrukhabad	1885.204	2101.275	2217.752	2308.522	2376.832	2426.893	2462.867	2488.354
160	Kannauj	1656.616	1843.532	1932.406	2000.342	2050.729	2087.269	2113.337	2131.717
161	Etawah	1581.810	1747.702	1818.824	1872.241	1911.260	1939.183	1958.874	1972.617
162	Auraiya	1379.545	1529.512	1587.505	1630.386	1661.286	1683.140	1698.392	1708.939
163	Kanpur Dehat	1796.184	1999.785	2075.469	2131.685	2172.361	2201.239	2221.464	2235.495
164	Kanpur Nagar	4581.268	5188.404	5389.957	5539.463	5647.671	5724.608	5778.622	5816.207
165	Jalaun	1689.974	1884.171	1965.217	2027.133	2073.140	2106.623	2130.625	2147.642
166	Jhansi	1998.603	2241.393	2332.986	2401.546	2451.579	2487.417	2512.748	2530.483
167	Lalitpur	1221.592	1365.811	1471.366	1557.111	1623.857	1674.100	1710.968	1737.518
168	Hamirpur	1104.285	1236.506	1275.119	1303.450	1323.789	1338.166	1348.217	1355.190
169	Mahoba	875.958	978.320	1020.002	1051.494	1074.641	1091.309	1103.135	1111.437
170	Banda	1799.410	1989.774	2081.127	2150.686	2202.159	2239.445	2266.039	2284.798
171	Chitrakoot	991.730	1103.222	1176.298	1234.046	1277.943	1310.339	1333.727	1350.347
172	Fatehpur	2632.733	2948.665	3068.153	3157.733	3223.078	3269.797	3302.723	3325.693
173	Pratapgarh	3209.141	3594.410	3792.913	3948.226	4065.546	4151.831	4214.053	4258.287

(continued)

Table 2.6 (continued)

District Code	Name	Total population							
		Census	Forecast						
		2011	2011	2016	2021	2026	2031	2036	2041
174	Kaushambi	1599.596	1781.870	1905.515	2003.828	2078.770	2134.079	2173.931	2202.154
175	Allahabad	5954.391	6680.868	7091.533	7413.143	7656.201	7834.983	7963.875	8055.457
176	Bara Banki	3260.699	3639.705	3893.497	4097.712	4254.961	4371.987	4456.884	4517.344
177	Faizabad	2470.996	2765.321	2920.088	3038.799	3126.418	3189.278	3233.468	3264.095
178	Ambedkar Nagar	2397.888	2686.665	2838.891	2957.607	3046.966	3112.444	3159.483	3192.796
179	Sultanpur	3797.117	4249.385	4506.749	4710.248	4865.178	4979.754	5062.669	5121.732
180	Bahraich	3487.731	3878.236	4225.726	4517.806	4751.692	4931.788	5066.310	5164.519
181	Shrawasti	1117.361	1240.360	1369.172	1482.807	1577.661	1653.197	1711.096	1754.170
182	Balrampur	2148.665	2387.343	2591.408	2763.558	2901.751	3008.286	3087.860	3145.893
183	Gonda	3433.919	3836.469	4148.224	4406.672	4610.720	4765.669	4879.879	4962.221
184	Siddharthnagar	2559.297	2846.635	3105.737	3327.994	3508.876	3649.844	3756.018	3833.937
185	Basti	2464.464	2747.039	2904.555	3028.534	3122.513	3191.729	3241.626	3277.042
186	Sant Kabir Nagar	1715.183	1920.351	2044.226	2140.906	2213.136	2265.360	2302.231	2327.829
187	Mahrajganj	2684.703	3003.278	3230.992	3416.822	3562.094	3671.884	3752.750	3811.189
188	Gorakhpur	4440.895	4971.698	5241.308	5451.130	5608.900	5724.486	5807.562	5866.454
189	Kushinagar	3564.544	3998.470	4297.445	4539.411	4726.974	4867.563	4970.312	5044.026
190	Deoria	3100.946	3498.574	3672.863	3806.773	3906.402	3978.768	4030.418	4066.826
191	Azamgarh	4613.913	5189.437	5489.633	5725.696	5904.892	6037.308	6133.225	6201.703

(continued)

Table 2.6 (continued)

District Code	Name	Total population							
		Census	Forecast						
		2011	2011	2016	2021	2026	2031	2036	2041
192	Mau	2205.968	2475.364	2632.819	2757.753	2853.218	2924.086	2975.574	3012.399
193	Ballia	3239.774	3608.284	3793.364	3937.092	4045.046	4124.116	4180.970	4221.308
194	Jaunpur	4494.204	5058.086	5299.414	5483.808	5620.423	5719.342	5789.781	5839.345
195	Ghazipur	3620.268	4045.272	4274.181	4450.510	4581.552	4676.378	4743.687	4790.817
196	Chandauli	1952.756	2193.813	2330.021	2437.407	2519.054	2579.435	2623.181	2654.403
197	Varanasi	3676.841	4134.020	4359.319	4534.387	4666.111	4762.833	4832.588	4882.242
198	Sant Ravidas Nagar	1578.213	1781.605	1865.559	1926.817	1970.259	2000.447	2021.127	2035.157
199	Mirzapur	2496.970	2802.434	2980.626	3120.963	3227.494	3306.120	3362.952	3403.412
200	Sonbhadra	1862.559	2097.856	2255.795	2382.011	2479.047	2551.443	2604.253	2642.142
201	Etah	1774.480	1986.412	2088.860	2168.646	2228.732	2272.844	2304.622	2327.204
202	Kanshiram Nagar	1436.719	1603.497	1687.354	1752.965	1802.563	1839.083	1865.451	1884.219
Bihar									
203	Pashchim Champaran	3935.042	4334.745	4683.809	4970.584	5195.752	5366.332	5492.082	5582.932
204	Purba Champaran	5099.371	5674.947	6188.595	6618.388	6960.325	7221.611	7415.183	7555.330
205	Sheohar	656.246	734.961	813.718	882.842	940.182	985.549	1020.111	1045.689
206	Sitamarhi	3423.574	3819.798	4180.324	4484.257	4727.280	4913.481	5051.516	5151.359
207	Madhubani	4487.379	4985.838	5354.959	5655.677	5890.226	6067.027	6196.896	6290.490
208	Supaul	2229.076	2443.068	2593.825	2711.716	2800.848	2866.533	2914.030	2947.908

(continued)

Table 2.6 (continued)

District Code	Name	Total population							
		Census	Forecast						
		2011	2011	2016	2021	2026	2031	2036	2041
209	Araria	2811.569	3101.505	3328.596	3508.499	3645.513	3746.766	3819.938	3871.967
210	Kishanganj	1690.400	1862.452	1996.109	2102.506	2183.997	2244.585	2288.645	2320.172
211	Purnia	3264.619	3613.475	3878.006	4087.830	4247.744	4365.953	4451.375	4512.098
212	Katihar	3071.029	3405.036	3660.652	3865.565	4023.196	4140.653	4226.112	4287.218
213	Madhepura	2001.762	2217.737	2401.309	2551.125	2667.801	2755.399	2819.372	2865.155
214	Saharsa	1900.661	2121.203	2293.334	2433.917	2543.687	2626.426	2687.147	2730.841
215	Darbhanga	3937.385	4426.955	4728.797	4971.214	5158.383	5298.543	5401.114	5474.920
216	Muzaffarpur	4801.062	5312.339	5733.805	6079.047	6349.344	6553.566	6703.754	6812.024
217	Gopalganj	2562.012	2867.433	3045.017	3186.047	3293.899	3374.012	3432.235	3473.879
218	Siwan	3330.464	3703.456	3947.093	4141.582	4290.702	4401.538	4482.029	4539.503
219	Saran	3951.862	4390.773	4677.177	4905.001	5079.076	5208.032	5301.388	5367.850
220	Vaishali	3495.021	3877.716	4192.244	4450.539	4653.241	4806.721	4919.809	5001.473
221	Samastipur	4261.566	4733.296	5070.004	5338.717	5544.161	5696.142	5805.854	5883.661
222	Begusarai	2970.541	3306.725	3559.479	3764.498	3923.769	4043.397	4130.993	4193.945
223	Khagaria	1666.886	1840.958	2001.743	2135.759	2242.302	2323.861	2384.504	2428.625
224	Bhagalpur	3037.766	3361.393	3580.452	3753.431	3884.790	3981.612	4051.423	4100.961
225	Banka	2034.763	2256.875	2417.190	2546.779	2647.307	2722.831	2778.212	2818.107
226	Munger	1367.765	1509.658	1583.673	1640.771	1683.413	1714.493	1736.744	1752.472

(continued)

Table 2.6 (continued)

District Code	Name	Total population							
		Census	Forecast						
		2011	2011	2016	2021	2026	2031	2036	2041
227	Lakhisarai	1000.912	1100.634	1170.916	1226.634	1269.091	1300.479	1323.166	1339.298
228	Sheikhpura	636.342	707.452	750.709	785.131	811.495	831.099	845.358	855.564
229	Nalanda	2877.653	3182.456	3338.446	3458.233	3547.364	3612.141	3658.419	3691.078
230	Patna	5838.465	6503.451	6961.186	7328.912	7612.647	7824.824	7979.795	8091.045
231	Bhojpur	2728.407	3034.696	3219.632	3365.549	3476.364	3558.105	3617.113	3659.051
232	Buxar	1706.352	1894.865	2018.138	2116.661	2192.341	2248.715	2289.753	2319.125
233	Kaimur (Bhabua)	1626.384	1808.021	1960.617	2087.577	2188.449	2265.692	2323.183	2365.073
234	Rohtas	2959.918	3299.078	3488.049	3635.489	3746.620	3828.219	3886.987	3928.726
235	Aurangabad	2540.073	2835.201	3066.538	3257.525	3408.237	3522.980	3607.978	3669.672
236	Gaya	4391.418	4882.772	5256.307	5557.922	5791.048	5965.222	6092.079	6182.769
237	Nawada	2219.146	2485.577	2671.860	2823.085	2940.733	3029.262	3094.225	3141.020
238	Jamui	1760.405	1961.790	2120.654	2251.556	2354.697	2433.131	2491.180	2533.283
239	Jehanabad	1125.313	1257.722	1346.508	1418.576	1474.673	1516.928	1547.977	1570.379
240	Arwal	700.843	781.033	827.595	864.554	892.829	913.855	929.163	940.138
Sikkim									
241	North District	43.709	49.099	49.807	50.218	50.454	50.589	50.665	50.709
242	West District	136.435	153.848	157.576	159.754	160.996	161.696	162.086	162.303
243	South District	146.850	166.396	171.163	173.978	175.598	176.517	177.034	177.324

(continued)

Table 2.6 (continued)

District Code	Name	Total population							
		Census	Forecast						
		2011	2011	2016	2021	2026	2031	2036	2041
244	East District	283.583	319.067	330.022	336.704	340.665	342.975	344.308	345.074
Arunachal Pradesh									
245	Tawang	49.977	56.150	61.311	65.370	68.376	70.504	71.960	72.935
246	West Kameng	83.947	95.077	98.123	100.024	101.194	101.912	102.351	102.621
247	East Kameng	78.690	85.140	91.093	95.650	98.956	101.259	102.818	103.853
248	Papum Pare	176.573	199.529	224.140	243.467	257.734	267.785	274.634	279.193
249	Upper Subansiri	83.448	89.814	97.999	104.486	109.321	112.760	115.123	116.708
250	West Siang	112.274	127.037	129.607	131.271	132.332	133.002	133.423	133.687
251	East Siang	99.214	111.272	114.172	115.987	117.096	117.765	118.166	118.403
252	Upper Siang	35.320	39.240	39.791	40.145	40.370	40.513	40.602	40.659
253	Changlang	148.226	166.535	173.333	177.972	181.061	183.085	184.396	185.239
254	Tirap	111.975	125.705	129.122	131.428	132.950	133.938	134.575	134.982
255	Lower Subansiri	83.030	90.984	102.694	112.578	120.309	126.001	130.005	132.731
256	Kurung Kumey	92.076	98.704	139.803	201.501	291.064	414.942	575.340	766.285
257	Dibang Valley	8.004	8.723	8.743	8.755	8.762	8.767	8.769	8.771
258	Lower Dibang Valley	54.080	61.921	63.283	64.050	64.475	64.708	64.836	64.905
259	Lohit	145.726	165.444	172.710	177.654	180.932	183.068	184.443	185.322
260	Anjaw	21.167	23.205	23.821	24.264	24.572	24.780	24.917	25.007

(continued)

Table 2.6 (continued)

District Code	Name	Total population							
		Census	Forecast						
		2011	2011	2016	2021	2026	2031	2036	2041
Nagaland									
261	Mon	250.260	297.695	300.506	301.575	301.977	302.128	302.185	302.206
262	Mokokchung	194.622	242.663	244.524	245.398	245.808	246.000	246.090	246.133
263	Zunheboto	140.757	170.221	172.632	173.811	174.380	174.653	174.783	174.845
264	Wokha	166.343	195.636	200.086	202.133	203.056	203.469	203.653	203.735
265	Dimapur	378.811	433.584	464.233	485.489	499.654	508.843	514.703	518.398
266	Phek	163.418	187.093	193.397	197.214	199.466	200.775	201.528	201.959
267	Tuensang	196.596	228.772	236.316	240.577	242.919	244.187	244.867	245.231
268	Longleng	50.484	96.425	96.447	96.452	96.453	96.453	96.453	96.453
269	Kiphire	74.004	102.470	102.500	102.504	102.505	102.505	102.505	102.505
270	Kohima	267.988	303.891	322.258	335.139	343.849	349.595	353.326	355.723
271	Peren	95.219	110.677	112.886	113.964	114.483	114.730	114.847	114.903
Manipur									
272	Senapati	479.148	527.924	618.382	703.576	778.883	841.809	891.961	930.448
273	Tamenglong	140.651	157.570	169.888	179.810	187.472	193.200	197.377	200.368
274	Churachandpur	274.143	305.833	321.307	333.022	341.650	347.878	352.306	355.423
275	Bishnupur	237.399	264.251	271.905	277.478	281.468	284.288	286.266	287.643
276	Thoubal	422.168	470.483	489.220	503.129	513.213	520.397	525.453	528.980

(continued)

Table 2.6 (continued)

District Code	Name	Total population							
		Census	Forecast						
		2011	2011	2016	2021	2026	2031	2036	2041
277	Imphal West	517.992	573.061	592.087	606.042	616.066	623.160	628.126	631.578
278	Imphal East	456.113	507.680	524.952	537.618	546.727	553.188	557.725	560.888
279	Ukhrul	183.998	204.833	222.416	236.964	248.475	257.263	263.790	268.538
280	Chandel	144.182	164.896	179.583	191.061	199.645	205.862	210.262	213.327
Mizoram									
281	Mamit	86.364	88.759	89.747	90.309	90.626	90.803	90.901	90.956
282	Kolasib	83.955	93.831	99.543	103.258	105.551	106.920	107.721	108.185
283	Aizawl	400.309	448.438	467.318	478.819	485.606	489.537	491.789	493.071
284	Champhai	125.745	141.545	146.892	150.198	152.172	153.325	153.991	154.372
285	Serchhip	64.937	72.490	75.556	77.454	78.571	79.209	79.568	79.768
286	Lunglei	161.428	179.176	184.141	187.094	188.805	189.782	190.335	190.647
287	Lawngtlai	117.894	131.248	140.190	146.039	149.666	151.841	153.119	153.862
Tripura									
288	Saiha	56.574	63.533	66.019	67.554	68.475	69.017	69.333	69.516
289	West Tripura	1725.739	1931.959	1990.292	2032.609	2062.783	2084.035	2098.874	2109.172
290	South Tripura	876.001	977.861	1000.094	1015.676	1026.460	1033.859	1038.905	1042.332
291	Dhalai	378.230	416.155	428.125	436.485	442.234	446.147	448.790	450.568
292	North Tripura	693.947	765.342	796.525	819.924	837.065	849.402	858.171	864.346

(continued)

Table 2.6 (continued)

District Code	Name	Total population							
		Census	Forecast						
		2011	2011	2016	2021	2026	2031	2036	2041
Meghalaya									
293	West Garo Hills	643.291	716.743	769.708	812.350	845.352	870.122	888.287	901.385
294	East Garo Hills	317.917	356.233	383.186	404.709	421.227	433.522	442.466	448.865
295	South Garo Hills	142.334	156.906	168.173	177.287	184.359	189.669	193.560	196.361
296	West Khasi Hills	383.461	428.429	464.295	493.438	516.138	533.245	545.817	554.887
297	Ribhoi	258.840	290.148	319.491	343.703	362.779	377.267	387.968	395.710
298	East Khasi Hills	825.922	912.721	967.026	1009.348	1041.212	1064.584	1081.400	1093.332
299	Jaintia Hills	395.124	442.495	483.262	516.747	543.036	562.953	577.638	588.250
Assam									
300	Kokrajhar	887.142	1009.694	1025.532	1036.028	1042.915	1047.404	1050.317	1052.202
301	Dhubri	1949.258	2146.414	2257.742	2343.448	2407.487	2454.275	2487.902	2511.787
302	Goalpara	1008.183	1103.287	1149.920	1184.782	1210.208	1228.420	1241.295	1250.315
303	Barpeta	1693.622	1847.651	1917.615	1969.819	2007.849	2035.072	2054.317	2067.801
304	Morigaon	957.423	1046.002	1092.074	1126.928	1152.615	1171.184	1184.421	1193.762
305	Nagaon	2823.768	3085.284	3209.126	3300.930	3367.338	3414.525	3447.632	3470.652
306	Sonitpur	1924.110	2132.330	2202.202	2253.568	2290.562	2316.815	2335.249	2348.098
307	Lakhimpur	1042.137	1143.172	1177.457	1202.549	1220.598	1233.420	1242.448	1248.764
308	Dhemaji	686.133	756.427	781.763	799.915	812.698	821.591	827.725	831.931

(continued)

Table 2.6 (continued)

District Code	Name	Total population							
		Census	Forecast						
		2011	2011	2016	2021	2026	2031	2036	2041
309	Tinsukia	1327.929	1462.594	1507.330	1539.968	1563.311	1579.770	1591.257	1599.218
310	Dibrugarh	1326.335	1464.737	1497.700	1521.639	1538.753	1550.850	1559.333	1565.247
311	Sivasagar	1151.050	1286.962	1319.052	1342.442	1359.187	1371.019	1379.304	1385.067
312	Jorhat	1092.256	1213.714	1239.566	1258.015	1270.959	1279.932	1286.101	1290.317
313	Golaghat	1066.888	1191.153	1224.378	1248.739	1266.295	1278.790	1287.604	1293.782
314	Karbi Anglong	956.313	1054.218	1089.614	1115.509	1134.119	1147.325	1156.611	1163.099
315	Dima Hasao	214.102	242.259	252.267	259.708	265.116	268.982	271.712	273.624
316	Cachar	1736.617	1914.335	1997.985	2061.790	2109.121	2143.513	2168.128	2185.557
317	Karimganj	1228.686	1351.801	1416.643	1466.496	1503.680	1530.793	1550.237	1564.016
318	Hailakandi	659.296	724.648	757.361	782.066	800.185	813.188	822.377	828.798
319	Bongaigaon	738.804	813.044	846.801	872.143	890.718	904.097	913.612	920.318
320	Chirang	482.162	537.117	542.621	546.230	548.581	550.108	551.096	551.735
321	Kamrup	1517.542	1669.518	1724.897	1766.119	1796.137	1817.649	1832.889	1843.598
322	Kamrup Metropolitan	1253.938	1422.192	1502.341	1562.885	1607.351	1639.331	1661.983	1677.854
323	Nalbari	771.639	846.884	863.475	875.331	883.698	889.550	893.618	896.433
324	Baksa	950.075	1058.262	1077.636	1091.243	1100.703	1107.231	1111.714	1114.783
325	Darrang	928.500	1020.357	1067.101	1102.194	1127.801	1146.102	1158.987	1167.963
326	Udalguri	831.668	927.018	941.555	951.743	958.824	963.714	967.079	969.387

(continued)

Table 2.6 (continued)

District Code	Name	Total population							
		Census	Forecast						
		2011	2011	2016	2021	2026	2031	2036	2041
West Bengal									
327	Darjiling	1846.823	2071.167	2159.075	2225.428	2274.271	2309.561	2334.718	2352.479
328	Jalpaiguri	3872.846	4323.777	4474.849	4586.559	4667.466	4725.188	4765.928	4794.464
329	Koch Bihar	2819.086	3104.602	3176.929	3229.081	3266.174	3292.300	3310.576	3323.298
330	Uttar Dinajpur	3007.134	3347.914	3543.774	3695.171	3808.591	3891.583	3951.270	3993.664
331	Dakshin Dinajpur	1676.276	1886.266	1950.308	1997.461	2031.474	2055.645	2072.642	2084.505
332	Maldah	3988.845	4428.140	4649.579	4815.900	4937.171	5023.705	5084.511	5126.783
333	Murshidabad	7103.807	7892.602	8305.537	8623.185	8860.192	9033.034	9156.990	9244.823
334	Birbhum	3502.404	3866.095	3998.426	4096.283	4167.021	4217.319	4252.666	4277.301
335	Barddhaman	7717.563	8588.373	8806.179	8963.895	9076.278	9155.443	9210.757	9249.188
336	Nadia	5167.600	5807.420	5996.736	6135.246	6234.678	6305.088	6354.466	6388.860
337	North Twenty	10009.781	11292.342	11696.224	11993.015	12206.647	12358.146	12464.455	12538.504
338	Hugli	5519.145	6172.348	6316.602	6420.263	6493.651	6545.062	6580.813	6605.548
339	Bankura	3596.674	3974.153	4076.081	4150.759	4204.494	4242.661	4269.520	4288.299
340	Puruliya	2930.115	3239.104	3353.536	3439.638	3502.781	3548.218	3580.473	3603.149
341	Haora	4850.029	5420.274	5599.652	5733.275	5830.677	5900.556	5950.120	5984.991
342	Kolkata	4496.694	5177.264	5198.594	5212.676	5221.921	5227.967	5231.914	5234.486
343	South Twenty Four Parganas	8161.961	9091.908	9495.174	9800.839	10026.493	10189.856	10306.456	10388.839

(continued)

Table 2.6 (continued)

District Code	Name	Total population							
		Census	Forecast						
		2011	2011	2016	2021	2026	2031	2036	2041
344	Paschim Medinipur	5913.457	6553.544	6735.629	6867.570	6961.425	7027.317	7073.151	7104.829
345	Purba Medinipur	5095.875	5657.471	5834.964	5963.773	6055.324	6119.436	6163.869	6194.443
Jharkhand									
346	Garhwa	1322.784	1473.013	1583.944	1672.791	1741.095	1791.969	1828.975	1855.432
347	Chatra	1042.886	1158.258	1251.498	1327.226	1386.121	1430.398	1462.843	1486.174
348	Kodarma	716.259	793.799	866.557	927.350	975.713	1012.703	1040.153	1060.067
349	Giridih	2445.474	2711.834	2918.392	3085.565	3215.249	3312.567	3383.782	3434.938
350	Deoghar	1492.073	1658.712	1792.833	1902.978	1989.405	2054.827	2103.017	2137.803
351	Godda	1313.551	1449.745	1548.241	1628.169	1690.366	1737.203	1771.609	1796.424
352	Sahibganj	1150.567	1270.608	1352.247	1417.342	1467.274	1504.447	1531.511	1550.897
353	Pakur	900.422	994.673	1069.544	1130.863	1178.896	1215.227	1241.985	1261.308
354	Dhanbad	2684.487	3027.373	3139.644	3222.434	3282.046	3324.234	3353.726	3374.165
355	Bokaro	2062.330	2318.853	2418.380	2492.795	2547.098	2586.022	2613.567	2632.881
356	Lohardaga	461.790	514.019	558.120	595.315	625.183	648.238	665.495	678.119
357	Purbi Singhbhum	2293.919	2546.982	2643.351	2714.803	2766.485	2803.199	2828.949	2846.845
358	Palamu	1939.869	2156.921	2312.216	2436.549	2532.150	2603.397	2655.269	2692.395
359	Latehar	726.978	804.612	864.088	911.971	948.896	976.436	996.470	1010.779
360	Hazaribagh	1734.495	1921.656	2040.458	2132.804	2202.099	2252.734	2289.021	2314.663

(continued)

Table 2.6 (continued)

District Code	Name	Total population							
		Census	Forecast						
		2011	2011	2016	2021	2026	2031	2036	2041
361	Ramgarh	949.443	1068.995	1105.074	1131.330	1150.076	1163.277	1172.483	1178.861
362	Dumka	1321.442	1460.652	1536.406	1595.913	1640.970	1674.140	1698.059	1715.048
363	Jamtara	791.042	882.465	938.006	982.447	1016.582	1041.987	1060.454	1073.650
364	Ranchi	2914.253	3229.311	3423.800	3574.180	3686.359	3767.826	3825.843	3866.590
365	Khunti	531.885	581.135	611.340	635.500	654.086	667.959	678.083	685.351
366	Gumla	1025.213	1118.978	1181.316	1230.787	1268.529	1296.464	1316.679	1331.070
367	Simdega	599.578	654.665	677.059	694.118	706.758	715.932	722.491	727.131
368	Pashchimi Singhbhum	1502.338	1641.006	1721.976	1785.583	1833.700	1869.068	1894.519	1912.554
369	Saraikela-Kharsawan	1065.056	1172.468	1247.788	1307.888	1353.934	1388.123	1412.920	1430.600
Odisha									
370	Bargarh	1481.255	1655.350	1692.636	1719.491	1738.516	1751.834	1761.082	1767.466
371	Jharsuguda	579.505	650.032	672.393	688.897	700.778	709.176	715.038	719.092
372	Sambalpur	1041.099	1159.566	1190.254	1212.661	1228.732	1240.112	1248.097	1253.664
373	Debagarh	312.520	345.680	355.497	362.671	367.818	371.462	374.016	375.796
374	Sundargarh	2093.437	2306.624	2363.989	2405.331	2434.674	2455.276	2469.630	2479.578
375	Kendujhar	1801.733	1986.366	2048.289	2094.011	2127.073	2150.621	2167.214	2178.817
376	Mayurbhanj	2519.738	2801.582	2906.341	2985.958	3044.835	3087.500	3117.964	3139.488
377	Baleshwar	2320.529	2583.081	2672.077	2737.350	2784.189	2817.274	2840.388	2856.410

(continued)

Table 2.6 (continued)

District Code	Name	Total population							
		Census	Forecast						
		2011	2011	2016	2021	2026	2031	2036	2041
378	Bhadrak	1506.337	1684.193	1742.580	1785.777	1817.033	1839.287	1854.950	1865.885
379	Kendrapara	1440.361	1594.721	1627.189	1650.280	1666.467	1677.700	1685.443	1690.754
380	Jagatsinghapur	1136.971	1277.400	1302.821	1320.706	1333.105	1341.611	1347.406	1351.334
381	Cuttack	2624.470	2913.966	2986.320	3038.189	3074.722	3100.134	3117.658	3129.670
382	Jajapur	1827.192	2042.304	2106.388	2153.423	2187.214	2211.116	2227.841	2239.453
383	Dhenkanal	1192.811	1323.495	1355.011	1377.982	1394.456	1406.134	1414.343	1420.081
384	Anugul	1273.821	1428.988	1476.054	1510.989	1536.380	1554.552	1567.416	1576.451
385	Nayagarh	962.789	1063.451	1084.414	1099.558	1110.342	1117.941	1123.257	1126.957
386	Khordha	2251.673	2517.190	2648.433	2747.063	2818.891	2870.017	2905.822	2930.614
387	Puri	1698.730	1888.832	1944.196	1984.977	2014.412	2035.347	2050.082	2060.376
388	Ganjam	3529.031	3948.028	4072.434	4164.603	4231.371	4278.953	4312.469	4335.883
389	Gajapati	577.817	640.332	656.885	669.010	677.711	683.862	688.166	691.154
390	Kandhamal	733.110	820.374	853.304	878.335	896.828	910.205	919.734	926.447
391	Baudh	441.162	488.399	510.263	527.073	539.601	548.726	555.259	559.880
392	Subarnapur	610.183	682.182	703.346	719.041	730.429	738.560	744.300	748.320
393	Balangir	1648.997	1776.771	1837.103	1882.403	1915.546	1939.337	1956.182	1967.994
394	Nuapada	610.382	670.627	688.617	701.695	711.022	717.585	722.158	725.324
395	Kalahandi	1576.869	1740.186	1817.124	1876.194	1920.182	1952.202	1975.123	1991.337

(continued)

Table 2.6 (continued)

District Code	Name	Total population							
		Census	Forecast						
		2011	2011	2016	2021	2026	2031	2036	2041
396	Rayagada	967.911	1066.137	1101.952	1128.566	1147.888	1161.681	1171.407	1178.208
397	Nabarangapur	1220.946	1354.518	1415.144	1461.162	1495.152	1519.753	1537.300	1549.683
398	Koraput	1379.647	1532.596	1595.127	1642.925	1678.436	1704.263	1722.758	1735.855
399	Malkangiri	613.192	666.470	689.237	706.045	718.227	726.937	733.104	737.442
Chhattisgarh									
400	Koriya	658.917	740.494	764.158	781.422	793.772	802.484	808.569	812.789
401	Surguja	2359.886	2631.435	2773.232	2883.052	2965.396	3025.652	3068.964	3099.696
402	Jashpur	851.669	932.636	955.294	971.718	983.416	991.645	997.383	1001.360
403	Raigarh	1493.984	1650.632	1722.688	1777.702	1818.460	1847.989	1869.038	1883.868
404	Korba	1206.640	1356.252	1425.471	1478.283	1517.447	1545.881	1566.208	1580.579
405	Janjgir-Champa	1619.707	1803.488	1938.974	2050.626	2138.237	2204.380	2252.875	2287.673
406	Bilaspur	2663.629	2906.073	3100.117	3254.582	3372.742	3460.406	3523.980	3569.325
407	Kabeerdham	822.526	890.764	955.838	1008.296	1048.886	1079.307	1101.566	1117.569
408	Rajnandgaon	1537.133	1695.450	1768.476	1823.510	1863.799	1892.674	1913.054	1927.284
409	Durg	3343.872	3686.279	3820.843	3920.105	3991.721	4042.568	4078.258	4103.110
410	Raipur	4063.872	4408.303	4698.945	4927.530	5100.658	5228.075	5319.885	5385.032
411	Mahasamund	1032.754	1124.046	1157.187	1181.601	1199.210	1211.716	1220.502	1226.627
412	Dhamtari	799.781	897.336	933.119	960.207	980.179	994.620	1004.915	1012.180

(continued)

Table 2.6 (continued)

District Code	Name	Total population							
		Census	Forecast						
		2011	2011	2016	2021	2026	2031	2036	2041
413	Uttar Bastar Kanker	748.941	827.229	852.558	871.084	884.380	893.795	900.397	904.996
414	Bastar	1413.199	1563.801	1626.440	1673.627	1708.281	1733.259	1751.022	1763.531
415	Narayanpur	139.820	156.145	164.135	170.264	174.825	178.143	180.518	182.198
416	Dakshin Bastar Dantewada	533.638	588.050	600.921	610.204	616.800	621.437	624.671	626.916
417	Bijapur	255.230	289.970	298.600	304.722	308.981	311.905	313.895	315.240
Madhya Pradesh									
418	Sheopur	687.861	770.128	821.225	861.427	891.927	914.433	930.704	942.294
419	Morena	1965.970	2189.255	2318.746	2418.704	2493.233	2547.376	2585.972	2613.113
420	Bhind	1703.005	1885.542	1965.009	2024.640	2068.168	2099.306	2121.260	2136.581
421	Gwalior	2032.036	2258.005	2400.011	2511.720	2596.550	2659.254	2704.686	2737.127
422	Datia	786.754	875.535	917.772	950.087	974.082	991.503	1003.946	1012.731
423	Shivpuri	1726.050	1933.917	2065.888	2170.959	2251.420	2311.215	2354.667	2385.731
424	Tikamgarh	1445.166	1617.592	1712.484	1786.322	1841.853	1882.554	1911.826	1932.593
425	Chhatarpur	1762.375	1976.235	2095.539	2188.450	2258.247	2309.270	2345.831	2371.654
426	Panna	1016.520	1134.543	1190.057	1232.249	1263.392	1285.886	1301.880	1313.124
427	Sagar	2378.458	2646.910	2762.303	2849.235	2913.010	2958.893	2991.440	3014.298
428	Damoh	1264.219	1407.195	1465.550	1509.603	1542.030	1565.457	1582.156	1593.945
429	Satna	2228.935	2492.262	2627.436	2731.538	2809.214	2865.814	2906.348	2935.017

(continued)

Table 2.6 (continued)

District Code	Name	Total population							
		Census	Forecast						
		2011	2011	2016	2021	2026	2031	2036	2041
430	Rewa	2365.106	2645.943	2803.843	2927.771	3021.727	3091.104	3141.343	3177.210
431	Umaria	644.758	714.816	758.331	792.378	818.114	837.064	850.751	860.500
432	Neemuch	826.067	925.951	960.283	985.812	1004.369	1017.638	1027.014	1033.584
433	Mandsaur	1340.411	1506.620	1566.436	1611.292	1644.132	1667.753	1684.527	1696.333
434	Ratlam	1455.069	1615.183	1695.333	1756.401	1801.544	1834.175	1857.381	1873.695
435	Ujjain	1986.864	2221.345	2319.137	2393.258	2447.975	2487.587	2515.859	2535.833
436	Shajapur	1512.681	1690.789	1773.028	1836.044	1882.916	1917.016	1941.425	1958.695
437	Dewas	1563.715	1746.914	1836.832	1905.216	1955.721	1992.221	2018.186	2036.452
438	Dhar	2185.793	2432.630	2595.319	2723.651	2821.132	2893.075	2945.050	2982.022
439	Indore	3276.697	3643.247	3967.808	4234.036	4442.788	4600.759	4717.119	4801.135
440	Khargone (West Nimar)	1873.046	2072.411	2187.250	2276.058	2342.580	2391.225	2426.176	2450.969
441	Barwani	1385.881	1540.377	1657.193	1751.522	1824.589	1879.384	1919.486	1948.315
442	Rajgarh	1545.814	1715.149	1824.415	1910.973	1977.012	2025.966	2061.483	2086.852
443	Vidisha	1458.875	1617.147	1697.847	1759.460	1805.090	1838.126	1861.654	1878.213
444	Bhopal	2371.061	2661.776	2865.575	3025.754	3147.097	3236.480	3300.968	3346.801
445	Sehore	1311.332	1460.863	1543.448	1607.220	1654.918	1689.752	1714.748	1732.462
446	Raisen	1331.597	1485.485	1560.330	1617.703	1660.423	1691.551	1713.876	1729.706
447	Betul	1575.362	1765.748	1825.300	1869.268	1901.060	1923.705	1939.661	1950.819

(continued)

Table 2.6 (continued)

District Code	Name	Total population							
		Census	Forecast						
		2011	2011	2016	2021	2026	2031	2036	2041
448	Harda	570.465	636.377	671.402	698.571	718.975	733.930	744.698	752.350
449	Hoshangabad	1241.350	1391.829	1448.021	1489.905	1520.336	1542.035	1557.302	1567.943
450	Katni	1292.042	1430.265	1500.914	1554.757	1594.533	1623.246	1643.629	1657.928
451	Jabalpur	2463.289	2740.433	2834.877	2904.862	2955.659	2991.977	3017.662	3035.690
452	Narsimhapur	1091.854	1221.068	1267.841	1302.898	1328.541	1346.967	1360.036	1369.223
453	Dindori	704.524	769.921	797.291	817.652	832.463	843.065	850.564	855.825
454	Mandla	1054.905	1164.329	1208.391	1241.482	1265.712	1283.125	1295.473	1304.145
455	Chhindwara	2090.922	2347.354	2433.087	2497.254	2544.196	2577.966	2601.969	2618.883
456	Seoni	1379.131	1521.555	1577.667	1619.072	1648.878	1669.955	1684.672	1694.858
457	Balaghat	1701.698	1885.418	1930.668	1963.336	1986.498	2002.713	2013.963	2021.721
458	Guna	1241.519	1386.522	1491.827	1576.563	1642.067	1691.150	1727.073	1752.915
459	Ashoknagar	845.071	936.197	986.583	1025.322	1054.157	1075.108	1090.067	1100.614
460	Shahdol	1066.063	1184.502	1231.128	1265.983	1291.410	1309.631	1322.521	1331.557
461	Anuppur	749.237	846.477	876.454	898.015	913.176	923.668	930.849	935.727
462	Sidhi	1127.033	1258.023	1343.460	1411.481	1463.642	1502.500	1530.829	1551.157
463	Singrauli	1178.273	1326.939	1444.680	1540.647	1615.515	1671.952	1713.405	1743.274
464	Jhabua	1025.048	1132.665	1214.560	1280.080	1330.512	1368.187	1395.708	1415.484
465	Alirajpur	728.999	818.425	867.284	905.454	934.280	955.498	970.822	981.738

(continued)

Table 2.6 (continued)

District Code	Name	Total population							
		Census	Forecast						
		2011	2011	2016	2021	2026	2031	2036	2041
466	Khandwa (East Nimar)	1310.061	1442.280	1505.983	1553.899	1588.969	1614.127	1631.917	1644.367
467	Burhanpur	757.847	837.423	870.242	894.654	912.400	925.086	934.047	940.323
Gujarat									
468	Kachchh	2092.371	2288.572	2463.478	2605.902	2716.931	2800.577	2861.984	2906.211
469	Banas Kantha	3120.506	3467.075	3685.391	3856.660	3986.416	4082.147	4151.401	4200.792
470	Patan	1343.734	1474.797	1506.802	1529.793	1546.063	1557.453	1565.368	1570.839
471	Mahesana	2035.064	2270.320	2319.512	2355.101	2380.470	2398.362	2410.887	2419.608
472	Sabar Kantha	2428.589	2669.142	2753.799	2816.458	2861.925	2894.445	2917.466	2933.643
473	Gandhinagar	1391.753	1560.185	1613.076	1652.045	1680.160	1700.138	1714.182	1723.979
474	Ahmadabad	7214.225	7996.604	8424.808	8752.930	8996.910	9174.297	9301.173	9390.861
475	Surendranagar	1756.268	1949.791	2030.026	2090.556	2135.049	2167.135	2189.953	2206.022
476	Rajkot	3804.558	4215.793	4418.444	4573.270	4688.281	4771.945	4831.882	4874.354
477	Jamnagar	2160.119	2373.188	2435.647	2480.822	2512.948	2535.519	2551.244	2562.135
478	Porbandar	585.449	647.003	657.747	665.411	670.815	674.597	677.227	679.050
479	Junagadh	2743.082	3035.994	3110.019	3163.592	3201.771	3228.680	3247.499	3260.589
480	Amreli	1514.190	1689.496	1711.425	1726.519	1736.815	1743.795	1748.507	1751.680
481	Bhavnagar	2880.365	3206.798	3327.924	3417.577	3482.396	3528.471	3560.828	3583.358
482	Anand	2092.745	2310.972	2363.707	2401.910	2429.143	2448.335	2461.749	2471.074

(continued)

Table 2.6 (continued)

District Code	Name	Total population							
		Census	Forecast						
		2011	2011	2016	2021	2026	2031	2036	2041
483	Kheda	2299.885	2536.512	2592.890	2633.519	2662.357	2682.605	2696.716	2706.498
484	Panch Mahals	2390.776	2646.516	2754.440	2835.001	2893.588	2935.393	2964.821	2985.339
485	Dohad	2127.086	2365.581	2553.116	2704.671	2822.103	2910.162	2974.588	3020.877
486	Vadodara	4165.626	4636.380	4795.560	4913.362	4998.707	5059.588	5102.540	5132.607
487	Narmada	590.297	650.513	668.080	680.872	690.022	696.483	701.003	704.146
488	Bharuch	1551.019	1726.543	1786.204	1830.767	1863.262	1886.543	1903.014	1914.563
489	The Dangs	228.291	252.899	266.910	277.739	285.851	291.788	296.058	299.091
490	Navsari	1329.672	1497.284	1531.805	1556.876	1574.803	1587.477	1596.367	1602.568
491	Valsad	1705.678	1907.540	2028.967	2125.263	2198.915	2253.702	2293.617	2322.259
492	Surat	6081.322	6839.848	7817.812	8697.043	9442.370	10043.232	10508.249	10856.847
493	Tapi	807.022	886.562	904.476	917.057	925.772	931.753	935.832	938.603
Daman & Diu									
494	Diu	52.074	57.873	59.954	61.482	62.578	63.351	63.889	64.262
495	Daman	191.173	214.045	266.523	320.309	370.797	414.404	449.443	476.010
Dadra & Nagar Haveli									
496	Dadra & Nagar Haveli	343.709	384.978	459.074	533.485	603.841	666.587	719.617	762.385
Maharashtra									
497	Nandurbar	1648.295	1815.909	1925.417	2011.009	2075.527	2122.846	2156.856	2180.947

(continued)

Table 2.6 (continued)

District Code	Name	Total population							
		Census	Forecast						
		2011	2011	2016	2021	2026	2031	2036	2041
498	Dhule	2050.862	2253.405	2340.498	2405.718	2453.361	2487.536	2511.733	2528.707
499	Jalgaon	4229.917	4701.441	4858.424	4974.091	5057.375	5116.350	5157.622	5186.266
500	Buldana	2586.258	2883.377	3001.803	3091.057	3156.504	3203.535	3236.845	3260.194
501	Akola	1813.906	2029.233	2093.114	2140.027	2173.780	2197.708	2214.493	2226.181
502	Washim	1197.160	1324.034	1376.354	1415.718	1444.571	1465.317	1480.029	1490.361
503	Amravati	2888.445	3233.619	3327.817	3396.574	3445.747	3480.403	3504.578	3521.319
504	Wardha	1300.774	1468.380	1496.776	1516.909	1530.977	1540.708	1547.391	1551.959
505	Nagpur	4653.570	5243.393	5471.288	5643.610	5770.625	5862.493	5928.034	5974.337
506	Bhandara	1200.334	1356.894	1382.056	1399.942	1412.481	1421.185	1427.185	1431.303
507	Gondiya	1322.507	1452.392	1472.686	1486.900	1496.747	1503.516	1508.146	1511.302
508	Gadchiroli	1072.942	1207.525	1250.065	1281.661	1304.618	1321.032	1332.633	1340.766
509	Chandrapur	2204.307	2507.206	2566.142	2607.438	2635.908	2655.317	2668.449	2677.289
510	Yavatmal	2772.348	3093.656	3193.116	3266.670	3319.922	3357.887	3384.656	3403.385
511	Nanded	3361.292	3780.598	3968.048	4111.160	4217.278	4294.276	4349.268	4388.101
512	Hingoli	1177.345	1307.476	1368.856	1415.700	1450.414	1475.583	1493.543	1506.214
513	Parbhani	1836.086	2041.244	2141.610	2218.406	2275.409	2316.772	2346.294	2367.119
514	Jalna	1959.046	2173.428	2287.035	2375.018	2441.010	2489.329	2524.086	2548.771
515	Aurangabad	3701.282	4135.823	4458.474	4717.836	4917.973	5067.590	5176.811	5255.166

(continued)

Table 2.6 (continued)

District Code	Name	Total population							
		Census	Forecast						
		2011	2011	2016	2021	2026	2031	2036	2041
516	Nashik	6107.187	6819.979	7250.901	7587.938	7842.185	8028.797	8163.031	8258.193
517	Thane	11060.148	12513.254	14017.619	15282.058	16288.945	17056.902	17623.591	18031.637
518	Mumbai Suburban	9356.962	10664.288	10952.328	11146.966	11276.493	11361.815	11417.639	11454.004
519	Mumbai	3085.411	3685.124	3686.813	3687.804	3688.384	3688.723	3688.922	3689.038
520	Raigarh	2634.200	2930.039	3093.055	3222.060	3320.581	3393.787	3447.080	3485.301
521	Pune	9429.408	10500.313	11390.150	12118.033	12687.712	13118.352	13435.421	13664.370
522	Ahmadnagar	4543.159	5086.491	5261.244	5390.614	5484.282	5551.017	5598.021	5630.860
523	Bid	2585.049	2865.457	2998.598	3099.899	3174.719	3228.777	3267.214	3294.236
524	Latur	2454.196	2748.652	2890.268	2999.403	3080.981	3140.582	3183.402	3213.798
525	Osmanabad	1657.576	1849.467	1905.835	1947.497	1977.645	1999.127	2014.268	2024.857
526	Solapur	4317.756	4843.077	5020.121	5152.501	5249.064	5318.235	5367.143	5401.406
527	Satara	3003.741	3384.962	3460.573	3514.687	3552.786	3579.304	3597.615	3610.191
528	Ratnagiri	1615.069	1884.510	1895.468	1902.093	1906.061	1908.423	1909.824	1910.655
529	Sindhudurg	849.651	978.251	984.136	988.245	991.072	992.996	994.297	995.172
530	Kolhapur	3876.001	4348.895	4472.300	4562.570	4627.389	4673.314	4705.547	4728.019
531	Sangli	2822.143	3172.991	3259.680	3323.320	3369.184	3401.799	3424.774	3440.850
Andhra Pradesh									
532	Adilabad	2741.239	3103.854	3209.277	3287.272	3343.765	3384.059	3412.483	3432.380

(continued)

Table 2.6 (continued)

District Code	Name	Total population							
		Census	Forecast						
		2011	2011	2016	2021	2026	2031	2036	2041
533	Nizamabad	2551.335	2873.392	2943.455	2993.978	3029.836	3055.000	3072.520	3084.650
534	Karimnagar	3776.269	4275.483	4384.016	4460.449	4513.269	4549.300	4573.660	4590.032
535	Medak	3033.288	3406.480	3539.578	3638.621	3710.352	3761.291	3796.962	3821.698
536	Hyderabad	3943.323	4535.369	4606.491	4648.313	4672.587	4686.572	4694.595	4699.187
537	Rangareddy	5296.741	5939.369	6824.269	7624.274	8304.350	8852.875	9276.847	9593.887
538	Mahbubnagar	4053.028	4535.082	4704.017	4830.177	4922.086	4987.841	5034.278	5066.772
539	Nalgonda	3488.809	3955.501	4047.449	4111.747	4155.984	4186.085	4206.419	4220.092
540	Warangal	3512.576	3966.507	4066.803	4138.374	4188.516	4223.195	4246.967	4263.163
541	Khammam	2797.370	3161.786	3241.723	3298.681	3338.602	3366.259	3385.266	3398.255
542	Srikakulam	2703.114	3048.449	3109.882	3154.021	3185.173	3206.882	3221.879	3232.176
543	Vizianagaram	2344.474	2654.776	2685.830	2706.063	2719.076	2727.376	2732.641	2735.971
544	Visakhapatnam	4290.589	4834.000	4996.200	5114.193	5198.076	5256.736	5297.289	5325.101
545	East Godavari	5154.296	5836.927	5912.758	5961.716	5992.943	6012.708	6025.156	6032.972
546	West Godavari	3936.966	4480.564	4529.914	4561.195	4580.821	4593.056	4600.652	4605.357
547	Krishna	4517.398	5093.542	5211.777	5296.865	5357.092	5399.226	5428.460	5448.629
548	Guntur	4887.813	5463.792	5571.596	5648.770	5703.225	5741.260	5767.637	5785.839
549	Prakasam	3397.448	3798.863	3897.374	3969.415	4021.073	4057.598	4083.164	4100.936
550	Sri Potti Sriramulu Nellore	2963.557	3309.508	3395.691	3458.516	3503.436	3535.112	3557.231	3572.572

(continued)

Table 2.6 (continued)

District Code	Name	Total population							
		Census	Forecast						
		2011	2011	2016	2021	2026	2031	2036	2041
551	Y. S. R	2882.469	3226.593	3317.731	3384.763	3433.060	3467.345	3491.427	3508.216
552	Kurnool	4053.463	4521.855	4692.554	4820.282	4913.527	4980.379	5027.691	5060.867
553	Anantapur	4081.148	4577.975	4724.895	4833.704	4912.530	4968.729	5008.340	5036.034
554	Chittoor	4174.064	4677.859	4818.313	4922.052	4997.037	5050.394	5087.939	5114.149
Karnataka									
555	Belgaum	4779.661	5329.572	5507.987	5640.180	5735.935	5804.163	5852.210	5885.767
556	Bagalkot	1889.752	2107.976	2189.245	2250.638	2295.856	2328.541	2351.849	2368.310
557	Bijapur	2177.331	2407.873	2523.683	2612.552	2678.593	2726.510	2760.676	2784.736
558	Bidar	1703.300	1912.140	1988.127	2045.365	2087.394	2117.682	2139.214	2154.374
559	Raichur	1928.812	2172.264	2273.210	2349.415	2405.345	2445.552	2474.027	2493.979
560	Koppal	1389.920	1559.909	1638.510	1699.097	1744.403	1777.519	1801.323	1818.229
561	Gadag	1064.570	1187.466	1213.833	1232.900	1246.462	1255.994	1262.638	1267.243
562	Dharwad	1847.023	2047.462	2111.603	2158.776	2192.785	2216.953	2233.950	2245.819
563	Uttara Kannada	1437.169	1614.654	1637.921	1654.256	1665.613	1673.454	1678.843	1682.534
564	Haveri	1597.668	1785.608	1832.569	1866.602	1890.815	1907.816	1919.643	1927.820
565	Bellary	2452.595	2728.950	2870.695	2979.253	3059.859	3118.345	3160.074	3189.491
566	Chitradurga	1659.456	1863.437	1910.587	1944.744	1969.063	1986.163	1998.084	2006.344
567	Davanagere	1945.497	2194.813	2245.409	2281.171	2306.067	2323.217	2334.944	2342.923

(continued)

Table 2.6 (continued)

District Code	Name	Total population							
		Census	Forecast						
		2011	2011	2016	2021	2026	2031	2036	2041
568	Shimoga	1752.753	1951.431	1976.331	1993.311	2004.799	2012.533	2017.721	2021.194
569	Udupi	1177.361	1316.700	1329.961	1338.905	1344.892	1348.880	1351.528	1353.285
570	Chikmagalur	1137.961	1302.986	1316.127	1325.236	1331.510	1335.810	1338.749	1340.753
571	Tumkur	2678.980	3052.476	3103.460	3139.434	3164.500	3181.813	3193.699	3201.825
572	Bangalore	9621.551	10656.865	11926.975	13013.298	13893.738	14576.701	15088.677	15462.724
573	Mandya	1805.769	2056.867	2075.916	2088.551	2096.873	2102.330	2105.898	2108.228
574	Hassan	1776.421	2022.880	2049.303	2067.663	2080.302	2088.945	2094.829	2098.823
575	Dakshina Kannada	2089.649	2324.358	2375.493	2412.287	2438.365	2456.651	2469.377	2478.189
576	Kodagu	554.519	626.512	631.852	635.576	638.158	639.941	641.168	642.011
577	Mysore	3001.127	3352.979	3462.049	3542.274	3600.041	3641.006	3669.741	3689.746
578	Chamarajanagar	1020.791	1152.386	1169.493	1181.339	1189.448	1194.956	1198.678	1201.185
579	Gulbarga	2566.326	2855.347	2988.377	3089.297	3163.568	3217.014	3254.855	3281.341
580	Yadgir	1174.271	1306.891	1386.262	1448.225	1494.699	1528.521	1552.601	1569.478
581	Kolar	1536.401	1721.801	1773.065	1811.143	1838.821	1858.625	1872.636	1882.470
582	Chikkaballapura	1255.104	1405.812	1438.628	1462.382	1479.297	1491.203	1499.514	1505.282
583	Bangalore Rural	990.923	1100.674	1145.395	1179.248	1204.182	1222.178	1234.977	1243.986
584	Ramanagara	1082.636	1222.570	1238.435	1249.316	1256.699	1261.672	1265.006	1267.235
Goa									

(continued)

Table 2.6 (continued)

District Code	Name	Total population							
		Census	Forecast						
		2011	2011	2016	2021	2026	2031	2036	2041
585	North Goa	818.008	921.019	940.888	954.973	964.799	971.578	976.219	979.379
586	South Goa	640.537	719.341	735.757	747.581	755.972	761.864	765.971	768.818
Lakshadweep									
587	Lakshadweep	64.473	73.418	75.062	76.174	76.914	77.401	77.719	77.926
Kerala									
588	Kasaragod	1307.375	1472.914	1505.410	1528.807	1545.420	1557.097	1565.249	1570.911
589	Kannur	2523.003	2851.938	2883.788	2905.734	2920.761	2931.009	2937.980	2942.717
590	Wayanad	817.420	926.773	942.518	953.464	961.000	966.154	969.663	972.045
591	Kozhikode	3086.293	3462.131	3517.410	3556.590	3584.057	3603.166	3616.389	3625.505
592	Malappuram	4112.920	4615.516	4772.947	4888.835	4972.373	5031.684	5073.343	5102.383
593	Palakkad	2809.934	3159.450	3216.493	3256.986	3285.340	3305.001	3318.545	3327.832
594	Thrissur	3121.200	3506.469	3546.349	3574.284	3593.683	3607.075	3616.282	3622.593
595	Ernakulam	3282.388	3678.080	3722.864	3754.586	3776.864	3792.416	3803.227	3810.720
596	Idukki	1108.974	1283.132	1290.295	1295.024	1298.135	1300.178	1301.517	1302.394
597	Kottayam	1974.551	2232.636	2247.327	2257.444	2264.370	2269.091	2272.302	2274.481
598	Alappuzha	2127.789	2408.740	2421.248	2429.731	2435.460	2439.317	2441.909	2443.649
599	Pathanamthitta	1197.412	1383.143	1386.691	1389.033	1390.577	1391.595	1392.266	1392.708
600	Kollam	2635.375	2989.783	3013.941	3030.531	3041.858	3049.562	3054.788	3058.328

(continued)

Table 2.6 (continued)

District Code	Name	Total population							
		Census	Forecast						
		2011	2011	2016	2021	2026	2031	2036	2041
601	Thiruvananthapuram	3301.427	3750.603	3791.131	3819.537	3839.301	3852.981	3862.415	3868.906
Tamil Nadu									
602	Thiruvallur	3728.104	4106.102	4485.429	4813.086	5084.748	5302.483	5472.309	5601.972
603	Chennai	4646.732	5239.076	5337.523	5409.352	5461.322	5498.714	5525.519	5544.691
604	Kancheepuram	3998.252	4363.600	4777.169	5139.129	5442.985	5689.273	5883.245	6032.543
605	Vellore	3936.331	4374.802	4517.619	4627.404	4710.049	4771.296	4816.164	4848.761
606	Tiruvannamalai	2464.875	2727.534	2795.274	2846.439	2884.385	2912.148	2932.264	2946.735
607	Viluppuram	3458.873	3807.867	3936.240	4036.715	4113.606	4171.448	4214.407	4246.013
608	Salem	3482.056	3822.273	3951.960	4051.755	4126.927	4182.664	4223.520	4253.222
609	Namakkal	1726.601	1900.431	1967.782	2020.107	2059.800	2089.382	2111.149	2127.019
610	Erode	2251.744	2495.677	2559.872	2608.524	2644.771	2671.435	2690.867	2704.935
611	The Nilgiris	735.394	860.447	866.140	870.015	872.658	874.463	875.699	876.547
612	Dindigul	2159.775	2377.561	2435.566	2479.795	2512.931	2537.432	2555.375	2568.423
613	Karur	1064.493	1158.753	1194.007	1221.299	1241.914	1257.209	1268.411	1276.542
614	Tiruchirappalli	2722.290	3004.939	3075.631	3128.517	3167.498	3195.923	3216.491	3231.293
615	Perambalur	565.223	623.995	644.084	659.727	671.646	680.581	687.198	692.056
616	Ariyalur	754.894	837.955	853.223	864.588	872.928	878.987	883.355	886.488
617	Cuddalore	2605.914	2868.168	2936.170	2987.105	3024.655	3052.018	3071.793	3086.000

(continued)

Table 2.6 (continued)

District Code	Name	Total population							
		Census	Forecast						
		2011	2011	2016	2021	2026	2031	2036	2041
618	Nagapattinam	1616.450	1799.545	1832.148	1856.414	1874.237	1887.203	1896.571	1903.307
619	Thiruvarur	1264.277	1404.514	1426.017	1441.896	1453.493	1461.897	1467.954	1472.304
620	Thanjavur	2405.890	2678.656	2726.315	2761.899	2788.157	2807.370	2821.342	2831.460
621	Pudukkottai	1618.345	1807.779	1848.976	1879.365	1901.431	1917.273	1928.554	1936.541
622	Sivaganga	1339.101	1452.892	1480.996	1501.844	1517.073	1528.073	1535.956	1541.573
623	Madurai	3038.252	3311.471	3396.818	3461.031	3508.594	3543.428	3568.734	3587.013
624	Theni	1245.899	1358.677	1378.360	1392.401	1402.304	1409.234	1414.056	1417.399
625	Virudhunagar	1942.288	2163.367	2222.749	2268.152	2302.258	2327.545	2346.113	2359.654
626	Ramanathapuram	1353.445	1478.753	1505.182	1524.653	1538.847	1549.117	1556.511	1561.815
627	Thoothukkudi	1750.176	1920.522	1948.805	1969.552	1984.595	1995.413	2003.147	2008.655
628	Tirunelveli	3077.233	3383.236	3475.672	3546.679	3600.146	3639.815	3668.933	3690.145
629	Kanniyakumari	1870.374	2054.477	2082.646	2103.181	2118.020	2128.676	2136.296	2141.729
630	Dharmapuri	1506.843	1660.327	1715.947	1758.483	1790.396	1814.002	1831.286	1843.848
631	Krishnagiri	1879.809	2083.073	2193.176	2281.347	2349.939	2402.121	2441.154	2469.990
632	Coimbatore	3458.045	3776.736	3910.758	4014.256	4092.466	4150.606	4193.304	4224.384
633	Tiruppur	2479.052	2733.943	2966.445	3168.085	3335.324	3469.030	3572.860	3651.699
Puducherry									
634	Yanam	55.626	62.483	81.263	103.726	129.276	156.763	184.625	211.191

(continued)

Table 2.6 (continued)

District Code	Name	Total population							
		Census	Forecast						
		2011	2011	2016	2021	2026	2031	2036	2041
635	Puducherry	950.289	1053.591	1127.557	1186.678	1232.155	1266.111	1290.904	1308.711
636	Mahe	41.816	46.530	47.865	48.854	49.570	50.082	50.443	50.696
637	Karaikal	200.222	222.093	231.505	238.712	244.075	247.981	250.783	252.770
Andaman & Nicobar Islands									
638	Nicobars	36.842	45.400	45.481	45.519	45.538	45.546	45.551	45.553
639	North & Middle Andaman	105.597	123.191	125.341	126.687	127.521	128.034	128.348	128.540
640	South Andaman	238.142	271.353	282.864	290.944	296.489	300.236	302.742	304.408

Source Author

References

Arriaga EE, Johnson PD, Jemison E (1994) Population analysis with microcomputers. Census Bureau, International Programs Center, Washington DC, United States

Dhar M (2022) Projection of district-level annual population by quinquennial age-group and sex from 2012 to 2031 in India. International Institute for Population Sciences, Mumbai

Isserman AM (1977) The accuracy of population projections for subcounty areas. J Am Inst Plann 43(3):247–259

Mondol H, Mallick UK, Biswas Md HA (2018) Mathematical modeling and predicting the current trends of human population growth in Bangladesh. Adv Model Anal 55(2):62–69

Openshaw S, Van Der Knaap G (1983) Small area population forecasting: some experience with British models. Tijdschr Econ Soc Geogr 74(4):291–304

Passos JR, de Pinho SZ, de Carvalho LR, Mischan MM (2012) Critical points in logistic growth curves and treatment comparisons. Scientia Agricola 69(5):308–312

Pearl R, Reed LJ (1920) On the rate of growth of the population of United States since 1790 and its mathematical representation. Proc Natl Acad Sci 6(6):275–288

Pflaumer P (2012) Forecasting the US population with the Gompertz growth curve. In: JSM proceedings, social statistics section. American Statistical Association, Alexandria, VA, pp 4967–4981

Pritchett HS (1891) A formula for predicting the population of the United States. Publ Am Stat Assoc 2(14):278–286

Siegel JS, Swanson DA (2004) The methods and materials of demography, 2nd edn. California, Elsevier Academic Press, San Diego

Schmitt RC, Crosetti AH (1951) Accuracy of the ratio method for forecasting city population. Land Econ 27(4):346–348

Smith SK (1987) Tests of forecast accuracy and bias for county population projections. J Am Stat Assoc 82(400):991–1003

Smith SK, Tayman J, Swanson DA (2013) A practitioner's guide to state and local population projections. Springer, New York

United Nations Population Fund (2009) District level population projections in selected states of India – 2006 to 2016. New Delhi, United Nations Population Fund

United Nations (2022) World population prospects 2022, Online. United Nations, Department of Economic and Social Affairs, Population Division, New York

White HR (1954) Empirical study of the accuracy of selected methods of projecting state populations. J Am Stat Assoc 49(267):480–498

Wilson T, Grossman I, Alexander M, Rees P, Temple J (2022) Methods for small area population forecasts: state-of-the-art and research needs. Popul Res Policy Rev 41:865–898

Chapter 3
Population at Stabilisation

Abstract This chapter estimates the population at stabilisation in 640 districts of India as they existed at the 2011 population census. The estimation exercise is based on the modelling of the pattern of district population growth during the period 1951–2011 through the simple logistic growth model. The paper reveals that, in all but a few districts of the country, the simple logistic growth model provides very good fit to district population growth pattern during the period 1951–2011, as revealed through the population enumerated at different decennial population censuses since 1951. The very good fit of the logistic growth model suggests that the upper asymptote of the model may be taken as an estimate of the upper limit of population growth or the population at stabilisation in the district. The validity of our approach is justified by the observation that sum of the population at stabilisation in 640 districts closely approximates the most likely upper limit of population growth in the country projected by the United Nations. The time when the population of a district stabilises will, however, be different for different districts as different districts are currently at different stages of population transition. The paper also highlights the inter-district variation in the density of population and male-female imbalance in population at the time of stabilisation.

Keywords India · Districts · Population growth · Stabilisation · Population growth modelling

Background

This chapter estimates the size of the population at stabilisation in the 640 districts of the country. Estimating the size of the population at stabilisation is important for understanding its implications for economic and social well-being and safeguarding environment. It is well known that the environment that sustains the life on the planet Earth is being endangered primarily by human-driven processes and their synergistic interactions (Brook et al. 2008). It is, therefore, imperative that the size of the population at stabilisation is estimated and its impact explored to plan for a

A. R. Chaurasia, *Population and Human Development in Districts of India*,
https://doi.org/10.1007/978-981-95-8169-6_3

sustainable future society. Population stabilises when it stops growing. The upper limit of population growth, therefore, is the population at stabilisation. There has never been any attempt to estimate the population at stabilisation in the districts of the country. Although, the National Population Policy 2000 had envisioned stabilising the population of the country by the year 2045 at a level consistent with the requirements of sustainable economic growth, social development and environmental protection by achieving the replacement fertility by the year 2010 (Government of India 2000), yet the Policy is silent about the size of the population at stabilisation.

The paper is divided into five sections including this background. The next section describes the data source. We have used the male and female population enumerated in the districts of India, as they existed at 2011 population census, at different decennial population censuses since 1951, These data have been made available by the Registrar General and Census Commissioner of India. The third section outlines the approach adopted for estimating the population at stabilisation which is based on modelling district population growth in the district during the 70 years between 1951 and 2011. The model has then been used to estimate the upper limit of population growth in the district. The fourth section of the paper presents results of the analysis while the last section summarises main findings and discusses their policy and programme relevance.

The Data

The main problem in modelling population growth pattern in the districts of India has been the increase in the number of districts at different decennial population censuses due to administrative reasons. At the 1961 population census, there were 339 districts in the country. This number increased to 640 at the 2011 population census and continues to increase. Recently, the Registrar General and Census Commissioner of India has reconstructed the population enumerated at different decennial population census since 1901 according to the administrative boundaries of 640 districts, as they existed at the time of the 2011 population census. This reconstructed district population dataset, spanning over a period of more than 110 years, provides, for the first time, an opportunity to model district population growth in the country. We have used this reconstructed dataset to model population growth during 1951–2011 in each of the 640 districts of the country. We have not used the district population enumerated at different decennial population censuses from 1901 through 1951 for two reasons. First, the population enumerated at different decennial population census during the period 1901–2041 is not available for many districts of the country. Second, we found that the population growth pattern during the period 1901–1951 is essentially different from the population growth pattern during the period 1951–2011 in a large proportion of districts because of the discontinuity in population growth pattern during the period 1901–1951 as compared to the population growth pattern during the period 1951–2011. Since the objective of the present analysis is to estimate the population of the district at stabilisation or the upper limit of district population growth or the size of the district population before it stops increasing, we have modelled the district population growth during 1951–2011 and used the model so obtained for estimating the upper limit of district population growth.

Population enumerated at the population census is known to be associated with several errors including errors of omission and duplication. The Registrar General and Census Commissioner of India undertakes post-enumeration survey after every decennial population census to estimate errors of omission and duplication during enumeration. The post-enumeration survey conducted after the 2011 population census has revealed that there was a net omission of around 23 persons for every 1000 persons enumerated at the 2011 population census (Government of India 2014). There was an estimated undercount of 23.08 persons for every 1000 persons counted which was offset by an estimated duplication of 0.10 persons for every 1000 persons counted. The post-enumeration survey, however, revealed no statistically significant difference in the net omission rate by sex but the net omission rate in females was estimated to be comparatively higher than the net omission rate in males. The post-enumeration survey has also revealed that the net omission rate was markedly higher in urban (29 persons per 1000 persons enumerated) as compared to the rural areas (20 persons per 1000 persons enumerated). There has also been little change in the net omission rate in the 2001 and the 2011 population censuses.

The estimates of the net omission rate are not available for the states/Union Territories and districts of the country, but it varies across administrative zones, being the lowest in the eastern but the highest in the central zone (Government of India 2014). In the absence of any information about the net omission rate in the districts, we have not made any adjustment in the enumerated population of districts and used the data provided by the Registrar General and Census Commissioner of India for population growth modelling without any adjustment.

There are also some gaps in the data made available by the Registrar General and Census Commissioner of India. In 42 districts, population enumerated at the 1951 population census was not available whereas in 10 districts, population enumerated at both 1951 and 1961 population censuses was missing. Therefore, modelling in 32 districts was based on population growth during 1961–2011 whereas, in 10 districts, it was based on population growth during 1971–2011. Moreover, population enumerated at 1981 population census was not available for 27 districts of Assam. We have, therefore taken the average of population enumerated at the 1971 population census and at the 1991 population census to estimate the population of the 27 districts of Assam at the 1981 population census.

The Method

The method that we have followed to estimate the upper limit of population growth or the population at stabilisation assumes that population growth follows an S-shaped growth trajectory which has three phases—an initial phase in which population increases exponentially; a middle phase in which population increases linearly; and a final phase in which population growth slows down and approaches an upper limit. The S-shaped trajectory of population growth may be characterised by different mathematical models including the logistic growth model, the Gompertz growth model and the generalised logistic growth model. The logistic growth model is the simplest description of an *S*-shaped growth pattern. It was first developed by Verhulst in 1838 (Verhlust 1838) and later re-invented independently by Pearl and Reed in 1920 (Pearl and Reed 1920). The model assumes that population grows exponentially under the constraint of an upper limit (Lotka 1956). Application of the logistic growth model to describe population growth has a long history. It was a popular method of population forecasting in the past. Several studies have shown that the application of the logistic growth curve may often provide reasonably accurate forecast of population growth (Dorn 1950; Leach 1981). In recent years, there is a renewed interest in the logistic growth model to describe and forecast population growth (Hrytsiuk et al. 2023; Burg and Ausubel 2023; Mondol et al. 2018; Sharif Ullah et al. 2019). Bhat (1999) had used this model to forecast the population of Delhi, the capital city of India. If the population growth in a district follows the logistic growth curve, then the upper asymptote of the logistic growth model gives the estimate of the upper limit of district population growth which may be taken as the population at stabilisation.

The upper asymptote of the logistic growth model can be estimated using the critical points separating three critical phases of logistic growth. There are several methods available to determine the critical points of the logistic growth curve (Passos et al. 2012). These include: i) accelerating growth function method, ii) tangent at the inflexion point method, iii) segmented regression method; iv) modified segmented regression method; v) non-significant difference method; and vi) non-significant difference by simulation method. Details of these methods are given elsewhere (Passos et al. 2012). The tangent at the point of inflexion method gives the size of the population at the point of interaction of the tangent at the inflexion point with the upper asymptote of the model. If K is the size of the population at the interaction of the tangent at the inflexion point of the population growth trajectory and the upper asymptote and U is the upper asymptote, then it can be shown for the logistic curve (Passos et al. 2012) that.

$$K = \frac{U}{(1 + \exp(-2))} \text{ or } U = K \times (1 + \exp(-2))$$

If it is assumed that population growth is maximal at the point of inflexion, then K can be calculated as

$$\mathrm{K} = \frac{4 \times \mathrm{s}}{\mathrm{r}}$$

where s is the slope of population growth trajectory when population growth rate is the highest and r is the intrinsic population growth rate or the 'width' or 'steepness' of population growth trajectory.

We have used the open source Loglet software package (Yung et al. 1999) to fit the logistic growth model to the pattern of population growth during the period 1951–2011 in each of the 640 districts of the country as they existed at the 2011 population census. The software provides estimate of the parameter *K* from which the upper asymptote of the model can be calculated. The software also provides estimates of mean absolute percentage error (MPAE), goodness-of-fit statistics (RSS and RMS) and estimates of R^2 to test the appropriateness of the logistic growth model in describing the district population growth pattern. Although, R^2 is not regarded as an appropriate measure for ascertaining the goodness-of-fit in case of nonlinear models (Spiess and Neumeyer 2010), yet it can be used for testing the goodness-of-fit as the logistic growth model can be transformed into a linear growth model through the Fisher-Pry transformation (Fisher and Pry 1971). If the model provides a good fit to population growth, then the upper asymptote of the model is the upper limit of district population growth or population at stabilisation.

Modelling Population Growth in Districts 1951–2011

Table 3.1 summarises results of the goodness-of-fit of the logistic growth model to population growth in 640 districts of the country during the period 1951–2011. In all but a few districts of the country, logistic growth model provides very good fit to population growth in the districts during the period 1951–2011. In 578 or more than 90 per cent districts, the MAPE is estimated to be less than 5 per cent in case of the male population whereas in 575 districts, the MAPE is estimated to be less than 5 per cent in case of female population. Similarly, in 620 or almost 97 per cent districts, R^2 is estimated to be at least 0.90 for the male population whereas, in case of the female population, it is estimated to be at least 0.90 in 611 districts. It is obvious from Table 3.1 that the logistic growth model provides good fit to both male and female population growth during the period 1951–2011 in all but a few districts. This also implies that the upper asymptote of the logistic growth model can be taken as an estimate of the upper limit of population growth or the size of the population at stabilisation in all but a few districts. We have, therefore, assumed that the logistic growth model describes the pattern of male and female population growth in the districts of the country during the period 1951–2011 (Fig. 3.1).

Table 3.1 suggests that there are only a few districts where the measures of the goodness-of-fit of the logistic growth model are comparatively lower. There are only 3 districts where fitting of the logistic growth curve to the pattern of growth in the male population during 1951–2011 is not found to be satisfactory as the R^2 is estimated to be less than 0.80. Similarly, there are 4 districts where fitting of the logistic growth curve to the pattern of female population growth is not found to be satisfactory because R^2 is estimated to be less than 0.80. Out of these 7 districts, 4 are in the north-eastern part of the country and 3 are in Nagaland alone. District Diu in the Union Territory of Daman and Diu is the only district in the country where the fitting of the logistic growth curve to the pattern of growth in both male population

Table 3.1 Results of fitting of the logistic growth model to population growth in districts, 1951–2011. Variation in MAPE and R^2 across districts

Mape			R^2		
Range	Males	Females	Range	Males	Females
<0.02	206	202	>= 0.95	558	520
0.02-0.04	315	305	0.90-0.95	62	91
0.04-0.06	80	94	0.85-0.90	10	19
0.06-0.08	21	22	0.80-0.85	5	4
0.08-0.10	8	7	<0.80	3	4
>= 0.10	10	8			
No data	2	2	No data	2	2
Total	640	640	Total	640	640

Source Author

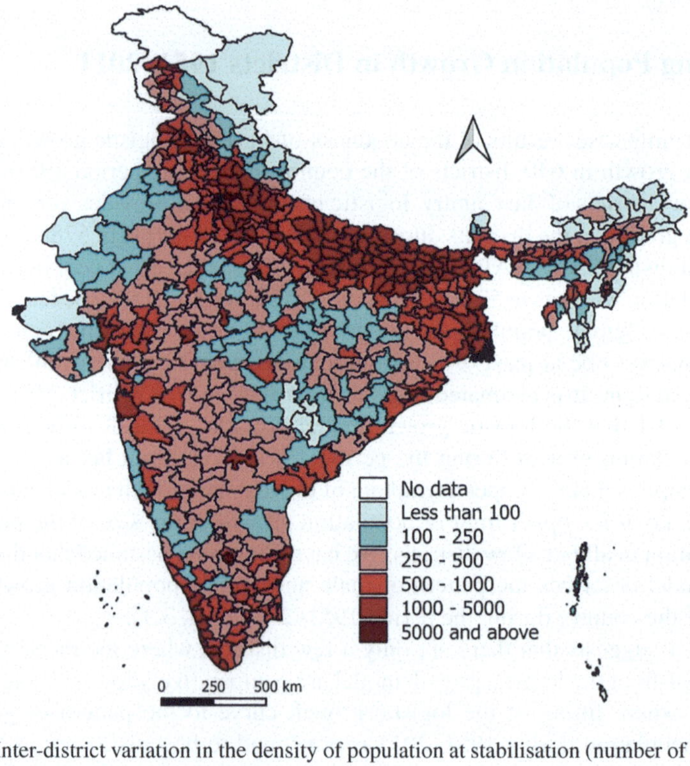

Fig. 3.1 Inter-district variation in the density of population at stabilisation (number of persons per square kilometre) in India. *Source* Author

and female population has not been found satisfactory. As regards the mean absolute percentage error (MAPE), there are only 10 districts where MAPE is estimated to be at least 10 per cent in case of male population whereas MAPE is estimated to be at least 10 per cent in only 8 districts in case of female population. One reason of relatively low measures of goodness-of-fit of the logistic population growth in these districts, during the period 1951–2011, is the rapid increase or decrease in the enumerated population in a decennial population census as compared to the immediately previous decennial population census. For example, in district Kiphire in Nagaland, the enumerated population more than doubled between the 1991 and the 2011 decennial population but, during the 2011 population census, the population of the district decreased relative to the population enumerated at the 2001 decennial population census. On the other hand, there are two districts in the National Capital Territory of Delhi—Central district and district New Delhi—where population has remained virtually unchanged or even decreased during the 60 years between 1951 and 2011. Reasons for the typical pattern of growth in either the male or the female population or both in these few districts of the country are not known at present but this typical pattern may be responsible for poor fit of logistic growth curve.

District Population at Stabilisation

Estimates of the population at stabilisation in the 640 districts are presented in the appendix table, along with the estimates of the increase in population before stabilisation along with the population density and the population sex ratio at the time of stabilisation. In district Thane in Maharashtra, the population is likely to stabilise at more than 18 million, the largest among the 640 districts whereas population will stabilise at less than 9 thousand in district Dibang Valley of Arunachal Pradesh, the smallest among the districts. There are 8 districts where population at stabilisation will be more than ten million. On the other hand, there are 15 districts where population will stabilise at less than 100 thousand. The proportionate increase in the population at stabilisation relative to population in 2011 is estimated to be the highest in district Kurung Kumey of Arunachal Pradesh but the lowest in district Mamit of Mizoram. In majority of the districts, the proportionate increase in population at stabilisation from the population in 2011 will range between 20 and 40 per cent. There are only 24 districts where the population at stabilisation will be more than 60 per cent higher compared to population enumerated at the 2011 population census whereas, in 7 districts, population at stabilisation will be more than two times the population enumerated at the 2011 population census.

Adding the population at stabilisation in 640 districts, the population of the country at stabilisation is estimated to be around 1620 million—around 828 million males and 792 million females (Table 3.2). This estimate of population at stabilisation in India is very close to the estimate of population at stabilisation or the size of the population of the country when it stops increasing according to the medium variant of population projection for India prepared by the United Nations (2022) and as projected by Kulkarni (2021). According to the most likely medium variant of the population projections prepared by the United Nations, the population of India is projected to peak to around 1697 million (United Nations 2022). On the other hand, Kulkarni (2021) has projected that the population of the country will peak to around 1660 million. The closeness of our estimate of the population of India at the time of stabilisation derived through the bottom-up approach of forecasting population provides the credence to the method adopted for estimating the population at stabilisation. The advantage of the approach employed in this chapter is that it provides estimates of the population at stabilisation for each of the 640 districts of the country as they existed at the 2011 population census. United Nations does not project population of the states/Union Territories or districts of India. There is no other source which provides estimates of the population at stabilisation in the districts of the country. Our estimates suggest that the population of the country will increase by around 409 million or by more than 37 per cent from the population enumerated at the 2011 population census before it stops increasing.

The estimation of population at stabilisation in the districts also permits estimating population at stabilisation in states and Union Territories of India. These estimates are given in Table 3.2. The population at stabilisation in Uttar Pradesh, the most populous state of the country, is estimated to be around 280 million whereas the population at stabilisation in Bihar will be around 150 million. On the other hand, population at stabilisation in Lakshadweep is estimated to be around 78 thousand. In Dadra and Nagar Haveli, Arunachal Pradesh and Daman and Diu, the population at stabilisation is expected to increase by around 150 per cent from the population enumerated at the 2011 population census. In 6 states/Union Territories, the increase in the population before it stops increasing, is expected to be more than 40 per cent relative to the population enumerated at the 2011 population census. It is also estimated that more than 40 per cent of the increase in population of the country before it stops increasing would be confined to only three states—Uttar Pradesh (20 per cent), Bihar (11 per cent) and Maharashtra (9 per cent). By comparison, the increase in population before it stops increasing in Kerala and Jammu and Kashmir is estimated to account for less than 1.5 per cent of the increase in the population of the country before it stops increasing. The differential contribution of the increase in the population of different states and Union Territories to the increase in the population of the country before it stops increasing is expected as different states/Union Territories are at distinct stages of population transition.

The distribution of population of the country at stabilisation across states and Union Territories will be different from the distribution of population at the 2011 population census. The share of the population at stabilisation of 19 states and Union Territories to the population at stabilisation of the country will be lower than the share of the population of these states and Union Territories to the population of the country enumerated at the 2011 population census. Uttar Pradesh, alone, will account for more than 17 per cent of the population at stabilisation of the country whereas Bihar will account for more than 9 per cent so that the two states will account for more than 26 per cent of the population at stabilisation of the country. The share of population at stabilisation of the Empowered Action Group (EAG) States to the

Table 3.2 Population at stabilisation in the states and Union Territories of India

Country/State/Union territory	Male (000)	Female (000)	Person (000)	Increase since 2011	
				Number (000)	%
Jammu and Kashmir	9731	8568	18299	5757	45.9
Himachal Pradesh	4372	4231	8603	1738	25.3
Punjab	18642	16677	35319	7576	27.3
Chandigarh	768	637	1405	350	33.2
Uttarakhand	6761	6780	13541	3455	34.2
Haryana	19458	17126	36584	11233	44.3
National Capital Territory of Delhi	11849	10603	22452	5665	33.7
Rajasthan	48831	45809	94640	26092	38.1
Uttar Pradesh	143462	136910	280372	80560	40.3
Bihar	77872	72306	150178	46078	44.3
Sikkim	388	349	737	127	20.7
Arunachal Pradesh	1825	1718	3543	2159	156.1
Nagaland	1333	1241	2574	595	30.1
Manipur	2067	2068	4135	1279	44.8
Mizoram	669	676	1345	248	22.6
Tripura	2304	2210	4514	840	22.9
Meghalaya	2143	2180	4323	1356	45.7
Assam	19634	19151	38785	7579	24.3
West Bengal	58455	55814	114269	22993	25.2
Jharkhand	23123	22094	45217	12229	37.1
Odisha	26202	25741	51943	9968	23.7
Chhattisgarh	16778	16715	33493	7947	31.1
Madhya Pradesh	50124	47396	97520	24893	34.3
Gujarat	42875	38164	81039	20600	34.1
Daman and Diu	415	187	602	359	147.6
Dadra and Nagar Haveli	597	286	883	540	157.1
Maharashtra	77684	72676	150360	37986	33.8
Andhra Pradesh	53758	53574	107332	22751	26.9
Karnataka	40524	39600	80124	19029	31.1
Goa	887	874	1761	303	20.7
Lakshadweep	40	38	78	14	21.5
Kerala	18898	20612	39510	6104	18.3
Tamil Nadu	44423	47155	91578	19431	26.9
Puducherry	894	1084	1978	729	58.4

(continued)

Table 3.2 (continued)

Country/State/Union territory	Male (000)	Female (000)	Person (000)	Increase since 2011	
				Number (000)	%
Andaman and Nicobar Islands	257	225	482	101	26.7
India	828043	791476	1619519	408664	33.7

Source Author

population at stabilisation of the country will be higher as compared to the share of the population of these states at the 2011 population census. On the other hand, the share of population at stabilisation of the four southern states—Andhra Pradesh, Karnataka, Kerala and Tamil Nadu—to the population at stabilisation of the country will be lower than the share of the population of these states to the population of the country enumerated at the 2011 population census. West Bengal accounted for more than 7.5 per cent of the population of the country enumerated at the 2011 population census. The share will decrease to around 7 per cent when the population stabilises.

Density of Population at Stabilisation

The population density at stabilisation will also vary across districts. The North East district in the National Capital Territory of Delhi will have the highest population density at the time of stabilisation. There will be more than 46 thousand persons per square Kilometre in this district at the time of stabilisation compared to 36 thousand persons per square Kilometre at the time of 2011 population census. In 14 districts, population density will be more than 10 thousand persons per square Kilometre at the time of stablisarion. Population density will also be exceptionally high in Yanam in Puducherry; Chandigarh in Chandigarh, Hyderabad in Andhra Pradesh; Mumbai and Mumbai Suburban in Maharashtra; Kolkata in West Bengal; and Chennai in Tamil Nadu at the time of stabilisation. The exceptionally high density of population at the time of stabilisation in these districts will have implications for the demand of resources and development of necessary infrastructure and facilities, putting enormous pressure on the environment and affecting the quality of life. On the other hand, population density at stabilisation will be less than 100 persons per square Kilometre in 47 districts. The density of population at the time of stabilisation will be the lowest—just around 1 person per square Kilometre—in district Dibang Valley of Arunachal Pradesh. Population density will also be less than 10 per sons per square Kilometre in district Anjaw of Arunachal Pradesh, district Leh (Ladakh) of Jammu & Kashmir and district Lahul & Spiti of Himachal Pradesh.

Sex Ratio of Population at Stabilisation

There will also be marked variation in the sex ratio of the population at stabilisation across districts. In 118 districts, there will be very high male advantage in the sex composition of the population at the time of stabilisation whereas in 18 districts, there will be very high female advantage. In majority of the districts, however, the sex ratio of population at the time of stabilisation will vary within the normal range of 95–105 males for every 100 females so that the sex ratio of the population of the country at the time of stabilisation will be around 105 males for every 100 females, compared to 107 males for every 100 females that prevailed in the country at the time of the 2011 population census. The sex ratio of the population at the time of stabilisation will also vary markedly across states and Union Territories. The Union Territories of Daman & Diu and Dadra & Nagar Haveli is likely to have sex ratio of the population at the time of stabilisation that will be very highly favourable to males whereas the Union Territory of Puducherry is likely to have sex ratio of population at the time of stabilisation that will be very highly favourable to females. Regional pattern of the sex ratio of the population at the time of stabilisation will be very marked. It will be The sex ratio will be favourable to females in southern states and Union Territories but favourable to males in northern states and Union Territories.

More specifically, in 45 of the 63 districts of Jammu and Kashmir, Punjab and Haryana, the sex ratio of population at stabilisation will be highly favourable to males whereas in 14 districts of these states, the sex ratio of the population at stabilisation will be moderately favourable to males. There is no district in these three states where the population sex ratio at stabilisation will be favourable to females. All these states are located in the north-west corner of the country. Similarly, in 12 districts of Uttar Pradesh, 7 districts of Madhya Pradesh and 6 districts of Rajasthan, the sex ratio of the population at stabilisation will be highly favourable to males. These 25 districts constitute a geographic continuity as may be seen from Fig. 3.2. The sex ratio of the population at stabilisation will also be highly favourable to males in 9 districts of Bihar which also constitute a geographical continuity. On the other hand, there are 13 states and Union Territories in the country where there is no district where the sex ratio of the population at stabilisation will be highly favourable to males. The 18 districts of the country where the sex ratio of the population at stabilisation will be highly favourable to females are in only 5 states and Union Territories and four of these states and Union Territories are in the southern part of the country. The sex ratio of the population at stabilisation will be either highly favourable or moderately favourable to females in all the four districts of Puducherry, in 10 of the 14 districts of Kerala and in 14 of the 32 districts of Tamil Nadu. The sex ratio of the population at stabilisation will also be highly favourable to females in 5 districts of Uttarakhand and in 4 districts of Arunachal Pradesh (Table 3.3).

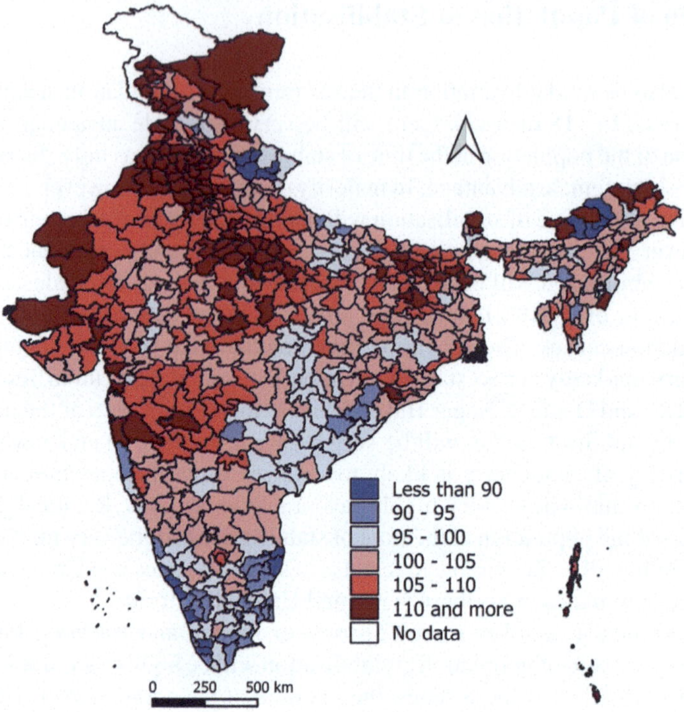

Fig. 3.2 The inter-district variation in the density of population at the time of stabilisation (number of persons per square kilometre) in India. *Source* Author

Conclusions

In this chapter, we have followed population growth modelling approach to estimate the population at the time of stabilisation or the size of the population when it stops growing in 640 districts of India. There has never been any attempt to estimate the size of the population at the time of stabilisation in the districts of the country. We have found that population growth during 1951–2011, as revealed through the population enumerated at different decennial censuses since 1951, can be modelled through the logistic growth model in all but a few districts. This means that the upper asymptote of the logistic growth model gives an estimate of the upper limit of population growth. The sum of the population at the time of stabilisation in 640 districts, then, gives an estimate of the population of the country at the time of stabilisation which is found to be very close to the upper limit of population growth of the country as revealed from the medium variant of the population projections prepared by the United Nations following the cohort-component method. The closeness of estimate of the upper limit of population growth in India obtained in this chapter by following a bottom-up approach with the estimate obtained through the industry standard cohort-component method of projection validates our approach

Table 3.3 Population sex ratio (males per 100 females) when population stops growing in India, states/Union Territories and districts

Country/State/ Union territory	Sex ratio M/F	Female advantage			Male advantage			Total
		High	Moderate	Average	Average	Moderate	High	
Andaman & Nicobar Islands	114	0	0	0	0	1	2	3
Andhra Pradesh	100	0	1	11	9	2	0	23
Arunachal Pradesh	106	4	0	1	4	2	5	16
Assam	103	0	0	2	23	2	0	27
Bihar	108	0	1	0	10	18	9	38
Chandigarh	121	0	0	0	0	0	1	1
Chhattisgarh	100	0	0	9	8	1	0	18
Dadra & Nagar Haveli	209	0	0	0	0	0	1	1
Daman & Diu	222	0	0	1	0	0	1	2
Delhi	112	0	0	0	0	2	7	9
Goa	102	0	0	1	1	0	0	2
Gujarat	112	0	0	1	7	14	4	26
Haryana	114	0	0	0	0	4	17	21
Himachal Pradesh	103	0	1	2	3	3	3	12
Jammu & Kashmir	114	0	0	1	1	6	14	22
Jharkhand	105	0	0	3	12	7	2	24
Karnataka	102	0	1	8	18	3	0	30
Kerala	92	5	5	4	0	0	0	14
Lakshadweep	106	0	0	0	0	1	0	1
Madhya Pradesh	106	0	0	6	18	19	7	50
Maharashtra	107	0	1	1	11	17	5	35
Manipur	100	0	0	5	3	0	1	9
Meghalaya	98	0	1	2	3	1	0	7
Mizoram	99	0	1	1	4	2	0	8
Nagaland	107	0	0	0	3	6	2	11
Odisha	102	0	3	6	15	5	1	30
Puducherry	82	2	2	0	0	0	0	4
Punjab	112	0	0	0	2	4	14	20
Rajasthan	107	0	0	1	10	16	6	33

(continued)

Table 3.3 (continued)

Country/State/ Union territory	Sex ratio M/F	Female advantage			Male advantage			Total
		High	Moderate	Average	Average	Moderate	High	
Sikkim	111	0	0	0	1	1	2	4
Tamil Nadu	94	2	12	15	3	0	0	32
Tripura	104	0	0	0	3	1	0	4
Uttar Pradesh	105	0	2	9	19	29	12	71
Uttarakhand	100	5	0	3	5	0	0	13
West Bengal	105	0	0	1	12	4	2	19
India	105	18	31	94	208	171	118	640

Source Author

of estimating population at the time of stabilisation in the districts of the country and hence in its states and Union Territories. Although, population stabilisation has been the key agenda in the development discourse of India right since independence, yet the present paper provides, for the first time, an estimate of population at the time of stabilisation in the districts. A district in India is the lowest administrative unit for development planning and programming for improving the quality of life of the people and for controlling population growth. In view of the social, cultural, economic and environmental diversity of the country India, there has always been emphasis on decentralised, district-based development planning and programming in India. Estimating the population of a district at the time of stabilisation, therefore, is expected to contribute, significantly, to estimating the long-term development and welfare needs of the people of the district and to analysing the long-term impact of population growth on development and environment at the local level.

The present analysis suggests that the long-term implications of population growth to social and economic development and the environment will be different in different districts of the country. For example, the population density, the single most important indicator of the population pressure on the environment at the local level, is likely to increase to more than 46 thousand persons per square Kilometre in district Northeast in the National Capital Territory of Delhi, the highest in the country. Similarly, even, in the long run, the male-female imbalance in population is likely to remain highly favourable to males in some geographically contiguous districts of the country despite all efforts of social and economic development including efforts to reduce and eliminate male-female disparities. At the national level, however, male-female imbalance is likely to normalise to around 105 males for every 100 females at the time of stabilisation.

Attempts to forecast district population in India are rare and there has never been any attempt to estimate population at the time of stabilisation in the districts or even in states and Union Territories. This is so when the size and the composition of population has always been an essential input to development planning and programming in India at either national or local (district) level. Population stabilisation has repeatedly been and continues to be stressed as necessary in the development discourse of the country for the accelerated social and economic progress, but there has never been any attempt to estimate the limit of population growth or population at stabilisation in districts which decide the population at stabilisation in states/Union Territories and in the country. Estimating the upper limit to population growth or population at stabilisation is a necessary input for estimating long-term development and welfare needs of the people of the district including the demand for water, energy and housing, and for assessing the long-term impact of population on natural resources at the local level. One reason to no attention to estimate population at stabilisation in the districts, of course, has been the paucity of district level data necessary for forecasting or projecting population growth. The present chapter makes the first attempt in this direction and shows that the long-term implications of population growth will be different in different districts.

Lastly, a caveat. The estimate of the population at stabilisation in the districts of India is derived following the data-driven approach which makes no assumption about the future trend in the determinants of population growth—fertility, mortality and migration. The approach is based on the characterisation of the pattern of population growth in the district observed during the period 1951–2011 as revealed through the population enumerated in the district in different decennial population census since 1951. This pattern of population growth may change with the arrival of new data which will then lead to new model describing the pattern of district population growth and hence new estimate of the upper limit of population growth or population at stabilisation. The arrival of new data may also lead to more than one population growth patterns in the same district, one confined to one sub-period and the other to another sub-period. There is also a possibility that one model depicts the increase in population while the other depicts the decrease in population as population starts decreasing when the upper limit of population is achieved. In the long run, forecasting district population growth may be more complex than the one which can be described by single wave logistic growth model.

Appendix

See Table 3.4.

Table 3.4 Size of the population at stabilisation, increase in population before population stops increasing, population sex ratio (males for every 100 females) and population density (persons per square Kilometre) in 640 districts of India as they existed at the 2011 population census

Country/State/Union Territory/District	Population at stabilisation			Increase in population since 2011		Population sex ratio at stabilisation	Population density at stabilisation
	Male (000)	Female (000)	Person (000)	Number (000)	%	Males per 100 females	Population per sq km^2
Jammu and Kashmir							
Kupwara	859	609	1468	598	68.7	141	617
Badgam	581	514	1095	341	45.2	113	804
Leh (Ladakh)	113	67	180	46	34.2	169	4
Kargil	105	83	188	47	33.5	126	13
Punch	400	352	752	275	57.6	114	449
Rajouri	539	446	985	343	53.5	121	375
Kathua	431	381	812	195	31.6	113	324
Baramula	761	671	1432	424	42.0	113	337
Bandipore	313	280	593	201	51.2	112	1718
Srinagar	914	838	1752	515	41.6	109	885
Ganderbal	268	221	489	192	64.6	121	1890
Pulwama	418	375	793	232	41.4	111	730
Shupiyan	197	198	395	128	48.1	99	1264
Anantnag	942	935	1877	798	73.9	101	525
Kulgam	274	260	534	110	25.9	105	1303
Doda	308	288	596	185	45.2	107	67
Ramban	219	205	424	140	49.3	107	319
Kishtwar	165	153	318	87	37.8	107	193
Udhampur	418	356	774	219	39.4	118	293
Reasi	225	207	432	117	37.1	109	251
Jammu	1064	942	2006	476	31.1	113	856
Samba	221	188	409	90	28.3	117	452
Himachal Pradesh							
Chamba	329	327	656	137	26.4	101	101
Kangra	921	929	1850	340	22.5	99	322
Lahul & Spiti	20	17	37	5	17.1	118	3
Kullu	304	290	594	156	35.7	105	108
Mandi	599	606	1205	206	20.6	99	305
Hamirpur	266	281	547	93	20.4	95	490
Una	345	325	670	148	28.4	106	435
Bilaspur	235	229	464	82	21.6	103	398

(continued)

Table 3.4 (continued)

Country/State/Union Territory/District	Population at stabilisation			Increase in population since 2011		Population sex ratio at stabilisation	Population density at stabilisation
	Male (000)	Female (000)	Person (000)	Number (000)	%	Males per 100 females	Population per sq km^2
Solan	415	364	779	199	34.2	114	402
Sirmaur	354	329	683	153	29.0	108	242
Shimla	529	488	1017	203	24.9	108	198
Kinnaur	55	45	100	16	18.7	122	16
Punjab							
Gurdaspur	1503	1324	2827	529	23.0	113	796
Kapurthala	525	469	994	179	22.0	112	609
Jalandhar	1441	1319	2760	566	25.8	109	1052
Hoshiarpur	985	958	1943	356	22.5	103	574
Shahid	365	358	723	111	18.1	102	564
Fatehgarh Sahib	394	342	736	136	22.6	115	624
Ludhiana	2445	2108	4553	1054	30.1	116	1273
Moga	649	572	1221	225	22.6	113	545
Firozpur	1393	1210	2603	575	28.3	115	491
Muktsar	594	533	1127	225	25.0	112	435
Faridkot	410	363	773	155	25.2	113	530
Bathinda	934	808	1742	353	25.4	116	519
Mansa	498	446	944	174	22.6	112	429
Patiala	1252	1110	2362	466	24.6	113	710
Amritsar	1766	1581	3347	857	34.4	112	1248
Tarn Taran	744	693	1437	317	28.3	107	595
Rupnagar	448	409	857	172	25.2	110	632
Sahibzada Ajit Singh Nagar	809	757	1566	571	57.4	107	1431
Sangrur	1097	969	2066	411	24.8	113	570
Barnala	390	348	738	143	24.0	112	498
Chandigarh							
Chandigarh	768	637	1405	350	33.2	121	12330
Uttarakhand							
Uttarkashi	215	207	422	92	27.8	104	53
Chamoli	232	242	474	82	21.0	96	59

(continued)

Table 3.4 (continued)

Country/State/Union Territory/District	Population at stabilisation			Increase in population since 2011		Population sex ratio at stabilisation	Population density at stabilisation
	Male (000)	Female (000)	Person (000)	Number (000)	%	Males per 100 females	Population per sq km^2
Rudraprayag	135	158	293	51	21.2	85	148
Tehri Garhwal	348	391	739	120	19.4	89	203
Dehradun	1253	1210	2463	766	45.1	104	797
Garhwal	376	420	796	109	15.8	90	149
Pithoragarh	285	290	575	92	19.1	98	81
Bageshwar	145	161	306	46	17.8	90	137
Almora	335	387	722	100	16.0	87	230
Champawat	158	161	319	59	22.7	98	180
Nainital	686	672	1358	403	42.3	102	319
Udham Singh Nagar	1184	1134	2318	669	40.6	104	912
Hardwar	1410	1345	2755	865	45.8	105	1168
Haryana							
Panchkula	397	365	762	200	35.6	109	848
Ambala	791	675	1466	337	29.9	117	931
Yamunanagar	855	750	1605	391	32.2	114	908
Kurukshetra	637	562	1199	234	24.3	113	784
Kaithal	706	622	1328	254	23.6	113	573
Karnal	1019	917	1936	430	28.6	111	768
Panipat	963	844	1807	602	49.9	114	1425
Sonipat	984	848	1832	382	26.3	116	863
Jind	888	769	1657	322	24.2	115	613
Fatehabad	607	548	1155	213	22.6	111	455
Sirsa	851	767	1618	323	24.9	111	378
Hisar	1195	1048	2243	499	28.6	114	563
Bhiwani	1100	977	2077	443	27.1	113	435
Rohtak	709	605	1314	253	23.8	117	753
Jhajjar	654	552	1206	248	25.9	118	658
Mahendragarh	625	551	1176	255	27.6	113	620
Rewari	644	554	1198	297	33.0	116	751
Gurgaon	2526	2085	4611	3096	204.4	121	3665
Mewat	1017	958	1975	886	81.3	106	1310

(continued)

Table 3.4 (continued)

Country/State/Union Territory/District	Population at stabilisation			Increase in population since 2011		Population sex ratio at stabilisation	Population density at stabilisation
	Male (000)	Female (000)	Person (000)	Number (000)	%	Males per 100 females	Population per sq km^2
Faridabad	1508	1406	2914	1105	61.0	107	3933
Palwal	782	724	1506	464	44.5	108	1109
National Capital Territory of Delhi							
Northwest	2650	2319	4969	1313	35.9	114	11218
North	635	586	1221	333	37.5	108	20020
Northeast	1507	1364	2871	629	28.1	110	46302
East	1159	1055	2214	505	29.6	110	35151
New Delhi	78	64	142	0	0.0	122	4057
Central	308	274	582	0	0.0	112	27730
West	1750	1590	3340	797	31.3	110	25693
Southwest	1822	1610	3432	1139	49.7	113	8152
South	1940	1741	3681	949	34.7	111	14901
Rajasthan							
Ganganagar	1344	1204	2548	579	29.4	112	232
Hanumangarh	1179	1067	2246	471	26.5	110	233
Bikaner	1831	1682	3513	1149	48.6	109	116
Churu	1399	1305	2704	664	32.6	107	195
Jhunjhunun	1387	1338	2725	587	27.5	104	460
Alwar	2737	2489	5226	1551	42.2	110	624
Bharatpur	1804	1642	3446	898	35.2	110	680
Dhaulpur	914	809	1723	516	42.8	113	568
Karauli	1099	967	2066	608	41.7	114	374
Sawai Madhopur	939	865	1804	469	35.1	109	401
Dausa	1243	1148	2391	757	46.3	108	697
Jaipur	5054	4662	9716	3090	46.6	108	872
Sikar	1843	1747	3590	913	34.1	106	464
Nagaur	2311	2172	4483	1176	35.5	106	253
Jodhpur	2717	2563	5280	1593	43.2	106	231
Jaisalmer	596	540	1136	467	69.7	110	30
Barmer	2128	1914	4042	1438	55.2	111	142
Jalor	1316	1251	2567	738	40.4	105	241

(continued)

Table 3.4 (continued)

Country/State/Union Territory/District	Population at stabilisation			Increase in population since 2011		Population sex ratio at stabilisation	Population density at stabilisation
	Male (000)	Female (000)	Person (000)	Number (000)	%	Males per 100 females	Population per sq km^2
Sirohi	750	709	1459	423	40.8	106	284
Pali	1280	1266	2546	509	25.0	101	206
Ajmer	1776	1719	3495	912	35.3	103	412
Tonk	954	929	1883	462	32.5	103	262
Bundi	741	708	1449	338	30.4	105	251
Bhilwara	1633	1625	3258	849	35.3	101	312
Rajsamand	748	736	1484	327	28.3	102	319
Dungarpur	955	945	1900	512	36.9	101	504
Banswara	1266	1198	2464	667	37.1	106	545
Chittaurgarh	991	975	1966	422	27.3	102	251
Kota	1380	1302	2682	732	37.5	106	514
Baran	834	807	1641	418	34.2	103	235
Jhalawar	945	895	1840	428	30.4	106	296
Udaipur	2153	2038	4191	1123	36.6	106	357
Pratapgarh	584	589	1173	304	35.1	99	263
Uttar Pradesh							
Saharanpur	2436	2255	4691	1224	35.3	108	1272
Muzaffarnagar	2842	2641	5483	1340	32.3	108	1368
Bijnor	2549	2452	5001	1318	35.8	104	1096
Moradabad	3531	3387	6918	2147	45.0	104	1861
Rampur	1646	1572	3218	882	37.8	105	1359
Jyotiba Phule Nagar	1348	1299	2647	807	43.8	104	1177
Meerut	2366	2148	4514	1071	31.1	110	1764
Baghpat	855	756	1611	308	23.6	113	1220
Ghaziabad	4553	4297	8850	4169	89.0	106	7507
Gautam Buddha Nagar	1932	1764	3696	2048	124.3	110	2883
Bulandshahr	2309	2145	4454	955	27.3	108	987
Aligarh	2678	2477	5155	1481	40.3	108	1412
Mahamaya Nagar	1057	956	2013	448	28.6	111	1094
Mathura	1868	1733	3601	1054	41.4	108	1078

(continued)

Table 3.4 (continued)

Country/State/Union Territory/District	Population at stabilisation			Increase in population since 2011		Population sex ratio at stabilisation	Population density at stabilisation
	Male (000)	Female (000)	Person (000)	Number (000)	%	Males per 100 females	Population per sq km^2
Agra	3295	2991	6286	1867	42.3	110	1556
Firozabad	1839	1674	3513	1015	40.6	110	1459
Mainpuri	1277	1169	2446	577	30.9	109	886
Budaun	2673	2533	5206	1524	41.4	106	1007
Bareilly	3233	3122	6355	1907	42.9	104	1543
Pilibhit	1443	1331	2774	744	36.6	108	753
Shahjahanpur	2241	2080	4321	1314	43.7	108	985
Kheri	3126	2914	6040	2018	50.2	107	786
Sitapur	3348	3079	6427	1943	43.3	109	1119
Hardoi	2923	2709	5632	1539	37.6	108	941
Unnao	2088	1930	4018	910	29.3	108	882
Lucknow	3489	3433	6922	2332	50.8	102	2738
Rae Bareli	2400	2295	4695	1289	37.9	105	1019
Farrukhabad	1317	1230	2547	661	35.1	107	1168
Kannauj	1135	1038	2173	517	31.2	109	1038
Etawah	1057	946	2003	421	26.6	112	867
Auraiya	916	815	1731	352	25.5	112	859
Kanpur Dehat	1208	1058	2266	470	26.2	114	750
Kanpur Nagar	3151	2748	5899	1318	28.8	115	1870
Jalaun	1153	1034	2187	497	29.4	111	479
Jhansi	1351	1220	2571	572	28.6	111	512
Lalitpur	908	892	1800	578	47.3	102	357
Hamirpur	734	636	1370	266	24.1	115	341
Mahoba	587	543	1130	254	29.0	108	359
Banda	1230	1098	2328	528	29.3	112	528
Chitrakoot	723	665	1388	397	40.0	109	432
Fatehpur	1752	1625	3377	744	28.2	108	813
Pratapgarh	2167	2193	4360	1151	35.9	99	1173
Kaushambi	1179	1087	2266	666	41.6	108	1273
Allahabad	4253	4013	8266	2311	38.8	106	1508
Bara Banki	2378	2277	4655	1394	42.8	104	1057

(continued)

Table 3.4 (continued)

Country/State/Union Territory/District	Population at stabilisation			Increase in population since 2011		Population sex ratio at stabilisation	Population density at stabilisation
	Male (000)	Female (000)	Person (000)	Number (000)	%	Males per 100 females	Population per sq km^2
Faizabad	1680	1649	3329	858	34.7	102	1422
Ambedkar Nagar	1614	1655	3269	871	36.3	98	1391
Sultanpur	2570	2688	5258	1460	38.5	96	1185
Bahraich	2759	2641	5400	1913	54.8	104	1031
Shrawasti	905	955	1860	743	66.5	95	1134
Balrampur	1611	1673	3284	1135	52.8	96	981
Gonda	2592	2561	5153	1719	50.1	101	1287
Siddharthnagar	1928	2093	4021	1462	57.1	92	1389
Basti	1652	1705	3357	893	36.2	97	1249
Sant Kabir Nagar	1198	1185	2383	667	38.9	101	1447
Mahrajganj	1982	1967	3949	1265	47.1	101	1338
Gorakhpur	3040	2961	6001	1560	35.1	103	1807
Kushinagar	2646	2570	5216	1651	46.3	103	1795
Deoria	2056	2093	4149	1048	33.8	98	1633
Azamgarh	3142	3219	6361	1747	37.9	98	1569
Mau	1571	1527	3098	892	40.4	103	1809
Ballia	2220	2094	4314	1074	33.2	106	1447
Jaunpur	2908	3043	5951	1457	32.4	96	1474
Ghazipur	2461	2433	4894	1274	35.2	101	1449
Chandauli	1407	1319	2726	774	39.6	107	1073
Varanasi	2564	2434	4998	1321	35.9	105	3256
Sant Ravidas Nagar	1008	1055	2063	486	30.8	96	2033
Mirzapur	1799	1698	3497	1000	40.0	106	794
Sonbhadra	1379	1352	2731	868	46.6	102	395
Etah	1236	1143	2379	605	34.1	108	979
Kanshiram Nagar	992	936	1928	491	34.2	106	986
Bihar							
Pashchim Champaran	2962	2835	5797	1862	47.3	104	1109
Purba Champaran	4106	3777	7883	2784	54.6	109	1987
Sheohar	563	545	1108	452	68.8	103	3174

(continued)

Table 3.4 (continued)

Country/State/Union Territory/District	Population at stabilisation			Increase in population since 2011		Population sex ratio at stabilisation	Population density at stabilisation
	Male (000)	Female (000)	Person (000)	Number (000)	%	Males per 100 females	Population per sq km^2
Sitamarhi	2753	2629	5382	1959	57.2	105	2346
Madhubani	3354	3157	6511	2023	45.1	106	1860
Supaul	1560	1466	3026	798	35.8	106	1248
Araria	2082	1909	3991	1179	41.9	109	1410
Kishanganj	1221	1173	2394	703	41.6	104	1270
Purnia	2385	2265	4650	1386	42.5	105	1440
Katihar	2331	2097	4428	1357	44.2	111	1449
Madhepura	1548	1421	2969	968	48.4	109	1661
Saharsa	1488	1345	2833	932	49.0	111	1679
Darbhanga	2940	2709	5649	1712	43.5	109	2479
Muzaffarpur	3786	3280	7066	2265	47.2	115	2227
Gopalganj	1715	1855	3570	1009	39.4	92	1756
Siwan	2400	2272	4672	1341	40.3	106	2105
Saran	2804	2715	5519	1568	39.7	103	2090
Vaishali	2802	2392	5194	1699	48.6	117	2551
Samastipur	3183	2876	6059	1797	42.2	111	2086
Begusarai	2276	2065	4341	1370	46.1	110	2263
Khagaria	1324	1210	2534	867	52.0	109	1705
Bhagalpur	2230	1984	4214	1176	38.7	112	1640
Banka	1517	1395	2912	877	43.1	109	964
Munger	941	847	1788	420	30.7	111	1260
Lakhisarai	717	659	1376	375	37.5	109	1121
Sheikhpura	449	431	880	243	38.2	104	1276
Nalanda	1961	1804	3765	888	30.8	109	1599
Patna	4330	4021	8351	2513	43.0	108	2608
Bhojpur	1987	1768	3755	1026	37.6	112	1568
Buxar	1212	1175	2387	681	39.9	103	1402
Kaimur (Bhabua)	1253	1213	2466	839	51.6	103	740
Rohtas	2074	1951	4025	1065	36.0	106	1037
Aurangabad	1972	1844	3816	1277	50.3	107	1155
Gaya	3250	3141	6391	2000	45.5	103	1284

(continued)

Table 3.4 (continued)

Country/State/Union Territory/District	Population at stabilisation			Increase in population since 2011		Population sex ratio at stabilisation	Population density at stabilisation
	Male (000)	Female (000)	Person (000)	Number (000)	%	Males per 100 females	Population per sq km^2
Nawada	1698	1553	3251	1032	46.5	109	1304
Jamui	1345	1289	2634	873	49.6	104	850
Jehanabad	849	774	1623	498	44.3	110	1744
Arwal	503	463	966	265	37.8	109	1514
Sikkim							
North District	29	22	51	7	16.1	133	12
West District	83	80	163	26	19.2	104	139
South District	92	85	177	31	21.0	108	237
East District	184	162	346	63	22.0	114	363
Arunachal Pradesh							
Tawang	46	29	75	25	49.6	160	34
West Kameng	56	48	104	19	22.8	117	14
East Kameng	48	58	106	27	34.4	83	26
Papum Pare	135	153	288	111	62.9	88	83
Upper Subansiri	54	65	119	36	43.4	83	17
West Siang	68	66	134	22	19.5	104	16
East Siang	59	60	119	20	19.7	97	33
Upper Siang	22	19	41	5	15.4	115	6
Changlang	95	91	186	39	26.0	105	40
Tirap	69	67	136	24	21.2	104	57
Lower Subansiri	63	75	138	55	66.1	85	39
Kurung Kumey	962	850	1812	1721	1868.6	113	300
Dibang Valley	5	4	9	1	9.6	129	1
Lower Dibang Valley	33	32	65	11	20.2	106	17
Lohit	96	91	187	41	28.2	105	36
Anjaw	13	12	25	4	18.9	106	4
Nagaland							
Mon	160	142	302	52	20.8	112	169
Mokokchung	129	117	246	52	26.5	110	152
Zunheboto	89	86	175	34	24.3	104	139
Wokha	103	101	204	37	22.5	102	125

(continued)

Table 3.4 (continued)

Country/State/Union Territory/District	Population at stabilisation			Increase in population since 2011		Population sex ratio at stabilisation	Population density at stabilisation
	Male (000)	Female (000)	Person (000)	Number (000)	%	Males per 100 females	Population per sq km^2
Dimapur	270	254	524	146	38.5	106	566
Phek	103	100	203	39	23.9	104	100
Tuensang	128	118	246	49	25.0	108	97
Longleng	51	46	97	46	91.1	112	172
Kiphire	53	49	102	29	38.5	108	91
Kohima	187	173	360	92	34.3	108	246
Peren	60	55	115	20	20.7	108	70
Manipur							
Senapati	518	512	1030	551	115.0	101	315
Tamenglong	105	102	207	67	47.4	102	47
Churachandpur	182	181	363	88	32.2	101	79
Bishnupur	145	146	291	53	22.5	99	586
Thoubal	265	271	536	115	27.2	98	1044
Imphal West	313	326	639	121	23.4	96	1232
Imphal East	281	287	568	112	24.5	98	801
Ukhrul	137	143	280	96	52.1	96	62
Chandel	121	99	220	76	52.5	122	66
Mizoram							
Mamit	47	44	91	5	5.4	106	30
Kolasib	54	54	108	25	29.6	100	79
Aizawl	240	255	495	94	23.6	94	138
Champhai	78	77	155	29	23.2	100	49
Serchhip	40	40	80	15	23.2	101	56
Lunglei	98	93	191	30	18.3	105	42
Lawngtlai	78	77	155	37	31.4	101	61
Saiha	35	35	70	13	23.3	99	50
Tripura							
West Tripura	1091	1041	2132	406	23.5	105	712
South Tripura	534	516	1050	173	19.8	104	343
Dhalai	234	220	454	76	20.1	106	189
North Tripura	445	433	878	185	26.6	103	431

(continued)

Table 3.4 (continued)

Country/State/Union Territory/District	Population at stabilisation			Increase in population since 2011		Population sex ratio at stabilisation	Population density at stabilisation
	Male (000)	Female (000)	Person (000)	Number (000)	%	Males per 100 females	Population per sq km^2
Meghalaya							
West Garo Hills	467	466	933	289	45.0	100	254
East Garo Hills	234	230	464	146	45.9	102	178
South Garo Hills	106	97	203	61	42.6	109	108
West Khasi Hills	286	290	576	193	50.3	98	110
Ribhoi	208	206	414	155	59.9	101	169
East Khasi Hills	546	575	1121	295	35.7	95	408
Jaintia Hills	297	316	613	218	55.2	94	161
Assam							
Kokrajhar	538	518	1056	168	19.0	104	320
Dhubri	1316	1251	2567	618	31.7	105	1180
Goalpara	642	628	1270	262	26.0	102	697
Barpeta	1072	1026	2098	405	23.9	104	919
Morigaon	612	603	1215	258	26.9	102	783
Nagaon	1753	1768	3521	697	24.7	99	886
Sonitpur	1201	1176	2377	453	23.5	102	457
Lakhimpur	633	630	1263	221	21.2	100	555
Dhemaji	431	410	841	155	22.6	105	260
Tinsukia	812	805	1617	289	21.7	101	427
Dibrugarh	796	782	1578	252	19.0	102	467
Sivasagar	712	686	1398	247	21.4	104	524
Jorhat	656	643	1299	207	18.9	102	456
Golaghat	662	646	1308	241	22.6	103	373
Karbi Anglong	597	580	1177	221	23.2	103	113
Dima Hasao	141	137	278	64	29.8	103	57
Cachar	1129	1097	2226	489	28.2	103	588
Karimganj	816	780	1596	367	29.9	105	882
Hailakandi	434	409	843	184	27.9	106	635
Bongaigaon	474	462	936	197	26.6	103	856
Chirang	281	272	553	71	14.7	103	288
Kamrup	952	916	1868	350	23.1	104	602

(continued)

Table 3.4 (continued)

Country/State/Union Territory/District	Population at stabilisation			Increase in population since 2011		Population sex ratio at stabilisation	Population density at stabilisation
	Male (000)	Female (000)	Person (000)	Number (000)	%	Males per 100 females	Population per sq km^2
Kamrup Metropolitan	848	866	1714	459	36.6	98	1794
Nalbari	462	441	903	131	17.0	105	858
Baksa	564	557	1121	171	18.0	101	456
Darrang	607	581	1188	259	27.9	104	749
Udalguri	494	480	974	143	17.2	103	484
West Bengal							
Darjiling	1194	1200	2394	546	29.6	99	760
Jalpaiguri	2487	2372	4859	986	25.5	105	780
Koch Bihar	1727	1625	3352	533	18.9	106	990
Uttar Dinajpur	2098	1994	4092	1085	36.1	105	1303
Dakshin Dinajpur	1078	1033	2111	435	25.9	104	951
Maldah	2739	2480	5219	1230	30.8	110	1398
Murshidabad	4778	4669	9447	2343	33.0	102	1774
Birbhum	2217	2115	4332	830	23.7	105	953
Barddhaman	4751	4584	9335	1617	21.0	104	1329
Nadia	3325	3141	6466	1298	25.1	106	1646
North Twenty	6473	6231	12704	2693	26.9	104	3103
Hugli	3389	3271	6660	1141	20.7	104	2115
Bankura	2209	2122	4331	734	20.4	104	629
Puruliya	1865	1790	3655	724	24.7	104	584
Haora	3103	2962	6065	1215	25.0	105	4134
Kolkata	2767	2472	5239	743	16.5	112	28320
South 24 Parganas	5391	5187	10578	2416	29.6	104	1062
Paschim Medinipur	3652	3522	7174	1260	21.3	104	766
Purba Medinipur	3214	3046	6260	1164	22.8	106	1328
Jharkhand							
Garhwa	980	937	1917	594	44.9	105	468
Chatra	782	758	1540	498	47.7	103	414
Kodarma	583	525	1108	391	54.6	111	436
Giridih	1905	1649	3554	1109	45.3	116	716

(continued)

Table 3.4 (continued)

Country/State/Union Territory/District	Population at stabilisation			Increase in population since 2011		Population sex ratio at stabilisation	Population density at stabilisation
	Male (000)	Female (000)	Person (000)	Number (000)	%	Males per 100 females	Population per sq km^2
Deoghar	1152	1068	2220	727	48.8	108	896
Godda	949	906	1855	542	41.2	105	819
Sahibganj	812	784	1596	446	38.7	104	774
Pakur	654	653	1307	406	45.1	100	722
Dhanbad	1752	1667	3419	734	27.3	105	1676
Bokaro	1371	1305	2676	614	29.8	105	928
Lohardaga	359	350	709	247	53.4	103	472
Purbi Singhbhum	1462	1425	2887	592	25.8	103	810
Palamu	1436	1343	2779	839	43.2	107	632
Latehar	531	512	1043	317	43.6	104	243
Hazaribagh	1225	1148	2373	638	36.8	107	667
Ramgarh	605	588	1193	243	25.6	103	890
Dumka	869	885	1754	432	32.7	98	466
Jamtara	566	538	1104	313	39.6	105	610
Ranchi	2019	1938	3957	1043	35.8	104	776
Khunti	351	351	702	171	32.1	100	277
Gumla	703	661	1364	339	33.0	106	254
Simdega	372	366	738	138	23.1	102	195
Pashchimi Singhbhum	956	998	1954	451	30.0	96	270
Saraikela-Kharsawan	730	741	1471	406	38.1	99	554
Odisha							
Bargarh	898	884	1782	300	20.3	102	305
Jharsuguda	369	359	728	148	25.6	103	344
Sambalpur	637	629	1266	225	21.6	101	191
Debagarh	191	189	380	67	21.5	101	129
Sundargarh	1246	1256	2502	408	19.5	99	258
Kendujhar	1109	1096	2205	403	22.4	101	266
Mayurbhanj	1588	1601	3189	669	26.6	99	306
Baleshwar	1493	1399	2892	571	24.6	107	760
Bhadrak	953	937	1890	384	25.5	102	755

(continued)

Table 3.4 (continued)

Country/State/Union Territory/District	Population at stabilisation			Increase in population since 2011		Population sex ratio at stabilisation	Population density at stabilisation
	Male (000)	Female (000)	Person (000)	Number (000)	%	Males per 100 females	Population per sq km^2
Kendrapara	852	850	1702	262	18.2	100	644
Jagatsinghapur	694	666	1360	222	19.6	104	815
Cuttack	1612	1543	3155	531	20.2	104	802
Jajapur	1150	1115	2265	438	24.0	103	781
Dhenkanal	735	698	1433	240	20.1	105	322
Anugul	821	777	1598	323	25.4	106	251
Nayagarh	597	538	1135	172	17.9	111	292
Khordha	1558	1425	2983	732	32.5	109	1061
Puri	1063	1020	2083	385	22.6	104	599
Ganjam	2217	2171	4388	859	24.3	102	535
Gajapati	340	357	697	120	20.8	95	161
Kandhamal	454	488	942	209	28.4	93	117
Baudh	287	283	570	129	29.3	101	184
Subarnapur	388	369	757	147	24.1	105	324
Balangir	1000	994	1994	346	21.0	101	303
Nuapada	364	368	732	122	20.0	99	190
Kalahandi	1015	1013	2028	452	28.6	100	256
Rayagada	576	617	1193	226	23.3	93	169
Nabarangapur	774	804	1578	357	29.2	96	298
Koraput	858	908	1766	386	28.0	95	201
Malkangiri	364	383	747	134	21.9	95	129
Chhattisgarh							
Koriya	414	408	822	163	24.8	102	124
Surguja	1600	1570	3170	811	34.3	102	202
Jashpur	510	500	1010	158	18.6	102	173
Raigarh	959	958	1917	423	28.3	100	271
Korba	811	803	1614	407	33.7	101	245
Janjgir-Champa	1195	1173	2368	748	46.2	102	614
Bilaspur	1886	1787	3673	1010	37.9	106	444
Kabeerdham	571	584	1155	333	40.4	98	273
Rajnandgaon	959	1000	1959	422	27.4	96	243

(continued)

Table 3.4 (continued)

Country/State/Union Territory/District	Population at stabilisation			Increase in population since 2011		Population sex ratio at stabilisation	Population density at stabilisation
	Male (000)	Female (000)	Person (000)	Number (000)	%	Males per 100 females	Population per sq km^2
Durg	2056	2102	4158	814	24.4	98	487
Raipur	2793	2740	5533	1469	36.2	102	447
Mahasamund	616	624	1240	208	20.1	99	259
Dhamtari	508	521	1029	229	28.6	97	252
Uttar Bastar Kanker	455	461	916	166	22.2	99	128
Bastar	881	911	1792	379	26.8	97	171
Narayanpur	93	93	186	46	33.1	100	40
Dakshin Bastar Dantewada	310	322	632	98	18.4	96	76
Bijapur	159	159	318	63	24.6	100	37
Madhya Pradesh							
Sheopur	512	457	969	281	40.9	112	147
Morena	1442	1232	2674	708	36.0	117	536
Bhind	1175	995	2170	467	27.4	118	487
Gwalior	1483	1329	2812	781	38.4	112	617
Datia	540	493	1033	246	31.3	110	356
Shivpuri	1286	1172	2458	732	42.4	110	244
Tikamgarh	1032	949	1981	535	37.0	109	392
Chhatarpur	1264	1166	2430	667	37.9	108	280
Panna	695	643	1338	322	31.7	108	188
Sagar	1603	1463	3066	687	28.9	110	299
Damoh	849	772	1621	357	28.2	110	222
Satna	1548	1452	3000	772	34.6	107	400
Rewa	1691	1570	3261	896	37.9	108	516
Umaria	454	429	883	238	37.0	106	217
Neemuch	535	513	1048	222	26.9	104	246
Mandsaur	871	852	1723	383	28.6	102	311
Ratlam	960	950	1910	455	31.3	101	393
Ujjain	1310	1272	2582	595	29.9	103	424
Shajapur	1025	973	1998	486	32.1	105	323
Dewas	1057	1021	2078	514	32.9	103	296

(continued)

Table 3.4 (continued)

Country/State/Union Territory/District	Population at stabilisation			Increase in population since 2011		Population sex ratio at stabilisation	Population density at stabilisation
	Male (000)	Female (000)	Person (000)	Number (000)	%	Males per 100 females	Population per sq km^2
Dhar	1557	1510	3067	881	40.3	103	376
Indore	2579	2420	4999	1722	52.6	107	1282
Khargone (West Nimar)	1255	1253	2508	635	33.9	100	313
Barwani	1011	1005	2016	630	45.5	101	371
Rajgarh	1084	1062	2146	600	38.8	102	349
Vidisha	993	923	1916	457	31.3	108	260
Bhopal	1753	1699	3452	1081	45.6	103	1245
Sehore	922	851	1773	462	35.2	108	270
Raisen	927	839	1766	435	32.6	110	209
Betul	999	977	1976	401	25.4	102	197
Harda	396	374	770	200	35.0	106	231
Hoshangabad	822	769	1591	350	28.2	107	237
Katni	863	827	1690	398	30.8	104	341
Jabalpur	1589	1488	3077	613	24.9	107	590
Narsimhapur	723	667	1390	298	27.3	109	271
Dindori	431	436	867	163	23.2	99	116
Mandla	655	669	1324	269	25.5	98	228
Chhindwara	1355	1303	2658	567	27.1	104	225
Seoni	869	848	1717	338	24.5	102	196
Balaghat	1012	1027	2039	337	19.8	99	221
Guna	926	887	1813	572	46.1	104	284
Ashoknagar	590	535	1125	279	33.1	110	241
Shahdol	681	671	1352	286	26.8	101	218
Anuppur	471	475	946	197	26.2	99	252
Sidhi	812	787	1599	472	41.9	103	330
Singrauli	944	870	1814	635	53.9	108	320
Jhabua	728	734	1462	437	42.6	99	406
Alirajpur	499	508	1007	278	38.1	98	316
Khandwa (East Nimar)	856	816	1672	362	27.6	105	227
Burhanpur	490	464	954	197	25.9	106	279

(continued)

Table 3.4 (continued)

Country/State/Union Territory/District	Population at stabilisation			Increase in population since 2011		Population sex ratio at stabilisation	Population density at stabilisation
	Male (000)	Female (000)	Person (000)	Number (000)	%	Males per 100 females	Population per sq km^2
Gujarat							
Kachchh	1624	1386	3010	917	43.8	117	66
Banas Kantha	2231	2084	4315	1195	38.3	107	402
Patan	819	764	1583	239	17.8	107	273
Mahesana	1275	1165	2440	404	19.9	109	554
Sabar Kantha	1530	1440	2970	542	22.3	106	402
Gandhinagar	910	836	1746	354	25.4	109	816
Ahmadabad	5034	4562	9596	2382	33.0	110	1184
Surendranagar	1159	1084	2243	486	27.7	107	215
Rajkot	2585	2387	4972	1168	30.7	108	444
Jamnagar	1336	1250	2586	426	19.7	107	182
Porbandar	350	333	683	98	16.7	105	295
Junagadh	1684	1606	3290	547	19.9	105	373
Amreli	895	863	1758	244	16.1	104	238
Bhavnagar	1899	1734	3633	753	26.1	109	362
Anand	1297	1195	2492	399	19.1	109	778
Kheda	1415	1313	2728	428	18.6	108	690
Panch Mahals	1538	1493	3031	640	26.8	103	579
Dohad	1578	1551	3129	1002	47.1	102	859
Vadodara	2679	2522	5201	1035	24.8	106	689
Narmada	362	349	711	121	20.5	104	252
Bharuch	1011	930	1941	390	25.1	109	298
The Dangs	151	155	306	78	34.1	98	173
Navsari	824	792	1616	287	21.6	104	720
Valsad	1266	1124	2390	684	40.1	113	794
Surat	6952	4774	11726	5644	92.8	146	2578
Tapi	473	471	944	137	17.0	100	301
Daman and Diu							
Diu	32	33	65	13	25.0	99	1668
Daman	383	155	538	346	181.1	248	7462
Dadra and Nagar Haveli							

(continued)

Table 3.4 (continued)

Country/State/Union Territory/District	Population at stabilisation			Increase in population since 2011		Population sex ratio at stabilisation	Population density at stabilisation
	Male (000)	Female (000)	Person (000)	Number (000)	%	Males per 100 females	Population per sq km^2
Dadra & Nagar Haveli	597	286	883	540	157.1	209	1799
Maharashtra							
Nandurbar	1137	1099	2236	587	35.6	103	375
Dhule	1322	1245	2567	516	25.2	106	357
Jalgaon	2745	2504	5249	1019	24.1	110	446
Buldana	1713	1599	3312	726	28.1	107	343
Akola	1150	1102	2252	438	24.2	104	397
Washim	787	726	1513	317	26.4	108	309
Amravati	1811	1747	3558	669	23.2	104	291
Wardha	807	755	1562	261	20.1	107	248
Nagpur	3085	2996	6081	1427	30.7	103	615
Bhandara	725	715	1440	240	20.0	102	352
Gondiya	760	758	1518	195	14.8	100	290
Gadchiroli	684	675	1359	286	26.7	101	94
Chandrapur	1368	1327	2695	491	22.3	103	236
Yavatmal	1772	1673	3445	673	24.3	106	254
Nanded	2310	2167	4477	1115	33.2	107	425
Hingoli	798	737	1535	358	30.4	108	318
Parbhani	1244	1171	2415	578	31.5	106	389
Jalna	1352	1254	2606	647	33.0	108	339
Aurangabad	2816	2622	5438	1737	46.9	107	537
Nashik	4394	4081	8475	2368	38.8	108	546
Thane	9534	9431	18965	7905	71.5	101	1984
Mumbai Suburban	6190	5331	11521	2164	23.1	116	25832
Mumbai	2064	1625	3689	604	19.6	127	23498
Raigarh	1880	1696	3576	941	35.7	111	500
Pune	7477	6727	14204	4775	50.6	111	908
Ahmadnagar	2929	2776	5705	1161	25.6	106	335
Bid	1765	1590	3355	770	29.8	111	314
Latur	1715	1569	3284	830	33.8	109	459

(continued)

Table 3.4 (continued)

Country/State/Union Territory/District	Population at stabilisation			Increase in population since 2011		Population sex ratio at stabilisation	Population density at stabilisation
	Male (000)	Female (000)	Person (000)	Number (000)	%	Males per 100 females	Population per sq km^2
Osmanabad	1071	978	2049	391	23.6	109	271
Solapur	2830	2648	5478	1161	26.9	107	368
Satara	1835	1802	3637	633	21.1	102	347
Ratnagiri	908	1004	1912	297	18.4	90	233
Sindhudurg	494	503	997	147	17.3	98	191
Kolhapur	2449	2330	4779	902	23.3	105	622
Sangli	1764	1713	3477	655	23.2	103	406
Andhra Pradesh							
Adilabad	1730	1747	3477	736	26.8	99	216
Nizamabad	1515	1596	3111	560	21.9	95	391
Karimnagar	2287	2336	4623	847	22.4	98	391
Medak	1951	1925	3876	842	27.8	101	400
Hyderabad	2423	2283	4706	762	19.3	106	21684
Rangareddy	5319	5056	10375	5077	95.9	105	1385
Mahbubnagar	2607	2532	5139	1086	26.8	103	279
Nalgonda	2113	2135	4248	759	21.8	99	298
Warangal	2159	2138	4297	784	22.3	101	335
Khammam	1696	1730	3426	628	22.5	98	214
Srikakulam	1605	1650	3255	551	20.4	97	558
Vizianagaram	1361	1380	2741	397	16.9	99	419
Visakhapatnam	2656	2728	5384	1093	25.5	97	482
East Godavari	3013	3033	6046	892	17.3	99	559
West Godavari	2308	2305	4613	676	17.2	100	596
Krishna	2743	2749	5492	975	21.6	100	629
Guntur	2902	2924	5826	938	19.2	99	511
Prakasam	2085	2055	4140	743	21.9	101	235
Sri Potti Sriramulu Nellore	1814	1792	3606	643	21.7	101	276
Y. S. R.	1775	1771	3546	663	23.0	100	231
Kurnool	2580	2556	5136	1082	26.7	101	291
Anantapur	2569	2529	5098	1017	24.9	102	267

(continued)

Table 3.4 (continued)

Country/State/Union Territory/District	Population at stabilisation			Increase in population since 2011		Population sex ratio at stabilisation	Population density at stabilisation
	Male (000)	Female (000)	Person (000)	Number (000)	%	Males per 100 females	Population per sq km[2]
Chittoor	2548	2624	5172	999	23.9	97	341
Karnataka							
Belgaum	3006	2955	5961	1181	24.7	102	444
Bagalkot	1213	1193	2406	516	27.3	102	367
Bijapur	1437	1402	2839	662	30.4	103	270
Bidar	1128	1061	2189	486	28.5	106	402
Raichur	1261	1278	2539	610	31.6	99	301
Koppal	941	916	1857	467	33.6	103	333
Gadag	646	632	1278	213	20.0	102	274
Dharwad	1142	1131	2273	425	23.0	101	533
Uttara Kannada	854	836	1690	253	17.6	102	164
Haveri	986	960	1946	348	21.8	103	403
Bellary	1640	1617	3257	803	32.8	101	385
Chitradurga	1017	1007	2024	365	22.0	101	240
Davanagere	1190	1169	2359	414	21.3	102	398
Shimoga	1020	1008	2028	275	15.7	101	239
Udupi	649	707	1356	179	15.2	92	379
Chikmagalur	667	678	1345	207	18.2	98	187
Tumkur	1614	1605	3219	540	20.2	101	304
Bangalore	8473	7878	16351	6730	69.9	108	7446
Mandya	1053	1060	2113	307	17.0	99	426
Hassan	1046	1061	2107	331	18.6	99	309
Dakshina Kannada	1238	1260	2498	408	19.5	98	514
Kodagu	317	326	643	89	16.1	97	157
Mysore	1878	1856	3734	733	24.4	101	592
Chamarajanagar	600	606	1206	186	18.2	99	214
Gulbarga	1684	1656	3340	774	30.2	102	305
Yadgir	801	805	1606	432	36.8	100	305
Kolar	961	944	1905	368	24.0	102	479
Chikkaballapura	771	747	1518	263	21.0	103	358
Bangalore Rural	652	613	1265	273	27.6	106	550

(continued)

Table 3.4 (continued)

Country/State/Union Territory/District	Population at stabilisation			Increase in population since 2011		Population sex ratio at stabilisation	Population density at stabilisation
	Male (000)	Female (000)	Person (000)	Number (000)	%	Males per 100 females	Population per sq km^2
Ramanagara	638	634	1272	189	17.5	101	362
Goa							
North Goa	502	484	986	168	20.5	104	568
South Goa	385	390	775	135	21.0	99	394
Lakshadweep							
Lakshadweep	40	38	78	14	21.5	106	2610
Kerala							
Kasaragod	758	825	1583	276	21.1	92	796
Kannur	1365	1588	2953	430	17.0	86	997
Wayanad	483	494	977	160	19.5	98	459
Kozhikode	1726	1920	3646	559	18.1	90	1555
Malappuram	2454	2713	5167	1054	25.6	90	1454
Palakkad	1614	1734	3348	538	19.1	93	747
Thrissur	1721	1915	3636	515	16.5	90	1201
Ernakulam	1881	1947	3828	545	16.6	97	1250
Idukki	649	655	1304	195	17.6	99	299
Kottayam	1119	1161	2280	305	15.4	96	1033
Alappuzha	1167	1280	2447	319	15.0	91	1729
Pathanamthitta	657	736	1393	196	16.4	89	525
Kollam	1451	1614	3065	430	16.3	90	1235
Thiruvananthapuram	1852	2031	3883	582	17.6	91	1774
Tamil Nadu							
Thiruvallur	2759	3205	5964	2237	60.0	86	1757
Chennai	2748	2844	5592	946	20.4	97	31958
Kancheepuram	2891	3565	6456	2458	61.5	81	1440
Vellore	2398	2534	4932	995	25.3	95	812
Tiruvannamalai	1448	1535	2983	518	21.0	94	482
Viluppuram	2103	2227	4330	871	25.2	94	602
Salem	2179	2151	4330	847	24.3	101	827
Namakkal	1071	1096	2167	441	25.6	98	634
Erode	1341	1399	2740	489	21.7	96	476

(continued)

Table 3.4 (continued)

Country/State/Union Territory/District	Population at stabilisation			Increase in population since 2011		Population sex ratio at stabilisation	Population density at stabilisation
	Male (000)	Female (000)	Person (000)	Number (000)	%	Males per 100 females	Population per sq km^2
The Nilgiris	425	453	878	143	19.4	94	342
Dindigul	1276	1326	2602	442	20.5	96	431
Karur	633	664	1297	233	21.9	95	447
Tiruchirappalli	1574	1694	3268	546	20.1	93	725
Perambalur	345	360	705	140	24.7	96	401
Ariyalur	438	456	894	139	18.5	96	461
Cuddalore	1535	1587	3122	515	19.8	97	843
Nagapattinam	934	986	1920	304	18.8	95	747
Thiruvarur	722	761	1483	219	17.3	95	652
Thanjavur	1369	1489	2858	452	18.8	92	838
Pudukkottai	959	996	1955	337	20.8	96	421
Sivaganga	767	788	1555	216	16.1	97	367
Madurai	1785	1849	3634	595	19.6	97	979
Theni	721	704	1425	179	14.4	103	497
Virudhunagar	1165	1230	2395	453	23.3	95	565
Ramanathapuram	766	809	1575	222	16.4	95	384
Thoothukkudi	984	1038	2022	272	15.5	95	426
Tirunelveli	1833	1912	3745	668	21.7	96	560
Kanniyakumari	1055	1100	2155	285	15.2	96	1280
Dharmapuri	948	928	1876	369	24.5	102	417
Krishnagiri	1265	1281	2546	666	35.4	99	496
Coimbatore	2091	2212	4303	846	24.5	95	910
Tiruppur	1893	1975	3868	1389	56.0	96	746
Puducherry							
Yanam	104	214	318	262	471.7	48	10601
Puducherry	643	708	1351	400	42.1	91	4594
Mahe	23	28	51	9	22.6	84	5696
Karaikal	124	134	258	57	28.5	92	1639
Andaman and Nicobar Islands							
Nicobars	25	20	45	9	23.6	123	25

(continued)

Table 3.4 (continued)

Country/State/Union Territory/District	Population at stabilisation			Increase in population since 2011		Population sex ratio at stabilisation	Population density at stabilisation
	Male (000)	Female (000)	Person (000)	Number (000)	%	Males per 100 females	Population per sq km^2
North & Middle Andaman	67	62	129	23	22.0	109	34
South Andaman	165	143	308	69	29.2	115	115

Source Author

References

Bhat PNM (1999) Population projections for Delhi: dynamic logistic model versus cohort-component method. Demogr India 28(2):153–165

Brook BW, Sodhi NS, Bradshaw CJA (2008) Synergies among extinction drivers under global change. Trends Ecol & Evol 23(8):453–460

Burg D, Ausubel JH (2023) Jewish population trajectories between the Jordan River and the Mediterranean Sea. Isr Aff 29(3):653–668

Dorn HF (1950) Pitfalls in population forecasts and projections. J Am Stat Assoc 45(251):311–334

Fisher J, Pry R (1971) A simple substitution model of technological change. Technol Forecast Soc Chang 3:75–88

Government of India (2000) National Population Policy 2000. Ministry of Health and Family Welfare, New Delhi

Government of India (2014) Census of India 2011. Report of the post enumeration survey. Registrar General and Census Commissioner of India, New Delhi

Government of India (2020a) Sample registration system statistical report 2020. Registrar General and Census Commissioner, India, New Delhi

Hrytsiuk P, Parfeniuk O, Shevchenko I (2023) The logistic dynamics of population growth as a prerequisite for global sustainable development. IOP Conf Ser: Earth Environ Sci 1126:012030

Kulkarni PM (2021) Demographic and regional decomposition of perspective population growth for India, 2021-2101. Demogr India 50(2):1–15

Leach D (1981) Re-evaluation of the logistic curve for human populations. J R Stat Soc Ser (Gen) 144(1):94–103

Lotka AJ (1956) Elements of physical biology. Dover, New York

Mondol H, Mallick UK, Biswas Md HA (2018) Mathematical modeling and predicting the current trends of human population growth in Bangladesh. Adv Model Anal 55(2):62–69

Passos JR, de Pinho SZ, de Carvalho LR, Mischan MM (2012) Critical points in logistic growth curves and treatment comparisons. Sci Agric 69(5):308–312

Pearl R, Reed LJ (1920) On the rate of growth of the population of United States since 1790 and its mathematical representation. Proc Natl Acad Sci 6(6):275–288

Sharif Ullah M, Mostafa G, Jahan N, Khan AH (2019) Analyzing and projection of future Bangladesh population using logistic growth model. Int J Mod Nonlinear Theory Appl 8:53–61

Spiess AN, Neumeyer N (2010) An evaluation of R^2 as an inadequate measure for nonlinear models in pharmacological and biochemical research: a Monte Carlo approach. BMC Pharmacol 10(1):6–17

United Nations (2022) World population prospects 2022, Online edn. United Nations, Department of Economic and Social Affairs, Population Division, New York

Verhlust P-F (1838) A note on the law of population growth. *Correspondence Mathematique et physique Publiee par a.* Quetelet (BrussS) 10:113–117

Yung J, Meyer PS, Ausubel JH (1999) Loglet lab for windows (version 1.1) tutorial. Technol Forecast Soc Chang 61(3):273–295

Chapter 4
Profiles of Fertility

Abstract This chapter presents estimates of total fertility rate (TFR) for 707 districts of the country based on the data available from the National Family Health Survey, 2019–2021. The chapter also classifies 707 districts into six profiles of fertility depending upon the total marital fertility rate (TMFR) and proportion of reproductive age women who are married in a district relative to the TMFR and the proportion of married women of reproductive age in the country. The chapter highlights the variation in TFR across districts. In 326 of the 707 districts of the country, TFR is estimated to be below the replacement level, although there are 67 districts in which TFR is estimated to be very high, at least 3 births per woman of reproductive age. The chapter also reveals that 707 districts can be classified into six fertility profiles depending upon the level of TMFR and the proportion of reproductive age women who are married.

Keywords India · Districts · Total fertility rate · Fertility profile

Introduction

Direct estimates of fertility are not available for the districts of India. Although, registration of births in India is mandatory under the Registration of Births and Deaths Act of 1969 (Government of India 1969), yet according to the latest round of the National Family Health Survey (2019–2021), birth of only around 89 per cent children below 5 years of age was found to be registered under the official civil registration system and this proportion varies widely across districts (Government of India 2022). Under the Registration of Births and Deaths Act, 1969, births are registered on *de-facto* rather than *de-jure* basis and, therefore, it is not possible to estimate fertility from the births registered under the civil registration system. Estimation of fertility also requires estimates of population by age and sex which are not available through the civil registration system. The only source of information about the population of the district by age and sex is the decennial population census. The last decennial population census in India was conducted way back in 2011 so

A. R. Chaurasia, *Population and Human Development in Districts of India*,
https://doi.org/10.1007/978-981-95-8169-6_4

that recent estimates of population of the district by age and sex are not available. Direct estimation of fertility in the districts of the country is also not reliable from the data available from the National Family Health Survey because the sample of households surveyed are too small to provide statistically valid estimates of fertility in the district.

In the absence of direct estimates, attempts have been made to estimate district fertility through the application of indirect methods. Different indirect methods of fertility estimation have been proposed. The most common of these methods is the P/F ratio method first proposed by Brass and its many refinements (Brass 1968, 1975; Hobcraft et al. 1982; Maultrie et al. 2013). Cho et al. (1986), on the other hand, have proposed the own children method based on the reverse survival technique while Rele (1967) has used the stable population method which has been modified by Swamy et al. (1992). Regression-based methods have also been proposed (Mauldin and Ross 1991; Jain 1997; Singh et al. 2012). The mean duration of the interval between successive live births has been suggested to estimate fertility (Srinivasan 1980; Yadava and Kumar 2002). Yadava et al. (2009) have proposed a method based on the weighted average of the proportionate distribution of live births by birth order whereas Yadava et al. (2009) have used the proportion of women having a live birth during the five years preceding the survey. Tiwari et al. (2020) have used the proportion of childless women of reproductive age to explain the variation in TFR.

United Nations (1967) has suggested a simple approach to estimate total marital fertility rate (TMFR) from the average parity of currently married women. This method is based on the hypothesis that, in populations that employ little birth control, the ratio of the average parity of currently married women at the end of the child-bearing period to the average parity of currently married women of a younger age group is closely related to the relative average parity of currently married women early and late in their twenties. If thc average number of children ever born (average parity) to women aged 15-19 years is P_1; average parity of currently married women aged 20–24 years is P_2, and so on so that the average parity of currently married women aged 45–50 years is P_7, then this hypothesis means that

$$\frac{\text{TMFR}}{P_3} \approx \frac{P_3}{P_2} \tag{4.1}$$

or

$$\text{TMFR} \approx \frac{P_3^2}{P_2} \tag{4.2}$$

If the Eq. (4.1) holds empirically, then TMFR can be approximated as

$$\text{TMFR} = \alpha + \beta * \left(P_3^2/P_2\right) \tag{4.3}$$

where α and β are constants to be determined.

Yadava and Tiwari (2007) have modified the approach suggested by the United Nations (1967) by considering the prevalence of contraception as a predictor of TFR whereas Gupta et al. (2014), have argued that with the increase in the age at marriage, there is a shift in fertility towards higher ages. They have, therefore, suggested that

$$\frac{\text{TFR}}{Q_4} \approx \frac{Q_4}{Q_3} \tag{4.4}$$

where, Q denotes the average parity of all women in a given age group not the average parity of currently married women. The TFR may now be calculated as

$$\text{TFR} = \gamma + \delta * \left(Q_4^2/Q_3\right) \tag{4.5}$$

where γ andδ are constants to be determined. Singh et al. (2022), on the other hand, have suggested that

$$\frac{\text{TMFR}}{P_5} \approx \frac{P_5}{P_4} \tag{4.6}$$

which means that TMFR may be estimated as

$$\text{TMFR} = \mu + \rho * \left(P_5^2/P_4\right) \tag{4.7}$$

Singh et al. (2022) have also tested the stability or the robustness of the regression model (4.7) by estimating the shrinkage or the decrease in the coefficient of determination which is attributed to the application of the regression model to a new data set. It is well known in the regression analysis that a fitted relationship performs less well on a new data set than the data set used for fitting the model (Everitt 2002). The robustness of the regression model implies that the regression model can be applied to dataset other than the one that is used to establish the relationship between the dependent and the independent variables.

Using the indirect methods of fertility estimation, there have been attempts in the past to estimate fertility in the districts of the country. The Registrar General and Census Commissioner of India has produced estimates of different indicators of fertility for the districts of the country based on the children ever born data collected during the 1981 and 1991 decennial population censuses through the application of Brass PF Ratio method (Government of India 1988, 1997). Similar exercise has, however, not been carried out by the Registrar General and Census Commissioner of India based on the data collected at the 2001 and 2011 decennial population censuses. District level estimates of fertility using data from decennial population censuses have also been prepared by Mishra et al. (1994), Guilmoto and Rajan (2002, 2013) and Kumar and Sathyanarayana (2012) using different indirect methods of fertility estimation. There has, however, been no decennial population census in India after

2011 so that census-based estimates of fertility for the districts of the country are not available after 2011.

The Government of India had also instituted the Annual Health Survey Programme in 2010 to generate annual estimates of key demographic indicators for the districts (Government of India 2011). This survey, however, did not cover all districts of the country and was discontinued after 2013. The fourth round of the National Family Health Survey (2015–2016) provided district level data which have been used by many authors to estimate fertility for 640 districts of the country as they existed at the time of the 2011 decennial population census (Singh et al. 2022; Chatterjee and Mohanty 2021; Jayachandran and Ram 2019). There, however, appears to be little attempt to estimate fertility in the districts of the country from the data available from the fifth round of the National Family Health Survey (2015–2016).

This chapter presents estimates of total fertility rate (TFR) for the 707 districts of the country based on the data available from the fifth round of the National Family Health Survey, 2019–2021 following the approach adopted by Singh et al. (2022). The estimate of TFR for the country based on the method adopted in this chapter is found to be a close approximation of the TFR estimated on the basis of full birth history data collected during the survey. This chapter also attempts to classify districts into different fertility profiles which are characterised in terms of the level of the fertility of currently married women of reproductive age and the proportion of women of reproductive age who were married at the time of the survey relative to the fertility of currently married women and the proportion of women who were married in the country as a whole. The analysis reveals that 707 districts can be classified into six fertility profiles depending upon the direction of the difference in marital fertility and proportion of married women between the district and the country.

Methods

We have used estimates of annual age-specific marital fertility rates available from the official sample registration system of the country to establish the empirical relationship between TMFR and P_3^2/P_2, P_4^2/P_3 and P_5^2/P_4. Annual estimates of age-specific marital fertility rates are available from the sample registration system for the period 1985–2020 for the country and for 20 states of the country which served the basis for establishing the empirical relationship between TMFR and P_3^2/P_2, P_4^2/P_3 and P_5^2/P_4. The sample registration system is the only source in India that provides estimates of age-specific marital fertility rates for the country and for its selected states annually. It is a dual record system that was launched by the Government of India in 1969 on a pilot basis and that now covers the entire country (Government of India 2025).

Results of the regression of TMFR on P_3^2/P_2, P_4^2/P_3 and P_5^2/P_4 are presented in Table 4.1 and depicted in Figs. 4.1, 4.2, and 4.3. The table shows that the relationship between TMFR and P_5^2/P_4 is nearly linear as reflected from the standardised regression coefficient and the coefficient of determination. This means that an estimate of

Table 4.1 Results of the regression of TMFR on P_3^2/P_2, P_4^2/P_3 and P_5^2/P_4 in India, 1985–2020

Particulars	P_3^2/P_2	P_4^2/P_3	P_5^2/P_4
Constant	0.758033	−0.093214	0.155519
B	1.049218	0.976749	0.945270
SE	0.011753	0.003494	0.001643
β	0.959274	0.995608	0.998958
't'	89.268770	279.545367	575.3391
'p' > |t|	0.000000	0.000000	0.000000
R^2	0.920092	0.9912222	0.997914
N	692	692	692

Source Author

TMFR can be obtained with a high degree of accuracy if P_5^2/P_4 is known. In other words, TMFR can be estimated as

$$\text{TMFR} = 0.9409 * P_5^2/P_4 + 0.1738 \tag{4.8}$$

Application of the Eq. (4.8) to the data available from the National Family Health Survey, 2019–2021 suggests a TMFR of around 3 births per married woman of reproductive age for the country for the period 2019–2021. The data available from the survey also suggests that around 71 per cent women of reproductive age in India were currently married at the time of the survey. This implies a TFR of around 2.2

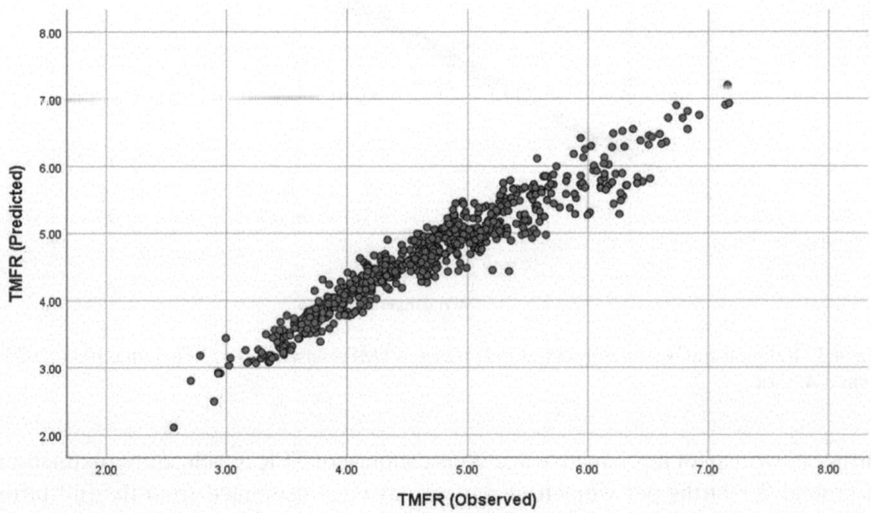

Fig. 4.1 Relationship between observed and predicted TMFR based on P_3^2/P_2 in India, 1985–2020. *Source* Author

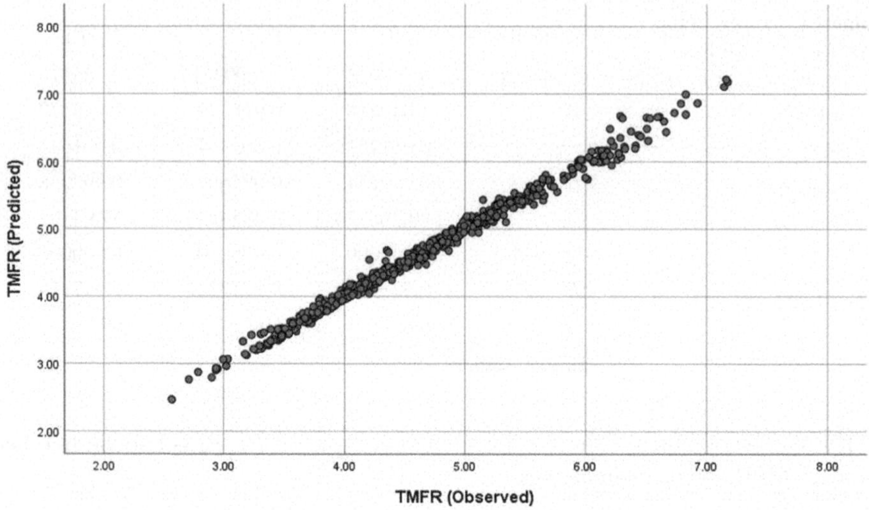

Fig. 4.2 Relationship between observed and predicted TMFR based on P_4^2/P_3 in India, 1985–2020. *Source* Author

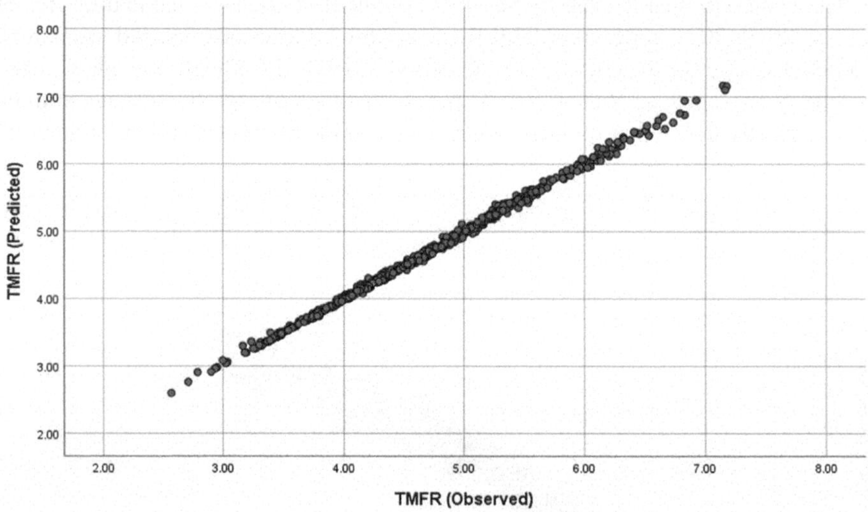

Fig. 4.3 Relationship between observed and predicted TMFR based on P_5^2/P_4 in India, 1985–2020. *Source* Author

births per woman of reproductive age. This estimate of TFR is a close approximation of around 2.1 births per women of reproductive age estimated from the full birth history data collected during the survey. This proximity of the two estimates of TFR provides credence to estimating district TFR using the Eq. (4.8).

If f denotes TFR, g denotes TMFR and m denotes proportion of currently married reproductive age women, then

$$f = g \times m \tag{4.9}$$

Let f_d denotes TFR of district d while f_c denotes TFR of the country. Then the difference between district TFR and country TFR can be decomposed as

$$\nabla f_d = f_d - f_c = (g_d \times m_d) - (g_c \times m_c) \tag{4.10}$$

We can write

$$\nabla f_d = \frac{f_d - f_c}{\ln\left(\frac{f_d}{f_c}\right)} \times \ln\left(\frac{f_d}{f_c}\right) = L_{dc} \times \ln\left(\frac{f_d}{f_c}\right) \tag{4.11}$$

where $L_{dc} = \frac{f_d - f_c}{\ln\left(\frac{f_d}{f_c}\right)}$ is the logarithmic mean of f_d and f_c. Now

$$\ln\left(\frac{f_d}{f_c}\right) = \ln\left(\frac{g_d}{g_c}\right) + \ln\left(\frac{m_d}{m_c}\right) \tag{4.12}$$

so that

$$\nabla f_d = \left(L_{dc} \times \ln\left(\frac{g_d}{g_c}\right)\right) + \left(L_{dc} \times \ln\left(\frac{m_d}{m_c}\right)\right) = \partial g_d + \partial m_d \tag{4.13}$$

where

$$\partial g_d = \left(L_{dc} \times \ln\left(\frac{g_d}{g_c}\right)\right) \tag{4.14}$$

and

$$\partial m_d = \left(L_{dc} \times \ln\left(\frac{m_d}{m_c}\right)\right) \tag{4.15}$$

Equation (4.13) shows that the difference between TFR of a district from the average TFR of the country can be decomposed into two components. One component of this difference is attributed to the difference in TMFR while the other component is attributed to the difference between the proportion of the women of reproductive age who are married. This decomposition provides a useful framework for constructing the profile of fertility in the district and in analysing how the profile of fertility varies across districts of the country. The characterisation of the profile of fertility of a district in terms of marital fertility and proportion of women of reproductive age who are married is important as it has implications for planning and programming for fertility regulation interventions in the district.

Based on Eq. (4.13), a district can be classified into one of the following mutually exclusive yet exhaustive 11 fertility profiles depending upon the direction of the difference in ∂g_d, ∂m_d and ∇f_d as defined above:

$$\begin{aligned}
\text{Profile 1}: \partial g_d > 0, \partial m_d > 0, \nabla f_d > 0\\
\text{Profile 2}: \partial g_d > 0, \partial m_d < 0, \nabla f_d > 0\\
\text{Profile 3}: \partial g_d > 0, \partial m_d < 0, \nabla f_d < 0\\
\text{Profile 4}: \partial g_d < 0, \partial m_d < 0, \nabla f_d < 0\\
\text{Profile 5}: \partial g_d < 0, \partial m_d > 0, \nabla f_d < 0\\
\text{Profile 6}: \partial g_d < 0, \partial m_d > 0, \nabla f_d > 0\\
\text{Profile 7}: \partial g_d = 0, \partial m_d = 0, \nabla f_d = 0\\
\text{Profile 8}: \partial g_d = 0, \partial m_d > 0, \nabla f_d > 0\\
\text{Profile 9}: \partial g_d = 0, \partial m_d < 0, \nabla f_d < 0\\
\text{Profile 10}: \partial g_d > 0, \partial m_d = 0, \nabla f_d > 0\\
\text{Profile 11}: \partial g_d < 0, \partial m_d = 0, \nabla f_d < 0
\end{aligned} \tag{4.16}$$

The 11 fertility profiles described above are mutually exclusive and exhaustive. Based on the difference in TMFR, proportion of married women of reproductive age and TFR of a district from the national average, a district can be classified into one and only one of the 11 fertility profiles. Each profile of fertility identified in Eq. (4.16) has a unique characterisation of fertility which has policy and programme implications. For example, fertility profile 2 suggests that TFR higher than the national average in districts having this fertility profile is due to TMFR higher than the national average. On the other hand, fertility profile 6 suggests that TFR higher than the national average in districts having this fertility profile is because of the higher proportion of women of reproductive age who are currently married in the district compared to the national average. The policy and programme implications for pursuing fertility transition in the two categories of districts are obviously different.

Inter-District Variation in TFR

Estimates of TFR for 707 districts of the country as they existed at the time of the National Family Health Survey, 2019–2021 are given in the appendix table along with estimates of TMFR and proportion of reproductive age women who were currently married at the time of the survey. The inter-district variation in TFR is depicted in Fig. 4.4 while the distribution of districts by the level of fertility in different states and Union Territories of the country are presented in Table 4.2. There are 328 (46.4 per cent) districts in the country where fertility was estimated to be below the replacement level (TFR of less than 2.1 births per woman of reproductive age) according to the information available from the National Family Health Survey, 2019–2021. In the remaining 279 districts, fertility was estimated to be above the replacement level, and, in 63 districts of the country, fertility was very high at the time of the survey as the TFR was at least 3 births per woman of reproductive age in these districts. There are 17 districts in which fertility was exceptionally high as the TFR, in these districts, was at least 3.5 births per woman of reproductive age. In 199 (28.1 per cent) districts fertility was moderately higher than the replacement level as TFR ranged between 2.1 and 2.5 births per woman of reproductive age in these districts. In 117 (16.5 per cent) districts fertility was markedly higher than the replacement level as TFR ranged between 2.5 and 3.0 births per woman of reproductive age. District South Goa in Goa had the lowest fertility in the country as TFR in the district was estimated to be 1.20 births per woman of reproductive age. On the other hand, TFR was estimated to be 4.71 births per woman of reproductive age in district West Khasi Hills of Meghalaya, which was the highest among the 707 districts.

Regional pattern in fertility in the districts is also very marked as may be seen from the Fig. 4.4. Most of the districts having above replacement fertility are located in the central part of the country comprising of the states of Rajasthan, Uttar Pradesh, Madhya Pradesh, Bihar and Jharkhand whereas in the north western and southern part of the country, fertility is below the replacement level with only a few exceptions. In the north-eastern part of the country, on the other hand, the scenario is mixed.

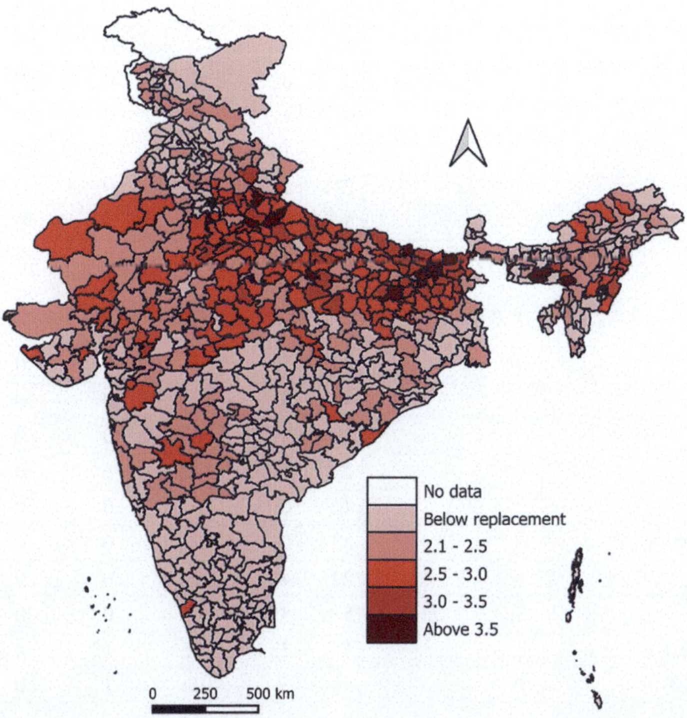

Fig. 4.4 Inter-district variation in total fertility rate (TFR) in India, 2019–2021. *Source* Author

Table 4.2 Variation in district TFR across states and Union Territories of India, 2019–2021

Country/State/Union Territory	Total fertility rate					Total
	<2.1	2.1–2.5	2.5–3.0	3.0–3.5	≥3.5	
Andaman and Nicobar Islands	3	0	0	0	0	3
Andhra Pradesh	13	0	0	0	0	13
Arunachal Pradesh	6	16	4	0	0	20
Assam	13	18	1	1	0	33
Bihar	0	1	8	21	8	38
Chandigarh	1	0	0	0	0	1
Chhattisgarh	15	11	1	0	0	27
Dadra and Nagar Haveli and Daman and Diu	3	0	0	0	0	3
Delhi	7	3	1	0	0	11
Goa	2	0	0	0	0	2
Gujarat	12	14	7	0	0	33
Haryana	11	9	1	0	1	22
Himachal Pradesh	9	3	0	0	0	12
Jammu and Kashmir	12	8	0	0	0	20
Jharkhand	3	6	14	1	0	24
Karnataka	23	7	0	0	0	30
Kerala	13	0	1	0	0	14
Ladakh	2	0	0	0	0	2
Lakshadweep	0	1	0	0	0	1
Madhya Pradesh	9	19	19	3	1	51
Maharashtra	21	12	3	0	0	36
Manipur	5	0	3	1	0	9
Meghalaya	5	1	0	2	3	11
Mizoram	6	2	0	0	0	8
Nagaland	3	3	3	2	0	11
Odisha	19	10	1	0	0	30
Puducherry	4	0	0	0	0	4
Punjab	20	2	0	0	0	22
Rajasthan	2	17	12	2	0	33
Sikkim	4	0	0	0	0	4
Tamil Nadu	32	0	0	0	0	32
Telangana	27	4	0	0	0	31
Tripura	5	3	0	0	0	8
Uttar Pradesh	1	23	35	12	4	75
Uttarakhand	5	5	2	1	0	13

(continued)

Table 4.2 (continued)

Country/State/Union Territory	Total fertility rate					Total
	<2.1	2.1–2.5	2.5–3.0	3.0–3.5	≥3.5	
West Bengal	12	7	1	0	0	20
India	328	199	117	46	17	707

Source *Author*

Among the 63 districts where fertility was very high at the time of the survey (TFR at least 3 births per woman of reproductive age), 29 are in Bihar alone while 16 are in Uttar Pradesh. Similarly, among the 17 districts where fertility is estimated to be exceptionally high (TFR equal to or more than 3.5 births per woman of reproductive age), 8 are in Bihar, 4 in Uttar Pradesh, 3 each in Meghalaya and 1 each in Haryana and Madhya Pradesh. There is no district in other states and Union Territories of the country where fertility was estimated to be equal to or more than 3.5 birth per woman of reproductive age at the time of the National Family Health Survey. There is also no district in 8 states/Union Territories where fertility was estimated to be above the replacement level at the time of the survey. In Tamil Nadu, fertility was estimated to be below the replacement level in 31 of the 32 districts. Similarly, in Kerala, fertility was estimated to be below the replacement level in 13 of the 14 districts. In Punjab, fertility was estimated to be below the replacement level in 20 of the 22 districts whereas in Himachal Pradesh fertility was estimated to be below the replacement level in 9 of the 12 districts. On the contrary, there is no district in Bihar where fertility was estimated to be below the replacement level. Bihar is the only state in the country where fertility was above the replacement district in all districts. In Uttar Pradesh, fertility was below the replacement level in only one of the 75 districts whereas, in Rajasthan, fertility was estimated to be below the replacement level in 2 of the 33 districts. Similarly, fertility was below the replacement level in three of the 24 districts of Jharkhand and in 3 of the 11 districts of Nagaland. Out of the 166 districts in the southern of the country, fertility was above the replacement level in only 28 districts out of which 15 are in Maharashtra alone. In contrast, in 311 districts of central India, fertility was below the replacement level in only 61 districts.

Profiles of Fertility

The TFR in any population is determined by two factors: (1) fertility of married women in the reproductive age (TMFR) and (2) proportion of reproductive age women who are married. The difference between the TFR of a district from the national average, therefore, is determined by the difference in TMFR and the difference in the proportion of reproductive age women who are married between the district and the national average. The relative contribution of the difference in TMFR and the difference in the proportion reproductive age women who are married of a district from the national average can be obtained through Eqs. (4.14) and (4.15) respectively. Based on the magnitude and the direction of these contributions, a district may be classified into one of the possible 11 mutually exclusive and exhaustive categories or profiles as defined by Eq. (4.16). This exercise suggests that 707 districts of the country can be classified into 6 of the 11 fertility profiles. There is no district which is classified in the remaining 5 fertility profiles.

The distribution of districts according to the fertility profile of the district and the level of fertility in the district is presented in Table 4.3. There are 328 districts where fertility was below the replacement level. The fertility profile of these districts has, however, not been the same. In 123 of these districts, both TMFR and the proportion of currently married women of reproductive age in the district was less than the national average so that TFR was less than the national average (Profile 4). On the other hand, in 181 of these districts, TMFR was below the national average, but the proportion of currently married reproductive age women was higher than the national average, but TFR was below the national average (Profile 5). In 24 districts, in contrast, TMFR was higher than the national average, but the proportion of currently married reproductive age women was lower than the national average and TFR was lower than the national average (Profile 3).

In 150 districts of the country, both TMFR and the proportion of currently married reproductive age women was higher than the national level so that TFR was higher than the national average (profile 1). On the other hand, in 137 districts, TMFR was above the national average, but the proportion of currently married reproductive age women was less than the national average but TFR was higher than the national average (Profile 2). In none of these 287 districts, fertility was below the replacement level and all districts with high or exceptionally high TFR are among these districts. Fertility was also not below the replacement level in any of the 29 districts in which TMFR was below the national average, but the proportion of currently married reproductive age women was above the national average so that TFR was above the national average (Profile 6). In all these districts, TFR was above the replacement level but less than 2.5 births per woman of reproductive age.

Table 4.3 Districts cross-classified by the level of TFR and the fertility profile, 2019–2021

Fertility profile	Total fertility rate					Total
	<2.1	2.1–2.5	2.5–3.0	3.0.3.5	≥3.5	
Profile 1: $\partial g_d > 0, \partial m_d > 0, \nabla f_d > 0$	0	46	67	27	10	150
Profile 2: $\partial g_d > 0, \partial m_d < 0, \nabla f_d > 0$	0	61	50	19	7	137
Profile 3: $\partial g_d > 0, \partial m_d < 0, \nabla f_d < 0$	24	20	0	0	0	44
Profile 4: $\partial g_d < 0, \partial m_d < 0, \nabla f_d < 0$	123	6	0	0	0	129
Profile 5: $\partial g_d < 0, \partial m_d > 0, \nabla f_d < 0$	181	37	0	0	0	218
Profile 6: $\partial g_d < 0, \partial m_d > 0, \nabla f_d > 0$	0	29	0	0	0	29
Total	328	199	117	46	17	707

Source Author

The distribution of districts by their fertility profile is different in different regions of the country as may be seen from Table 4.4 and Fig. 4.5. Out of 311 districts in central India, 213 have either fertility profile 1 or fertility profile 2. In contrast, there are only 7 districts out of 166 districts in the southern region of the country which have either fertility profile 1 or fertility profile 2. There are only 33 districts in this region which have fertility profile 4 which means that both TMFR and proportion of currently married reproductive age women in these districts were below the national average. The fertility profile in 117 districts in this region of the country is profile 5 which means that TMFR in these districts is below the national average, but the proportion of currently married reproductive age women is higher than the national average. The TFR in these districts is, however, below the national average. On the other hand, out of 311 districts in the central region of the country, only 42 districts have both TMFR and the proportion of currently married reproductive age women less than the national average (Profile 4). Out of the 38 districts in Bihar, 30 districts have fertility profile 1 while the remaining 8 districts have fertility profile 2. Similarly, out of 24 districts in Jharkhand, 13 districts have fertility profile 1 while 8 districts have fertility profile 2.

Table 4.4 Distribution of districts by states/Union Territories and fertility profiles

India/State/Union Territory	Fertility profile						Total
	1 $\partial g_d > 0$ $\partial m_d > 0$ $\nabla f_d > 0$	2 $\partial g_d > 0$ $\partial m_d < 0$ $\nabla f_d > 0$	3 $\partial g_d > 0$ $\partial m_d < 0$ $\nabla f_d < 0$	4 $\partial g_d < 0$ $\partial m_d < 0$ $\nabla f_d < 0$	5 $\partial g_d < 0$ $\partial m_d > 0$ $\nabla f_d < 0$	6 $\partial g_d < 0$ $\partial m_d > 0$ $\nabla f_d > 0$	
Andaman and Nicobar Islands	0	0	0	2	1	0	3
Andhra Pradesh	0	0	0	0	13	0	13
Arunachal Pradesh	4	7	2	5	1	1	20
Assam	11	3	1	4	13	1	33
Bihar	33	5	0	0	0	0	38
Chandigarh	0	0	0	1	0	0	1
Chhattisgarh	0	8	4	15	0	0	27
Dadra and Nagar Haveli and Daman and Diu	0	0	1	0	2	0	3
Delhi	1	1	4	4	1	0	11
Goa	0	0	0	2	0	0	2
Gujarat	9	2	0	6	12	4	33
Haryana	4	1	2	5	8	2	22

(continued)

Table 4.4 (continued)

India/State/Union Territory	Fertility profile						Total
	1 $\partial g_d > 0$ $\partial m_d > 0$ $\nabla f_d > 0$	2 $\partial g_d > 0$ $\partial m_d < 0$ $\nabla f_d > 0$	3 $\partial g_d > 0$ $\partial m_d < 0$ $\nabla f_d < 0$	4 $\partial g_d < 0$ $\partial m_d < 0$ $\nabla f_d < 0$	5 $\partial g_d < 0$ $\partial m_d > 0$ $\nabla f_d < 0$	6 $\partial g_d < 0$ $\partial m_d > 0$ $\nabla f_d > 0$	
Himachal Pradesh	1	0	0	1	10	0	12
Jammu and Kashmir	0	4	10	6	0	0	20
Jharkhand	13	7	1	1	1	1	24
Karnataka	2	0	0	7	21	0	30
Kerala	1	0	0	1	12	0	14
Ladakh	0	0	2	0	0	0	2
Lakshadweep	0	0	1	0	0	0	1
Madhya Pradesh	26	8	0	8	4	5	51
Maharashtra	3	0	1	6	20	6	36
Manipur	0	4	3	2	0	0	9
Meghalaya	0	6	3	2	0	0	11
Mizoram	0	1	2	5	0	0	8
Nagaland	1	7	3	0	0	0	11
Odisha	3	2	0	11	13	1	30
Puducherry	0	0	0	4	0	0	4
Punjab	0	1	0	10	11	0	22
Rajasthan	19	6	1	1	3	3	33
Sikkim	0	0	0	4	0	0	4
Tamil Nadu	0	0	0	9	23	0	32
Telangana	1	0	0	2	27	1	31
Tripura	1	0	0	0	5	2	8
Uttar Pradesh	13	59	3	0	0	0	75
Uttarakhand	1	5	0	3	4	0	13
West Bengal	3	0	0	2	13	2	20
India	150	137	44	129	218	29	707

Source Author

There are 29 districts in the country where TFR was higher than the national average not because TMFR was higher than the TMFR of the country but because the proportion of currently married reproductive age women was higher than the corresponding proportion at the national level (Profile 6). Among these 29 districts, 18 districts are in Maharashtra (6), Madhya Pradesh (5), Gujarat (4) and Rajasthan (3). The remaining 11 districts are in Arunachal Pradesh (1), Assam (1), Haryana (2), Jharkhand (2), Odisha (1) Telangana (1), Tripura (2) and West Bengal (2).

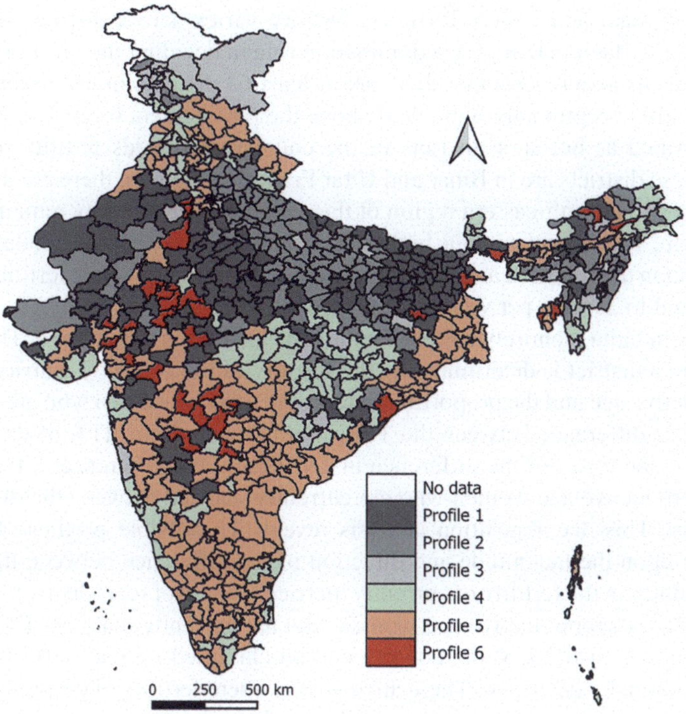

Fig. 4.5 Profiles of fertility in districts of India, 2019–2021. *Source* Author

Conclusions

This chapter highlights the variation in fertility, as measured by TFR, across the districts of the country. The data available from the latest round of the National Family Health Survey 2019–2021 suggest that fertility in around 46 per cent districts is already below the replacement level. However, in more than half of the districts of the country, fertility still remains above the replacement level. Nearly all but a few districts where fertility appears to be above the replacement level are located in the central region of the country comprising of the states of Bihar, Chhattisgarh, Gujarat, Jharkhand, Madhya Pradesh, Odisha, Rajasthan and Uttar Pradesh. Fertility also appears to be above the replacement level in many districts in the north-eastern region of the country. In the southern region of the country, comprising of the states of Andhra Pradesh, Goa, Karnataka, Kerala, Tamil Nadu and Telangana, and Union Territories of Andaman and Nicobar Islands, Lakshadweep and Puducherry, fertility appears to be below the replacement level in all but a few districts. In the northern region of the country comprising of the states of Delhi, Haryana, Himachal Pradesh, Punjab, Uttarakhand, Union Territories of Chandigarh, Jammu and Kashmir and Ladakh, fertility remains below the replacement level in majority of the districts.

Within each state and Union Territory, fertility varies across districts indicating district-specific factors that play a dominating role in deciding the level of fertility.

The analysis also reveals that there are at least 63 districts in the country where fertility is still exceptionally high, well above the replacement level. These districts may be termed as hot spot districts of the country as regards fertility transition. Most of these districts are in Bihar and Uttar Pradesh, although, there are districts in Meghalaya in the north-eastern region of the country where fertility remains exceptionally high. The district with the highest TFR in the country is in Meghalaya. There is no district in the southern and northern region of the country where fertility is more than or equal to 3 births per woman of reproductive age.

Fertility in India is entirely confined within the institution of marriage. This means that TFR in a district is determined by the fertility of the currently married women of reproductive age and the proportion of reproductive age women who are currently married. The difference between the TFR of a district and the TFR of the country, therefore, is the result of the difference in TMFR and the difference in the proportion of reproductive age women who are currently married between the district and the country. This decomposition analysis reveals 11 possible profiles of fertility depending upon the magnitude and direction of the difference between the district and the country in the fertility of currently married women of reproductive age and in the proportion of reproductive age women who are currently married. The analysis reveals that 707 districts of the country can be classified into six fertility profiles which are mutually exclusive. There are districts where fertility of currently married women of reproductive age is lower than the fertility of currently married women of reproductive age in the country but the total fertility rate in these districts is higher than the total fertility rate in the country because the proportion of reproductive age women who are married in the district is higher than the proportion of currently married reproductive age women in the country. Similarly, there are districts where fertility of married women of reproductive age is higher than the national average but the TFR in these districts is lower than TFR of the country because the proportion of reproductive age women who are married is lower than the proportion of reproductive age women who are currently married at the national level.

The profiling of fertility of district in terms of TMFR and the proportion of currently married reproductive age women relative to that in the country has implications for fertility transition, especially, in those districts where TFR remains above the replacement level. It is important to analyse whether TFR in the district in access to the replacement level fertility is due to high fertility of currently married women of reproductive age or is due to a higher proportion of reproductive age women who are currently married. This distinction is important as factors that influence fertility of married women are different from the factors that influence the proportion of reproductive age women who are married. For example, a reduction in maternal mortality may lead to an increase in the proportion of reproductive age women who are currently married but may not have any impact on the fertility of married women. Similarly, an increase in the prevalence of breastfeeding may contribute to a decrease in the fertility of currently married women of reproductive age but may not have any impact on the proportion of women who are currently married. In districts where fertility is above the replacement level because of a high proportion of the reproductive age women who are currently married, a different approach of fertility reduction is needed as compared to those districts where fertility is high because of high fertility of currently married reproductive age women.

Appendix

See Table 4.5.

Table 4.5 Total fertility rate (TFR), total marital fertility rate (TMFR), proportion of reproductive age women currently married and the profile of fertility in districts of India, 2019–2021

State/Union Territory	District	TMFR	Proportion married	TFR	Fertility profile
Andaman and Nicobar Islands					
	Nicobar	2.136	64.205	1.371	4
	North and Middle Andaman	2.400	74.074	1.778	5
	South Andaman	1.872	70.522	1.320	4
Andhra Pradesh					
	Anantapur	2.215	79.780	1.767	5
	Chittoor	2.128	76.331	1.624	5
	East Godavari	2.506	75.533	1.893	5
	Guntur	2.114	77.804	1.645	5
	Krishna	2.265	72.592	1.644	5
	Kurnool	2.533	76.205	1.930	5
	Prakasam	2.141	79.255	1.697	5
	Sri Potti Sriramulu Nellore	2.076	73.554	1.527	5
	Srikakulam	2.166	71.957	1.558	5
	Visakhapatnam	2.091	73.311	1.533	5
	Vizianagaram	2.458	75.110	1.846	5
	West Godavari	2.322	76.923	1.786	5
	Y.S.R	2.560	77.308	1.979	5
Arunachal Pradesh					
	Anjaw	2.574	70.525	1.815	4
	Changlang	2.983	69.282	2.067	4
	Dibang Valley	3.022	66.514	2.010	4
	East Kameng	3.531	71.528	2.525	1

(continued)

Table 4.5 (continued)

State/Union Territory	District	TMFR	Proportion married	TFR	Fertility profile
	East Siang	2.812	64.105	1.803	4
	Kra Daadi	2.945	77.757	2.290	6
	Kurung Kumey	3.550	66.052	2.345	2
	Lohit	3.415	64.736	2.210	2
	Longding	3.512	65.781	2.310	2
	Lower Dibang Valley	3.245	65.055	2.111	3
	Lower Subansiri	3.145	65.634	2.064	3
	Namsai	3.850	72.968	2.809	1
	Papum Pare	3.573	63.855	2.282	2
	Siang	3.929	63.750	2.505	2
	Tawang	3.503	66.943	2.345	2
	Tirap	3.488	70.958	2.475	1
	Upper Siang	3.150	70.865	2.232	2
	Upper Subansiri	3.640	75.045	2.732	1
	West Kameng	2.558	66.401	1.698	4
	West Siang	3.025	70.947	2.146	5
Assam					
	Baksa	2.448	75.334	1.844	5
	Barpeta	3.289	75.146	2.472	1
	Biswanath	2.947	74.575	2.197	5
	Bongaigaon	2.697	76.174	2.055	5
	Cachar	3.378	71.578	2.418	1
	Charaideo	2.880	67.711	1.950	4
	Chirang	2.695	75.273	2.028	5
	Darrang	3.186	77.022	2.454	1
	Dhemaji	3.044	77.809	2.368	6
	Dhubri	3.424	80.611	2.760	1
	Dibrugarh	3.010	69.476	2.091	4
	Dima Hasao	3.240	66.017	2.139	3
	Goalpara	3.007	73.015	2.195	5
	Golaghat	2.905	75.326	2.189	5
	Hailakandi	3.416	70.990	2.425	1
	Hojai	3.275	72.560	2.376	1
	Jorhat	2.573	72.940	1.877	5
	Kamrup	2.664	73.675	1.963	5
	Kamrup Metropolitan	2.039	68.884	1.405	4
	Karbi Anglong	3.438	68.869	2.367	2

(continued)

Table 4.5 (continued)

State/Union Territory	District	TMFR	Proportion married	TFR	Fertility profile
	Karimganj	3.518	70.270	2.472	2
	Kokrajhar	2.565	75.945	1.948	5
	Lakhimpur	2.487	75.730	1.884	5
	Majuli	3.318	73.604	2.443	1
	Morigaon	3.163	76.744	2.427	1
	Nagaon	3.352	74.077	2.483	1
	Nalbari	2.363	74.774	1.767	5
	Sivasagar	2.564	73.925	1.895	5
	Sonitpur	3.450	71.650	2.472	1
	South Salmara Mancachar	3.979	77.076	3.066	1
	Tinsukia	2.472	69.149	1.710	4
	Udalguri	3.007	71.696	2.156	5
	West Karbi Anglong	3.131	70.710	2.214	2
Bihar					
	Araria	4.691	79.011	3.706	1
	Arwal	3.985	73.105	2.913	1
	Aurangabad	4.624	71.791	3.320	1
	Banka	4.107	79.912	3.282	1
	Begusarai	4.925	76.106	3.748	1
	Bhagalpur	4.243	75.385	3.199	1
	Bhojpur	3.940	74.068	2.918	1
	Buxar	4.716	70.337	3.317	2
	Darbhanga	4.287	73.578	3.155	1
	Gaya	4.840	72.971	3.532	1
	Gopalganj	4.356	69.447	3.025	2
	Jamui	3.966	80.735	3.202	1
	Jehanabad	4.029	75.044	3.023	1
	Kaimur (Bhabua)	3.776	71.500	2.700	1
	Katihar	4.312	77.101	3.325	1
	Khagaria	4.788	79.774	3.819	1
	Kishanganj	5.157	67.797	3.497	2
	Lakhisarai	4.931	76.531	3.774	1
	Madhepura	4.326	81.651	3.532	1
	Madhubani	4.462	75.399	3.364	1
	Munger	3.735	75.044	2.803	1
	Muzaffarpur	4.205	75.042	3.155	1

(continued)

Table 4.5 (continued)

State/Union Territory	District	TMFR	Proportion married	TFR	Fertility profile
	Nalanda	4.402	73.942	3.255	1
	Nawada	4.673	72.490	3.388	1
	Pashchim Champaran	4.307	75.620	3.257	1
	Patna	3.377	73.454	2.481	1
	Purba Champaran	4.138	76.632	3.171	1
	Purnia	4.420	79.015	3.492	1
	Rohtas	4.256	69.474	2.957	2
	Saharsa	4.261	83.080	3.540	1
	Samastipur	4.207	79.530	3.346	1
	Saran	4.065	71.910	2.923	1
	Sheikhpura	4.645	74.732	3.471	1
	Sheohar	4.397	73.998	3.254	1
	Sitamarhi	4.990	74.497	3.718	1
	Siwan	4.164	65.108	2.711	2
	Supaul	3.669	80.602	2.958	1
	Vaishali	4.376	77.025	3.371	1
Chandigarh					
	Chandigarh	2.579	64.744	1.670	4
Chhattisgarh					
	Balod	2.348	66.216	1.555	4
	Baloda Bazar	3.365	66.807	2.248	2
	Balrampur	3.518	70.037	2.464	2
	Bastar	3.344	69.196	2.314	2
	Bemetara	3.415	68.790	2.349	2
	Bijapur	2.836	65.439	1.856	4
	Bilaspur	3.025	65.181	1.971	4
	Dantewada	3.014	65.753	1.982	4
	Dhamtari	2.631	66.442	1.748	4
	Durg	2.686	67.196	1.805	4
	Gariyaband	2.669	69.338	1.850	4
	Janjgir–Champa	2.791	66.937	1.868	4
	Jashpur	2.915	70.866	2.066	4
	Kabeerdham	3.209	67.989	2.182	3
	Kodagaon	3.625	61.172	2.218	2
	Korba	3.414	63.789	2.178	3
	Koriya	3.020	68.969	2.083	4
	Mahasamund	2.594	67.385	1.748	4

(continued)

Table 4.5 (continued)

State/Union Territory	District	TMFR	Proportion married	TFR	Fertility profile
	Mungeli	4.302	65.825	2.832	2
	Narayanpur	3.530	60.804	2.146	3
	Raigarh	2.941	64.251	1.890	4
	Raipur	3.040	66.860	2.033	4
	Rajnandgaon	2.711	66.228	1.795	4
	Sukma	3.332	64.839	2.160	3
	Surajpur	3.141	70.626	2.218	2
	Surguja	3.314	68.047	2.255	2
	Uttar Bastar Kanker	2.980	62.585	1.865	4
Dadra and Nagar Haveli and Daman and Diu					
	Dadra and Nagar Haveli	2.680	72.031	1.930	5
	Daman	2.755	73.724	2.031	5
	Diu	3.599	55.300	1.990	3
Delhi					
	Central	2.460	64.167	1.579	4
	East	2.727	66.966	1.826	4
	New Delhi	3.187	71.785	2.288	1
	North	3.712	69.661	2.586	2
	North East	3.190	66.007	2.105	3
	North West	2.892	69.077	1.998	4
	Shahdara	3.153	62.766	1.979	3
	South	3.167	67.081	2.124	3
	South East	2.862	63.278	1.811	4
	South West	2.176	71.353	1.553	5
	West	3.154	66.517	2.098	3
Goa					
	North Goa	1.860	66.239	1.232	4
	South Goa	1.996	60.094	1.200	4
Gujarat					
	Ahmadabad	2.716	72.824	1.978	5
	Amreli	3.083	69.495	2.143	4
	Anand	2.809	79.206	2.225	6
	Aravali	3.516	74.248	2.611	1
	Banas Kantha	3.687	74.712	2.755	1
	Bharuch	3.052	71.992	2.197	5
	Bhavnagar	2.975	69.192	2.058	4

(continued)

Table 4.5 (continued)

State/Union Territory	District	TMFR	Proportion married	TFR	Fertility profile
	Botad	3.600	70.761	2.548	2
	Chhota Udaipur	3.150	74.651	2.351	1
	Devbhumi Dwarka	3.556	71.324	2.536	1
	Dohad	3.992	73.409	2.930	1
	Gandhinagar	2.876	78.248	2.250	6
	Gir Somnath	2.990	67.221	2.010	4
	Jamnagar	2.653	67.901	1.801	4
	Junagadh	2.721	73.394	1.997	5
	Kachchh	3.387	68.904	2.334	2
	Kheda	2.846	76.964	2.190	5
	Mahesana	3.640	75.186	2.737	1
	Mahisagar	3.000	79.301	2.379	6
	Morbi	2.531	71.354	1.806	5
	Narmada	2.929	75.813	2.221	6
	Navsari	2.418	69.238	1.674	4
	Panch Mahals	2.905	74.394	2.161	5
	Patan	3.267	72.420	2.366	1
	Porbandar	2.735	70.242	1.921	4
	Rajkot	2.969	72.647	2.157	5
	Sabar Kantha	2.910	74.419	2.165	5
	Surat	2.820	72.169	2.035	5
	Surendranagar	3.230	73.077	2.360	1
	Tapi	2.514	72.163	1.814	5
	The Dangs	3.308	77.155	2.552	1
	Vadodara	2.494	75.026	1.871	5
	Valsad	2.107	73.229	1.543	5
Haryana					
	Ambala	2.550	68.622	1.750	4
	Bhiwani	3.246	71.635	2.325	1
	Charkhi Dadri	3.005	74.579	2.241	6
	Faridabad	2.975	69.686	2.073	4
	Fatehabad	2.580	70.151	1.810	4
	Gurgaon	2.679	74.708	2.002	5
	Hisar	2.803	72.203	2.024	5
	Jhajjar	2.681	71.923	1.928	5
	Jind	3.069	69.363	2.129	4

(continued)

Table 4.5 (continued)

State/Union Territory	District	TMFR	Proportion married	TFR	Fertility profile
	Kaithal	3.112	73.431	2.286	1
	Karnal	3.390	70.458	2.389	2
	Kurukshetra	2.632	74.210	1.953	5
	Mahendragarh	3.050	77.075	2.351	6
	Mewat	5.031	72.027	3.624	1
	Palwal	4.133	72.313	2.989	1
	Panchkula	3.175	68.862	2.187	3
	Panipat	2.941	74.304	2.185	5
	Rewari	2.686	75.402	2.025	5
	Rohtak	2.852	69.267	1.976	4
	Sirsa	3.117	69.894	2.179	3
	Sonipat	2.635	73.917	1.947	5
	Yamunanagar	2.679	71.878	1.926	5
Himachal Pradesh					
	Bilaspur	2.612	76.291	1.993	5
	Chamba	2.931	73.750	2.161	5
	Hamirpur	2.626	76.222	2.002	5
	Kangra	2.513	69.472	1.746	4
	Kinnaur	2.712	75.620	2.050	5
	Kullu	2.535	74.351	1.885	5
	Lahul & Spiti	2.641	79.708	2.105	5
	Mandi	2.501	78.603	1.966	5
	Shimla	2.495	71.032	1.772	5
	Sirmaur	3.458	71.956	2.488	1
	Solan	2.642	73.077	1.931	5
	Una	2.922	71.704	2.095	5
Jammu $ Kashmir					
	Anantnag	3.292	55.862	1.839	3
	Badgam	3.384	57.210	1.936	3
	Bandipore	3.535	53.187	1.880	3
	Baramula	3.483	56.003	1.951	3
	Doda	3.343	67.218	2.247	2
	Ganderbal	3.496	60.092	2.101	3
	Jammu	2.279	65.330	1.489	4
	Kathua	2.665	62.602	1.668	4
	Kishtwar	2.964	62.915	1.865	4

(continued)

Table 4.5 (continued)

State/Union Territory	District	TMFR	Proportion married	TFR	Fertility profile
	Kulgam	3.316	59.164	1.962	3
	Kupwara	3.921	57.870	2.269	2
	Pulwama	3.010	58.109	1.749	4
	Punch	3.474	61.482	2.136	3
	Rajouri	3.370	65.094	2.194	3
	Ramban	3.574	63.525	2.270	2
	Reasi	3.193	68.007	2.171	3
	Samba	2.795	67.607	1.890	4
	Shupiyan	3.496	59.522	2.081	3
	Srinagar	2.537	58.124	1.475	4
	Udhampur	3.357	66.303	2.226	2
Jharkhand					
	Bokaro	3.038	75.431	2.292	6
	Chatra	3.728	75.226	2.804	1
	Deoghar	3.507	82.184	2.882	1
	Dhanbad	3.164	72.593	2.297	1
	Dumka	3.349	79.568	2.665	1
	Garhwa	3.992	73.156	2.921	1
	Giridih	3.278	79.587	2.609	1
	Godda	3.442	79.123	2.723	1
	Gumla	3.830	67.828	2.598	2
	Hazaribagh	3.361	76.190	2.561	1
	Jamtara	3.114	80.538	2.508	1
	Khunti	3.301	67.081	2.215	2
	Kodarma	3.531	74.866	2.643	1
	Latehar	4.052	70.364	2.851	2
	Lohardaga	3.520	66.016	2.324	2
	Pakur	3.431	75.407	2.587	1
	Palamu	3.782	70.796	2.678	2
	Pashchimi Singhbhum	3.299	69.132	2.281	2
	Purbi Singhbhum	2.821	70.367	1.985	4
	Ramgarh	4.131	71.623	2.959	1
	Ranchi	3.123	64.784	2.023	3
	Sahibganj	4.068	77.902	3.169	1
	Saraikela-Kharsawan	2.719	72.095	1.960	5
	Simdega	3.551	64.384	2.287	2

(continued)

Table 4.5 (continued)

State/Union Territory	District	TMFR	Proportion married	TFR	Fertility profile
Karnataka					
	Bagalkot	3.038	72.185	2.193	5
	Bangalore	1.988	73.807	1.468	5
	Bangalore Rural	2.258	75.126	1.696	5
	Belgaum	3.014	72.638	2.189	5
	Bellary	2.639	68.527	1.808	4
	Bidar	3.101	69.687	2.161	4
	Bijapur	3.293	74.638	2.458	1
	Chamarajanagar	2.415	73.790	1.782	5
	Chikkaballapura	2.354	76.612	1.803	5
	Chikmagalur	2.167	73.744	1.598	5
	Chitradurga	2.198	76.694	1.686	5
	Dakshina Kannada	2.598	67.867	1.763	4
	Davanagere	2.338	72.579	1.697	5
	Dharwad	2.125	69.746	1.482	4
	Gadag	2.869	70.664	2.028	4
	Gulbarga	2.898	72.613	2.105	5
	Hassan	2.113	72.052	1.522	5
	Haveri	2.430	73.431	1.785	5
	Kodagu	2.242	73.434	1.647	5
	Kolar	2.395	75.000	1.796	5
	Koppal	2.595	73.255	1.901	5
	Mandya	2.222	72.500	1.611	5
	Mysore	2.489	72.682	1.809	5
	Raichur	3.242	72.340	2.345	1
	Ramanagara	2.284	71.604	1.635	5
	Shimoga	2.027	72.460	1.469	5
	Tumkur	2.236	77.014	1.722	5
	Udupi	2.178	68.750	1.497	4
	Uttara Kannada	2.344	68.358	1.603	4
	Yadgir	3.070	71.418	2.192	5
Kerala					
	Alappuzha	1.957	74.017	1.448	5
	Ernakulam	2.058	72.449	1.491	5
	Idukki	2.229	73.270	1.633	5
	Kannur	2.232	77.333	1.726	5
	Kasaragod	2.663	75.129	2.000	5
	Kollam	2.029	73.507	1.492	5

(continued)

Table 4.5 (continued)

State/Union Territory	District	TMFR	Proportion married	TFR	Fertility profile
	Kottayam	2.186	72.110	1.576	5
	Kozhikode	2.432	74.736	1.818	5
	Malappuram	3.249	77.800	2.528	1
	Palakkad	2.405	75.111	1.806	5
	Pathanamthitta	1.999	69.423	1.388	4
	Thiruvananthapuram	1.964	74.425	1.462	5
	Wayanad	2.720	73.174	1.990	5
Ladakh					
	Kargil	3.640	57.603	2.097	3
	Leh (Ladakh)	3.235	59.715	1.932	3
Lakshadweep					
	Lakshadweep	3.182	68.492	2.179	3
Madhya Pradesh					
	Agar Malwa	3.214	79.491	2.555	1
	Alirajpur	4.358	73.675	3.211	1
	Anuppur	3.252	70.734	2.301	2
	Ashoknagar	3.712	76.680	2.847	1
	Balaghat	2.513	68.433	1.720	4
	Barwani	3.311	74.537	2.468	1
	Betul	2.876	67.753	1.948	4
	Bhind	3.700	74.032	2.739	1
	Bhopal	2.706	66.017	1.787	4
	Burhanpur	3.214	70.370	2.262	2
	Chhatarpur	3.232	72.432	2.341	1
	Chhindwara	2.742	65.969	1.809	4
	Damoh	3.676	75.201	2.764	1
	Datia	3.646	72.594	2.647	1
	Dewas	3.048	75.676	2.306	6
	Dhar	2.652	76.406	2.026	5
	Dindori	3.594	71.012	2.552	1
	Guna	3.766	74.249	2.796	1
	Gwalior	3.380	70.971	2.399	1
	Harda	3.797	73.680	2.798	1
	Hoshangabad	3.798	70.535	2.679	2
	Indore	2.673	75.450	2.017	5
	Jabalpur	2.512	68.880	1.730	4
	Jhabua	4.646	76.000	3.531	1

(continued)

Table 4.5 (continued)

State/Union Territory	District	TMFR	Proportion married	TFR	Fertility profile
	Katni	3.250	71.311	2.318	1
	Khandwa (East Nimar)	3.604	71.611	2.581	1
	Khargone (West Nimar)	2.943	74.579	2.195	5
	Mandla	2.999	70.071	2.101	4
	Mandsaur	2.967	76.460	2.269	6
	Morena	3.935	75.680	2.978	1
	Narsinghpur	2.926	73.795	2.159	5
	Neemuch	3.014	74.531	2.246	6
	Panna	3.496	70.561	2.467	2
	Raisen	4.089	68.354	2.795	2
	Rajgarh	3.062	76.930	2.355	6
	Ratlam	3.043	78.335	2.384	6
	Rewa	4.439	69.254	3.074	2
	Sagar	3.909	74.808	2.925	1
	Satna	3.827	70.933	2.715	1
	Sehore	3.926	74.179	2.912	1
	Seoni	2.909	69.856	2.032	4
	Shahdol	2.736	70.460	1.927	4
	Shajapur	3.278	75.982	2.490	1
	Sheopur	3.323	75.185	2.499	1
	Shivpuri	3.682	75.164	2.767	1
	Sidhi	4.059	70.426	2.858	2
	Singrauli	4.615	72.418	3.342	1
	Tikamgarh	3.327	76.204	2.535	1
	Ujjain	3.144	76.361	2.400	1
	Umaria	3.526	69.417	2.448	2
	Vidisha	3.942	71.064	2.802	1
Maharashtra					
	Ahmadnagar	2.454	75.626	1.856	5
	Akola	2.725	74.176	2.022	5
	Amravati	2.376	71.442	1.698	5
	Aurangabad	2.662	78.508	2.090	5
	Bhandara	2.433	73.147	1.780	5
	Bid	2.789	78.361	2.185	5
	Buldana	2.485	75.468	1.875	5
	Chandrapur	2.514	72.987	1.835	5
	Dhule	3.059	78.535	2.402	6

(continued)

Table 4.5 (continued)

State/Union Territory	District	TMFR	Proportion married	TFR	Fertility profile
	Gadchiroli	2.275	73.197	1.665	5
	Gondiya	2.442	72.899	1.780	5
	Hingoli	2.766	76.603	2.119	5
	Jalgaon	2.668	77.193	2.060	5
	Jalna	2.924	79.943	2.338	6
	Kolhapur	2.347	76.205	1.788	5
	Latur	3.236	77.519	2.509	1
	Mumbai	2.170	67.742	1.470	4
	Mumbai Suburban	1.955	67.459	1.319	4
	Nagpur	2.371	68.785	1.631	4
	Nanded	2.951	75.489	2.228	6
	Nandurbar	2.944	73.784	2.172	5
	Nashik	3.389	76.742	2.601	1
	Osmanabad	2.861	80.704	2.309	6
	Palghar	2.678	69.371	1.858	4
	Parbhani	2.803	80.536	2.257	6
	Pune	2.223	73.749	1.639	5
	Raigarh	3.020	72.785	2.198	5
	Ratnagiri	3.148	69.378	2.184	3
	Sangli	1.758	77.387	1.361	5
	Satara	2.839	75.716	2.150	5
	Sindhudurg	1.917	68.997	1.323	4
	Solapur	3.577	76.644	2.742	1
	Thane	2.729	68.506	1.870	4
	Wardha	2.171	74.411	1.616	5
	Washim	2.965	78.537	2.328	6
	Yavatmal	2.363	73.849	1.745	5
Manipur					
	Bishnupur	3.047	65.319	1.990	4
	Chandel	4.185	66.248	2.773	2
	Churachandpur	3.252	61.806	2.010	3
	Imphal East	2.932	64.689	1.897	4
	Imphal West	3.131	64.540	2.021	3
	Senapati	4.258	64.883	2.763	2
	Tamenglong	3.688	69.899	2.578	2
	Thoubal	3.135	63.816	2.001	3
	Ukhrul	4.736	63.624	3.013	2

(continued)

Table 4.5 (continued)

State/Union Territory	District	TMFR	Proportion married	TFR	Fertility profile
Meghalaya					
	East Garo Hills	3.668	63.060	2.313	2
	East Jantia Hills	6.343	61.224	3.883	2
	East Khasi Hills	3.851	54.474	2.098	3
	North Garo Hills	3.138	59.420	1.864	3
	Ribhoi	5.224	58.941	3.079	2
	South Garo Hills	3.054	67.701	2.067	4
	South West Garo Hills	3.126	63.248	1.977	3
	South West Khasi Hills	5.543	65.566	3.634	2
	West Garo Hills	3.107	67.336	2.092	4
	West Jaintia Hills	5.771	60.269	3.478	2
	West Khasi Hills	6.766	69.601	4.709	2
Mizoram					
	Aizawl	3.433	48.818	1.676	3
	Champhai	2.827	60.073	1.698	4
	Kolasib	2.890	61.033	1.764	4
	Lawngtlai	3.402	62.743	2.135	3
	Lunglei	2.594	53.049	1.376	4
	Mamit	3.485	63.747	2.222	2
	Saiha	2.638	62.636	1.653	4
	Serchhip	2.751	55.724	1.533	4
Nagaland					
	Dimapur	3.185	57.372	1.828	3
	Kiphire	3.561	72.896	2.596	1
	Kohima	4.034	48.778	1.968	3
	Longleng	4.295	69.989	3.006	2
	Mokokchung	3.397	55.395	1.882	3
	Mon	3.642	60.958	2.220	2
	Peren	3.928	62.540	2.457	2
	Phek	4.626	55.032	2.546	2
	Tuensang	4.976	60.628	3.017	2
	Wokha	3.714	59.824	2.222	2
	Zunheboto	4.662	58.259	2.716	2
Odisha					
	Anugul	2.674	73.353	1.961	5
	Balangir	2.924	68.020	1.989	4
	Baleshwar	2.704	78.863	2.132	5

(continued)

Table 4.5 (continued)

State/Union Territory	District	TMFR	Proportion married	TFR	Fertility profile
	Bargarh	2.324	70.833	1.646	4
	Baudh	2.343	73.154	1.714	5
	Bhadrak	2.807	72.250	2.028	5
	Cuttack	2.036	72.598	1.478	5
	Debagarh	2.553	70.692	1.805	4
	Dhenkanal	2.933	73.333	2.151	5
	Gajapati	2.972	66.596	1.979	4
	Ganjam	3.107	73.001	2.268	6
	Jagatsinghapur	2.388	72.642	1.735	5
	Jajapur	2.898	73.812	2.139	5
	Jharsuguda	2.714	65.164	1.768	4
	Kalahandi	2.853	70.526	2.012	4
	Kandhamal	3.484	69.628	2.426	2
	Kendrapara	2.887	71.228	2.056	5
	Kendujhar	3.322	71.167	2.364	1
	Khordha	2.039	71.696	1.462	5
	Koraput	2.751	66.182	1.821	4
	Malkangiri	3.040	70.599	2.146	4
	Mayurbhanj	2.596	75.485	1.959	5
	Nabarangapur	3.748	72.857	2.731	1
	Nayagarh	2.755	78.712	2.169	5
	Nuapada	3.211	69.990	2.248	2
	Puri	2.337	73.327	1.714	5
	Rayagada	3.163	71.901	2.274	1
	Sambalpur	2.400	64.769	1.554	4
	Subarnapur	2.579	70.389	1.815	4
	Sundargarh	2.544	65.329	1.662	4
Puducherry					
	Karaikal	2.251	65.279	1.469	4
	Mahe	2.391	70.344	1.682	4
	Puducherry	2.414	66.667	1.609	4
	Yanam	2.382	69.940	1.666	4
Punjab					
	Amritsar	2.686	71.233	1.913	5
	Barnala	2.375	72.753	1.728	5
	Bathinda	2.688	72.885	1.959	5
	Faridkot	2.689	73.033	1.964	5

(continued)

Table 4.5 (continued)

State/Union Territory	District	TMFR	Proportion married	TFR	Fertility profile
	Fatehgarh Sahib	2.503	72.128	1.805	5
	Fazilka	3.005	70.416	2.116	4
	Firozpur	2.712	71.951	1.951	5
	Gurudaspur	2.466	71.884	1.773	5
	Hoshiarpur	2.762	66.917	1.848	4
	Jalandhar	2.910	67.517	1.965	4
	Kapurthala	2.778	66.865	1.857	4
	Ludhiana	3.230	69.141	2.233	2
	Mansa	2.937	69.409	2.039	4
	Moga	2.597	70.084	1.820	4
	Muktsar	2.767	71.795	1.987	5
	Pathankot	2.626	71.113	1.867	5
	Patiala	2.744	71.702	1.967	5
	Rupnagar	2.900	67.751	1.965	4
	Sahibzada Ajit Singh Nagar	2.334	75.426	1.761	5
	Sangrur	2.758	69.956	1.930	4
	Shahid Bhagat Singh Nagar	2.505	65.755	1.647	4
	Tarn Taran	2.951	69.265	2.044	4
Rajasthan					
	Ajmer	2.894	73.131	2.117	5
	Alwar	3.721	74.693	2.779	1
	Banswara	3.194	70.840	2.263	2
	Baran	3.127	72.288	2.261	1
	Barmer	3.566	68.761	2.452	2
	Bharatpur	4.218	70.136	2.958	2
	Bhilwara	3.600	75.198	2.707	1
	Bikaner	3.319	75.826	2.517	1
	Bundi	3.397	72.534	2.464	1
	Chittaurgarh	3.033	77.628	2.354	6
	Churu	3.438	73.177	2.516	1
	Dausa	3.811	72.117	2.748	1
	Dhaulpur	4.107	73.660	3.025	1
	Dungarpur	3.295	70.940	2.338	1
	Ganganagar	2.788	71.975	2.007	5
	Hanumangarh	3.356	72.586	2.436	1
	Jaipur	3.106	73.213	2.274	6

(continued)

Table 4.5 (continued)

State/Union Territory	District	TMFR	Proportion married	TFR	Fertility profile
	Jaisalmer	3.673	71.228	2.616	1
	Jalor	3.594	71.479	2.569	1
	Jhalawar	3.081	75.000	2.310	6
	Jhunjhunun	3.187	68.300	2.177	3
	Jodhpur	3.137	71.941	2.256	1
	Karauli	4.493	72.586	3.262	1
	Kota	2.947	68.604	2.022	4
	Nagaur	3.294	70.110	2.309	2
	Pali	3.462	67.705	2.344	2
	Pratapgarh	3.545	74.166	2.629	1
	Rajsamand	2.996	71.189	2.133	5
	Sawai Madhopur	4.007	74.295	2.977	1
	Sikar	3.320	72.273	2.400	1
	Sirohi	4.225	69.292	2.928	2
	Tonk	3.509	72.836	2.556	1
	Udaipur	3.357	72.319	2.428	1
Sikkim					
	East District	2.259	62.544	1.413	4
	North District	2.351	67.995	1.599	4
	South District	1.962	70.604	1.385	4
	West District	2.318	63.384	1.469	4
Tamil Nadu					
	Ariyalur	2.517	75.669	1.904	5
	Chennai	2.063	67.564	1.394	4
	Coimbatore	1.917	72.914	1.398	5
	Cuddalore	2.126	72.109	1.533	5
	Dharmapuri	2.307	73.950	1.706	5
	Dindigul	2.283	76.715	1.751	5
	Erode	2.256	71.574	1.614	5
	Kancheepuram	2.446	70.034	1.713	4
	Kanniyakumari	2.009	75.446	1.516	5
	Karur	2.118	71.567	1.516	5
	Krishnagiri	2.650	76.696	2.033	5
	Madurai	2.149	70.104	1.507	4
	Nagapattinam	2.389	68.010	1.625	4
	Namakkal	2.049	71.031	1.455	5

(continued)

Table 4.5 (continued)

State/Union Territory	District	TMFR	Proportion married	TFR	Fertility profile
	Perambalur	2.716	74.459	2.022	5
	Pudukkottai	2.768	72.277	2.001	5
	Ramanathapuram	2.283	75.216	1.717	5
	Salem	2.574	73.429	1.890	5
	Sivaganga	2.163	71.778	1.552	5
	Thanjavur	2.202	71.545	1.576	5
	The Nilgiris	2.252	69.803	1.572	4
	Theni	2.455	75.915	1.864	5
	Thiruvallur	2.261	74.235	1.678	5
	Thiruvannamalai	2.326	70.537	1.641	4
	Thiruvarur	2.485	69.212	1.720	4
	Thoothukkudi	2.763	68.608	1.896	4
	Thrissur	1.978	70.929	1.403	5
	Tiruchirappalli	2.335	71.226	1.663	5
	Tirunelveli	2.133	73.737	1.573	5
	Tiruppur	2.247	71.409	1.605	5
	Vellore	2.719	73.467	1.997	5
	Viluppuram	2.901	72.368	2.099	5
	Virudhunagar	2.235	69.381	1.551	4
Telangana					
	Adilabad	3.282	74.030	2.430	1
	Bhadradri Kothagudem	2.187	72.439	1.584	5
	Hyderabad	2.671	66.549	1.778	4
	Jagitial	2.329	75.163	1.750	5
	Jangoan	2.370	76.556	1.814	5
	Jayashankar Bhupalapally	2.266	76.637	1.737	5
	Jogulamba Gadwal	2.767	77.527	2.145	5
	Kamareddy	2.312	76.101	1.759	5
	Karimnagar	1.949	74.946	1.461	5
	Khammam	2.219	73.774	1.637	5
	Komaram Bheem Asifabad	2.400	73.140	1.755	5
	Mahabubabad	2.185	74.362	1.625	5
	Mahabubnagar	2.795	74.904	2.094	5
	Mancherial	2.080	71.544	1.488	5
	Medak	2.629	75.255	1.978	5
	Medchal-Malkajgiri	2.563	72.626	1.861	5

(continued)

Table 4.5 (continued)

State/Union Territory	District	TMFR	Proportion married	TFR	Fertility profile
	Nagarkurnool	2.475	72.860	1.803	5
	Nalgonda	2.335	76.138	1.778	5
	Nirmal	2.138	74.497	1.593	5
	Nizamabad	2.192	74.684	1.637	5
	Peddapalli	2.127	73.168	1.556	5
	Rajanna Sircilla	2.586	75.940	1.964	5
	Ranga Reddy	2.713	73.503	1.994	5
	Sangareddy	3.102	73.970	2.295	6
	Siddipet	2.361	77.111	1.820	5
	Suryapet	2.384	77.614	1.850	5
	Vikarabad	2.843	74.395	2.115	5
	Wanaparthy	2.726	70.629	1.925	4
	Warangal Rural	2.292	78.596	1.801	5
	Warangal Urban	2.387	74.252	1.772	5
	Yadadri Bhuvanagiri	2.495	71.821	1.792	5
Tripura					
	Dhalai	2.901	78.322	2.272	6
	Gomati	2.198	81.709	1.796	5
	Khowai	2.074	80.068	1.661	5
	North Tripura	2.635	72.643	1.914	5
	Sepahijala	2.717	81.481	2.214	6
	South Tripura	2.190	81.525	1.786	5
	Unakoti	3.245	75.879	2.462	1
	West Tripura	2.086	81.159	1.693	5
Uttar Pradesh					
	Agra	4.237	69.447	2.943	2
	Aligarh	3.491	70.858	2.474	2
	Ambedkar Nagar	4.819	60.455	2.913	2
	Amethi	4.831	65.598	3.169	2
	Auraiya	4.156	70.893	2.946	2
	Azamgarh	3.626	63.290	2.295	2
	Baghpat	3.764	67.960	2.558	2
	Bahraich	4.175	74.494	3.110	1
	Ballia	3.612	65.321	2.360	2
	Balrampur	4.030	72.253	2.912	1
	Banda	4.258	69.089	2.942	2
	Bara Banki	4.697	64.434	3.026	2

(continued)

Table 4.5 (continued)

State/Union Territory	District	TMFR	Proportion married	TFR	Fertility profile
	Bareilly	4.598	63.870	2.937	2
	Basti	3.529	65.621	2.316	2
	Bijnor	4.097	60.707	2.487	2
	Budaun	4.985	67.980	3.389	2
	Bulandshahr	3.640	71.715	2.610	1
	Chandauli	3.470	70.456	2.445	2
	Chitrakoot	3.588	69.376	2.489	2
	Deoria	3.154	67.243	2.121	3
	Etah	5.073	67.882	3.443	2
	Etawah	3.755	69.315	2.603	2
	Faizabad	3.503	66.552	2.331	2
	Farrukhabad	4.461	66.565	2.970	2
	Fatehpur	4.261	66.125	2.818	2
	Firozabad	4.145	71.234	2.953	1
	Gautam Buddha Nagar	3.173	69.263	2.198	3
	Ghaziabad	3.693	68.449	2.527	2
	Ghazipur	3.925	67.941	2.667	2
	Gonda	4.130	67.826	2.801	2
	Gorakhpur	3.942	66.251	2.612	2
	Hamirpur	3.695	67.661	2.500	2
	Hapur	3.711	69.842	2.592	2
	Hardoi	4.588	67.442	3.094	2
	Jalaun	3.421	71.967	2.462	1
	Jaunpur	3.395	66.585	2.261	2
	Jhansi	3.232	71.501	2.311	1
	Jyotiba Phule Nagar	4.838	62.953	3.046	2
	Kannauj	4.449	66.568	2.962	2
	Kanpur Dehat	3.612	69.165	2.498	2
	Kanpur Nagar	3.388	67.283	2.280	2
	Kanshiram Nagar	5.336	70.742	3.775	2
	Kaushambi	5.592	67.451	3.772	2
	Kheri	3.985	68.704	2.738	2
	Kushinagar	3.445	68.553	2.361	2
	Lalitpur	4.149	76.420	3.171	1
	Lucknow	3.132	65.742	2.059	3
	Mahamaya Nagar	3.677	68.818	2.531	2
	Maharajganj	3.422	71.018	2.431	1

(continued)

Table 4.5 (continued)

State/Union Territory	District	TMFR	Proportion married	TFR	Fertility profile
	Mahoba	4.171	66.714	2.783	2
	Mainpuri	4.448	66.463	2.956	2
	Mathura	4.151	72.252	2.999	1
	Mau	4.054	62.055	2.515	2
	Meerut	4.431	66.191	2.933	2
	Mirzapur	3.427	71.887	2.464	1
	Moradabad	4.042	63.008	2.546	2
	Muzaffarnagar	3.686	67.213	2.477	2
	Pilibhit	3.805	68.855	2.620	2
	Pratapgarh	4.257	66.145	2.816	2
	Prayagraj	4.739	67.259	3.187	2
	Rae Bareli	4.363	64.752	2.825	2
	Rampur	5.721	61.224	3.503	2
	Saharanpur	3.373	65.444	2.208	2
	Sambhal	4.182	67.299	2.814	2
	Sant Kabir Nagar	3.527	66.025	2.329	2
	Sant Ravidas Nagar (Bhadohi)	3.570	71.768	2.562	1
	Shahjahanpur	5.190	70.774	3.673	2
	Shamli	3.977	64.919	2.582	2
	Shrawasti	4.349	79.389	3.453	1
	Siddharthnagar	4.497	69.862	3.141	2
	Sitapur	3.758	69.842	2.625	2
	Sonbhadra	3.531	72.699	2.567	1
	Sultanpur	4.682	65.476	3.065	2
	Unnao	3.622	66.125	2.395	2
	Varanasi	3.385	67.819	2.296	2
Uttarakhand					
	Almora	2.480	67.213	1.667	4
	Bageshwar	2.760	73.044	2.016	5
	Chamoli	2.710	71.190	1.929	5
	Champawat	3.742	69.880	2.615	2
	Dehradun	2.894	66.488	1.924	4
	Garhwal	3.097	65.160	2.018	4
	Hardwar	4.708	68.366	3.218	2
	Nainital	3.227	68.921	2.224	2
	Pithoragarh	2.919	72.000	2.102	5

(continued)

Table 4.5 (continued)

State/Union Territory	District	TMFR	Proportion married	TFR	Fertility profile
	Rudraprayag	3.214	72.014	2.314	1
	Tehri Garhwal	3.190	70.231	2.240	2
	Udham Singh Nagar	3.631	69.921	2.539	2
	Uttarkashi	2.888	73.429	2.121	5
West Bengal					
	Bankura	2.558	79.539	2.035	5
	Birbhum	2.735	80.202	2.193	5
	Dakshin Dinajpur	2.504	77.787	1.948	5
	Darjiling	2.371	70.686	1.676	4
	Haora	2.011	74.608	1.500	5
	Hugli	1.827	78.585	1.436	5
	Jalpaiguri	3.161	72.297	2.285	1
	Koch Bihar	2.916	79.893	2.330	6
	Kolkata	1.866	68.483	1.278	4
	Maldah	3.046	77.257	2.353	6
	Murshidabad	2.689	79.304	2.133	5
	Nadia	2.344	80.805	1.894	5
	North Twenty Four Parganas	2.271	77.736	1.765	5
	Paschim Barddhaman	2.511	74.125	1.861	5
	Paschim Medinipur	2.423	82.673	2.003	5
	Purba Barddhaman	2.463	80.927	1.993	5
	Purba Medinipur	1.959	83.219	1.630	5
	Puruliya	3.275	75.135	2.461	1
	South Twenty Four Parganas	2.585	81.341	2.103	5
	Uttar Dinajpur	3.537	71.690	2.535	1

Source Author

References

Brass W (1968) Methods of analysis and estimation. In: Brass W (eds) The demography of Tropical Africa. Princeton University Press, Princeton

Brass W (1975) Methods for estimating fertility and mortality from limited and defective data. NC, University of North Carolina, Laboratories for Population Statistics, Chapel Hill

Chatterjee S, Mohanty SK (2021) Fertility transition and socioeconomic development in districts of India, 2001–2016. J Biosoc Sci 1–19

Cho L-J, Retherford RD, Choe MK (1986) The own-children method of fertility estimation. University of Hawaii Press, Honolulu

Everitt BS (2002) Cambridge dictionary of statistics. Oxford University Press, Oxford

Government of India (1969) Registration of births and deaths Act 1969. New Delhi, Government of India

Government of India (1988) Fertility in India. An analysis of 1981 census data. Occasional Paper No. 13 of 1988, New Delhi, Ministry of Home Affairs. Office of the Registrar General, India. Demography Division

Government of India (1997) District level estimates of fertility and child mortality for 1991 and their inter relations with other variables. Occasional Paper No. 1 of 1997, New Delhi, Registrar General and Census Commissioner of India

Government of India (2011) Annual health survey. New Delhi, Ministry of Home Affairs, Office of the Registrar General and Census Commissioner of India

Government of India (2022) National family health survey (NFHS-5) 2019–2021. India. New Delhi, Ministry of Health and Family Welfare. International Institute for Population Sciences

Government of India (2025) Sample registration system statistical report 2022. New Delhi, Ministry of Home Affairs, Office of the Registrar General and Census Commissioner of India

Guilmoto CZ, Rajan SI (2002) District level estimates of fertility from India's 2001 Census. Econ Polit Wkly XXXVII(7):665–672

Guilmoto CZ, Rajan SI (2013) Fertility at the district level in India. *Economic and Political Weekly* 48(23).

Gupta K, Singh BP, Singh KK (2014) Estimation of total fertility rates in India using indirect techniques. J Nat Acad Math 28:21–28

Hobcraft JN, Goldman N, Chidambaram VC (1982) Advances in the P/F ratio method for the analysis of birth histories. Popul Stud 36(2):291–316

Jain A (1997) Consistency between contraceptive use and fertility in India. Demogr India 26(1):19–36

Jayachandran AA, Ram F (2019) Estimation of district level TFR of eight EAG states and Assam from NFHS-4, 2015–16. Demogr India 48(1):63–73

Kumar S, Sathyanarayana KM (2012) District-level estimates of fertility and implies sex ratio at birth in India. Econ Polit Wkly 47(33)

Mauldin WP, Ross JA (1991) Family planning programmes: efforts and results, 1982–1989. Stud Fam Plann 22(6):350–367

Maultrie T, Dorrington R, Hill A, Hill K, Timæmus I, Zaba B (2013) Tools for demographic estimation. International Union for the Scientific Study of Population, Paris

Mishra VK, Palmore JA, Sinha SK (1994) Indirect estimates of fertility and mortality at the district level, 1981. Occasional Paper No. 4 of 1994, New Delhi, Office of the Registrar General India

Rele JR (1967) Fertility analysis through extension of stable population concepts. PhD Dissertation, Berkeley, University of California

Singh KK, Singh BP, Gupta K (2012) Estimation of total fertility rate and birth averted due to contraception: regression approach. Int J Stat Appl 2(5):47–55

Singh BP, Chaurasia AR, Tiwari AK (2022) A simple approach to estimate fertility at the district level. In: Chaurasia AR (ed) India 2021: population health. Bhopal, Madhya Pradesh, India, MLC Foundation and 'Shyam' Institute

Srinivasan K (1980) Birth interval analysis in fertility surveys. International Statistical Institute, The Hague

Swamy VS, Saxena AK, Palmore JA, Mishra V, Rele JR, Luther NY (1992) Evaluating the sample registration system using indirect estimates of fertility and mortality, New Delhi, Registrar General of India. Occasional Paper

Tiwari AK, Singh BP, Patel V (2020) Retrospective study of investigation of possible predictors for total fertility rate in India. J Sci Res Rep 26(9):111–119

United Nations (1967) Manuals on methods of estimating population. Manual 4. Methods of estimating basic demographic measures from incomplete data. New York, United Nations, Department of Economics and Social Affairs

Yadava RC, Kumar A (2002) On an indirect estimation of total fertility rate from open birth interval. Demogr India 31(2):211–222

Yadava RC, Tiwari AK (2007) A modified procedure to estimate total fertility rate. J Empir Res Soc Sci 2(1):82–87

Yadava RC, Tiwari AK, Sharma SS (2009) Indirect measurements of total fertility rate. J Fam Welf 55(2):70–73

Chapter 5
Differentials in Longevity

Abstract This chapter analyses the differentials in longevity across the districts of India based on inter-district variation in the life expectancy at birth and inequality adjusted life expectancy at birth. District level estimates of life expectancy at birth and inequality adjusted life expectancy at birth are based on the life tables constructed for 639 districts of the country as they existed at the 2011 population census following a non-parametric approach of modelling the transition in age-specific probabilities of death at the country level and the strong linear relationship between the geometric mean of the age-specific probabilities of death and the probability of death in the first five years of life. The analysis reveals that both male and female life expectancy at birth and inequality adjusted life expectancy at birth vary widely across the districts of the country and inter-district variation in female life expectancy at birth and inequality adjusted life expectancy at birth is relatively more marked than inter-district variation in male life expectancy at birth and inequality adjusted life expectancy at birth. Both female life expectancy at birth and inequality adjusted life expectancy at birth have also been found to be higher than male life expectancy at birth and inequality adjusted female life expectancy at birth in all districts of the country but the female-male gap in both life expectancy at birth and in inequality adjusted life expectancy at birth vary widely across districts. This chapter also discusses the inequality in the distribution of the age at death or the distribution of the length of life across the districts of the country.

Keywords India · Districts · Life expectancy at birth · Inequality adjusted life expectancy at birth

Introduction

A long and healthy life is the cherished goal of all societies and all social and economic development efforts. The life expectancy, or the expected number of years of life remaining at a specified age is universally the most widely used indicator of longevity. A high life expectancy at birth is commonly viewed as a matter of

A. R. Chaurasia, *Population and Human Development in Districts of India*,
https://doi.org/10.1007/978-981-95-8169-6_5

aggrandisement while a low life expectancy at birth reflects the poor quality of the life of the people across countries and within countries. Improvement in the life expectancy at birth also reflects the progress in human well-being. The life expectancy at birth is used to measure the individual capacity dimension of the human development index that has been developed and used by the United Nations since 1990 to measure the progress in human well-being and to rank countries in terms of human development (United Nations 2022). Life expectancy at birth is also widely used to assess potential health inequalities across countries and populations, and across administrative areas within countries for evaluating the effectiveness of public health programmes and interventions. Changes in the life expectancy at birth contributes to the development of future healthcare policies and programmes and is central to health needs profiling at the local level.

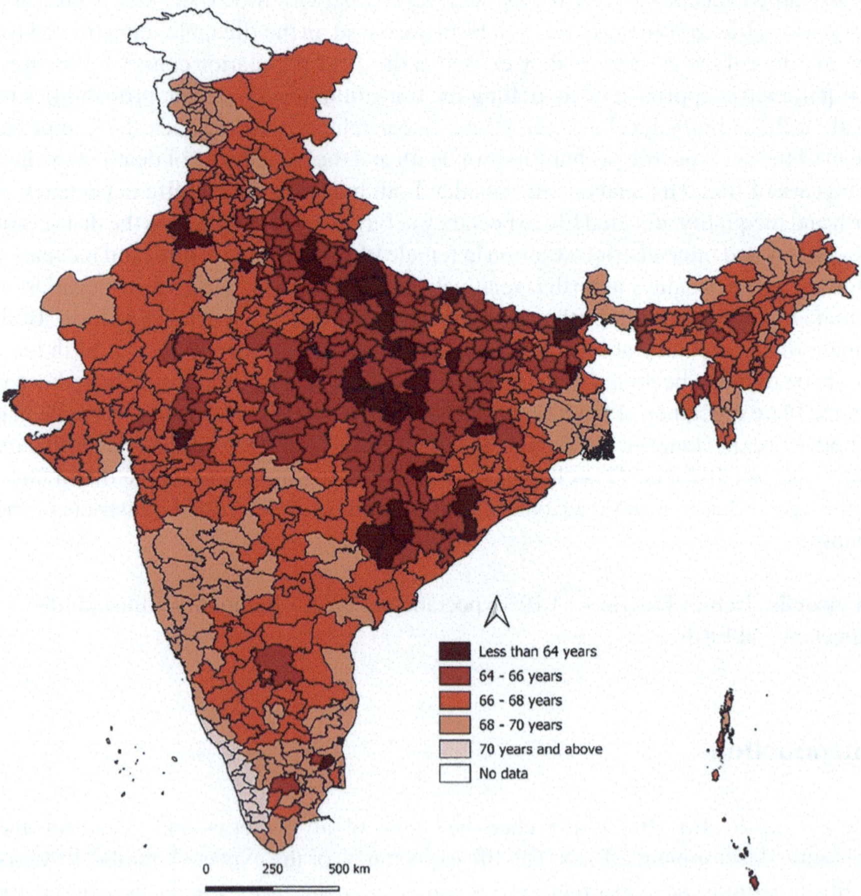

Fig. 5.1 Inter-district variation in male life expectancy at birth in India. *Source* Author

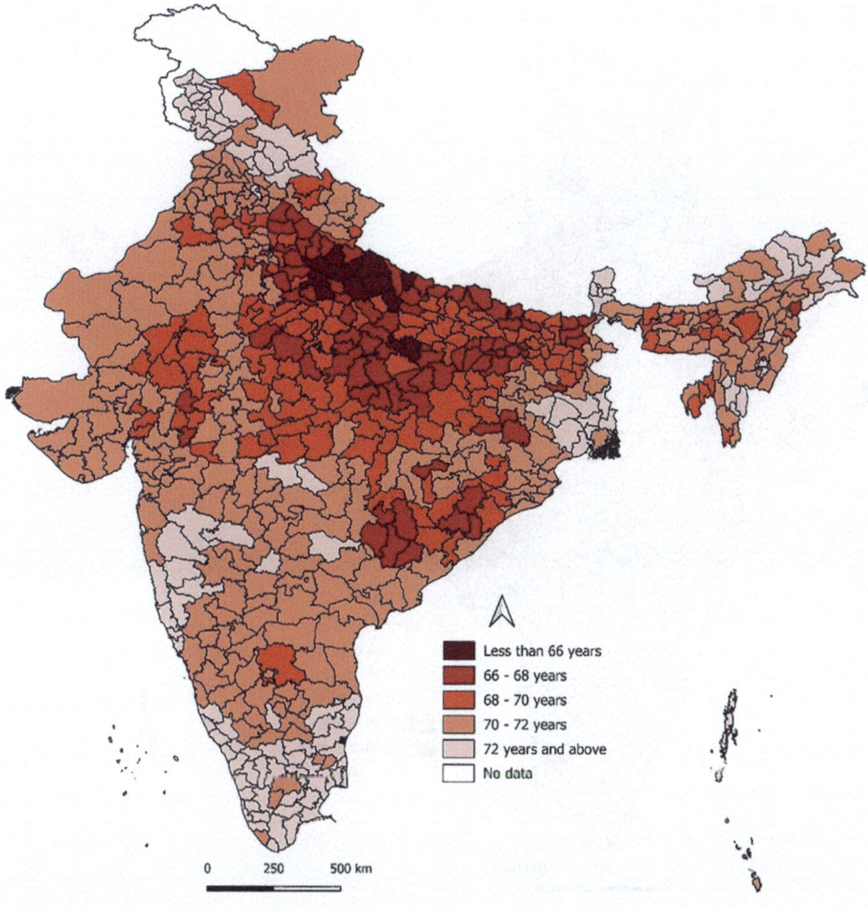

Fig. 5.2 Inter-district variation in female life expectancy at birth in India. *Source* Author

The life expectancy at birth is conventionally obtained from the life table which is an empirical model that synthesises the mortality experience of the population in a clear and concise manner (Namboodari and Suchindran 1987), although there are other indirect methods of its estimation. Construction of the life table requires age-specific death rates as the basic input. The estimation of age-specific death rates requires population and death counts by age and sex. The registration of births and deaths in India is mandatory under the Registration of Births and Deaths Act 1969 (Government of India 1969) which has recently been amended (Government of India 2023). However, a significant proportion of deaths in the country remain unregistered and there is marked variation in the completeness of death registration within the country. There are also errors in the classification of deaths by age and sex (Singh 2023). Population count by age and sex in the country is available through the decennial population census only. The civil registration system of the country,

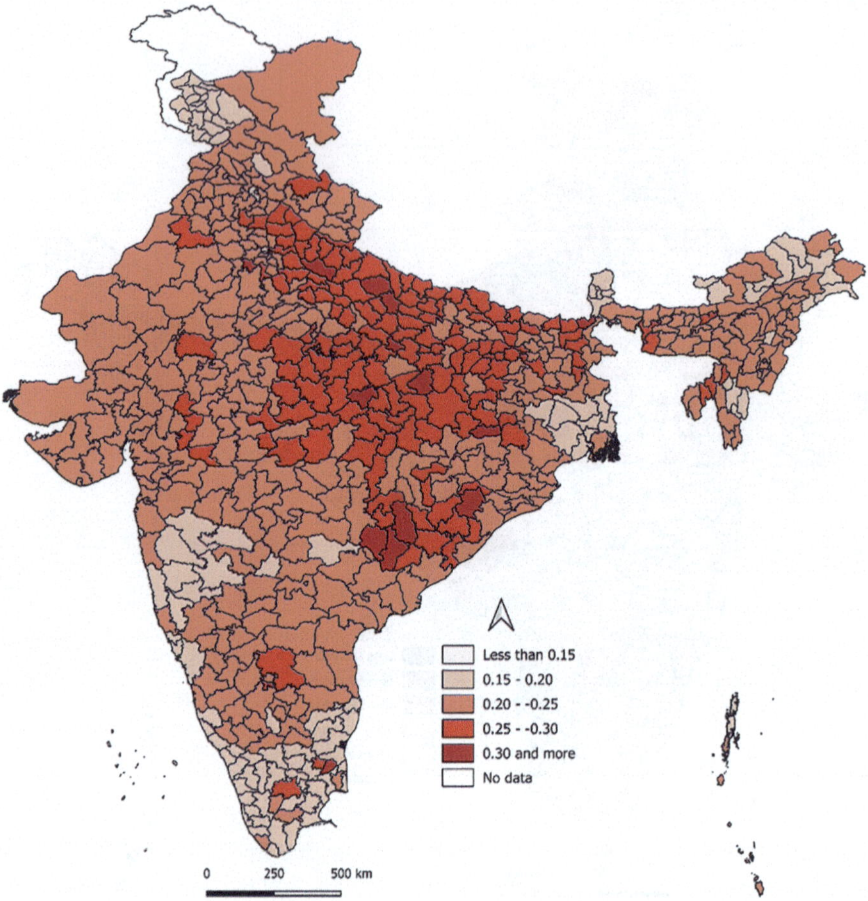

Fig. 5.3 Inequality in the age at death distribution in males (index *A*) across districts in India. Source Author

therefore, does not provide estimates of age-specific death rate for either the country or its constituent states/Union Territories and districts. Because of the imperfect civil registration system, the Government of India launched the sample registration system in 1967 to provide annual estimates of key demographic indicators including age-specific death rates by sex (Government of India 2022b). The system, however, provides estimates of age-specific death rates for selected states of the country only and life tables for these states are available (Government of India, 2022). The sample registration system does not provide estimates of age-specific death rates for the districts of the country. The Government of India also launched the National Family Health Survey (NFHS) Programme in 1992 that provides information related to demographic and health situation in the country (Government of India 2022c). The first three rounds of the NFHS provided estimates of key population and health related indicators up to the state level only. However, the fourth and the fifth rounds of NFHS

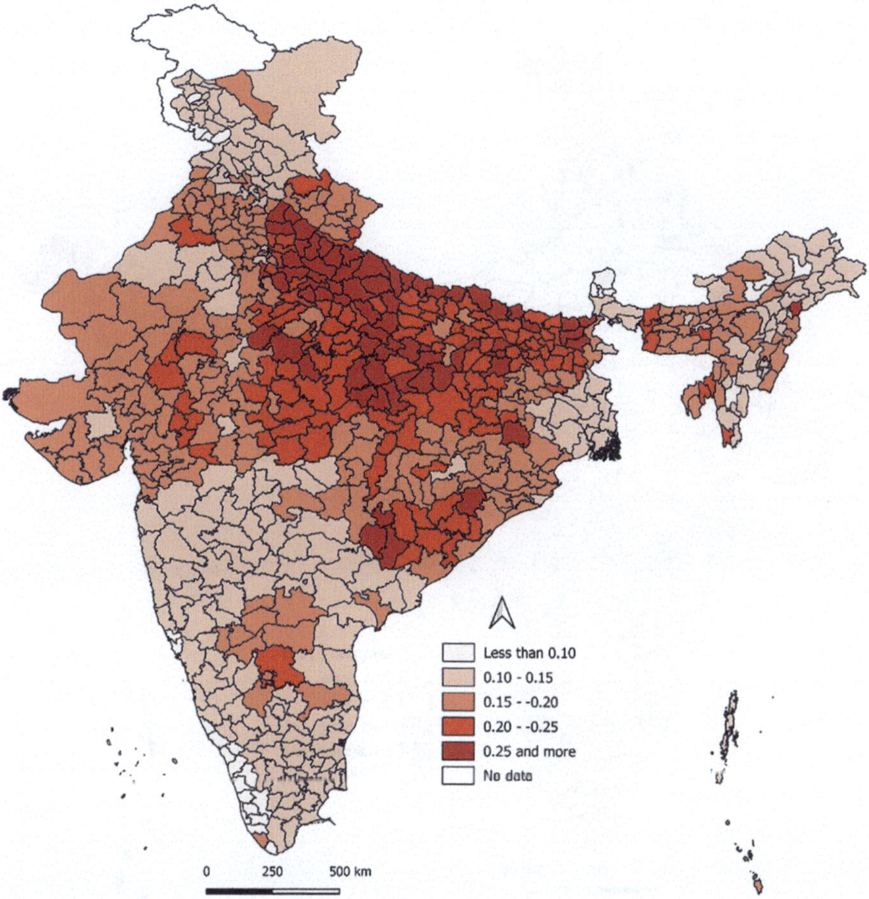

Fig. 5.4 Inequality in the age at death distribution in females (index A) across districts in India. *Source* Author

have provided estimates of selected population and health related indicators up to the district level but because of sample size limitations, the data available through the NFHS are not adequate enough to construct life tables for the districts.

In the absence of the estimates of age-specific death rates, there have been attempts to construct district life tables through indirect approaches. Ponnapalli (2011) constructed life tables for the districts of Punjab using an indirect estimate of life expectancy at birth and the model life tables. Kesarwani (2015) has constructed life tables for all districts of the country based on national and state level life tables available through the sample registration system and district level estimates of infant mortality rate derived from the summary birth history data available from the decennial population census. The method followed by Kesarwani (2015) uses the linear relationship that exists between life expectancy at birth and the probability of death in the first year of life and the relationship between the life expectancy and birth

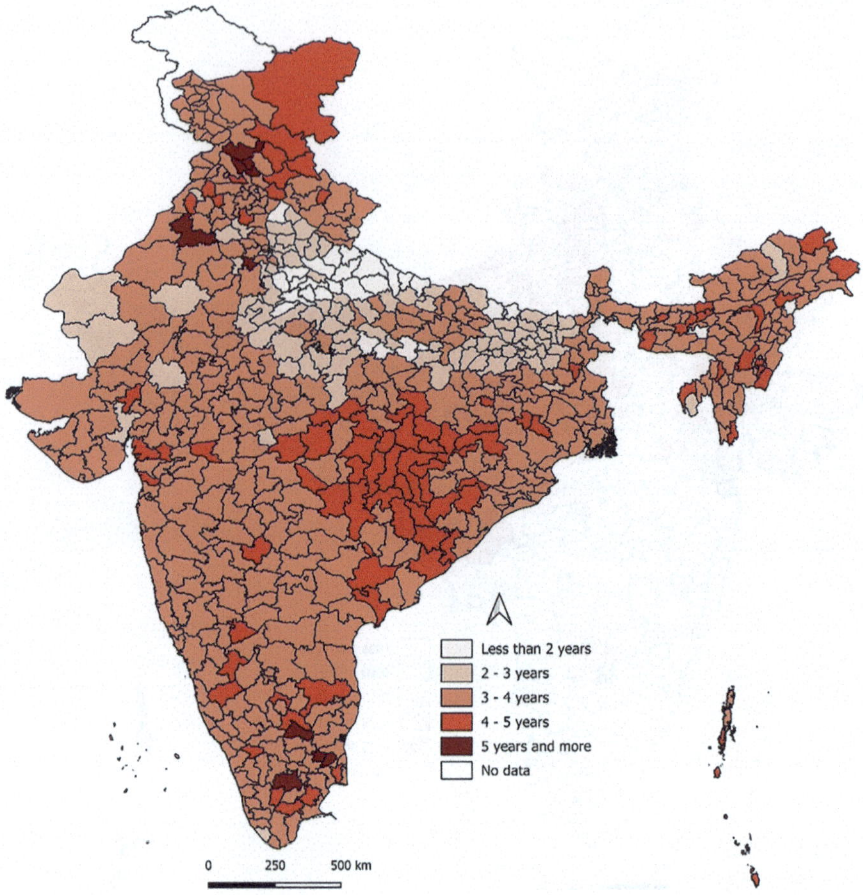

Fig. 5.5 Female-male difference in the life expectancy at birth in districts of India. *Source* Author

and the life expectancy at successive ages. The same approach has also been used by Saikia and Borah (2016) to construct life tables for 23 districts of Assam. These studies have first estimated the life expectancy at birth based on the strong linear relationship between the life expectancy at birth and child mortality, especially, infant mortality, and then constructed life tables by either using the model life tables or using the association between the life expectancy at birth and life expectancy at successive ages. Chaurasia (2023), on the other hand, has proposed a non-parametric approach for constructing life tables and has used this approach to construct life tables for the districts of the country (Chaurasia 2024). This approach involves first estimating the age-specific probabilities of death in the district and then constructing life tables using estimated age-specific probabilities of death.

The non-availability of district life tables is a major impediment to measuring and monitoring population health at the district level in the country and to analyse

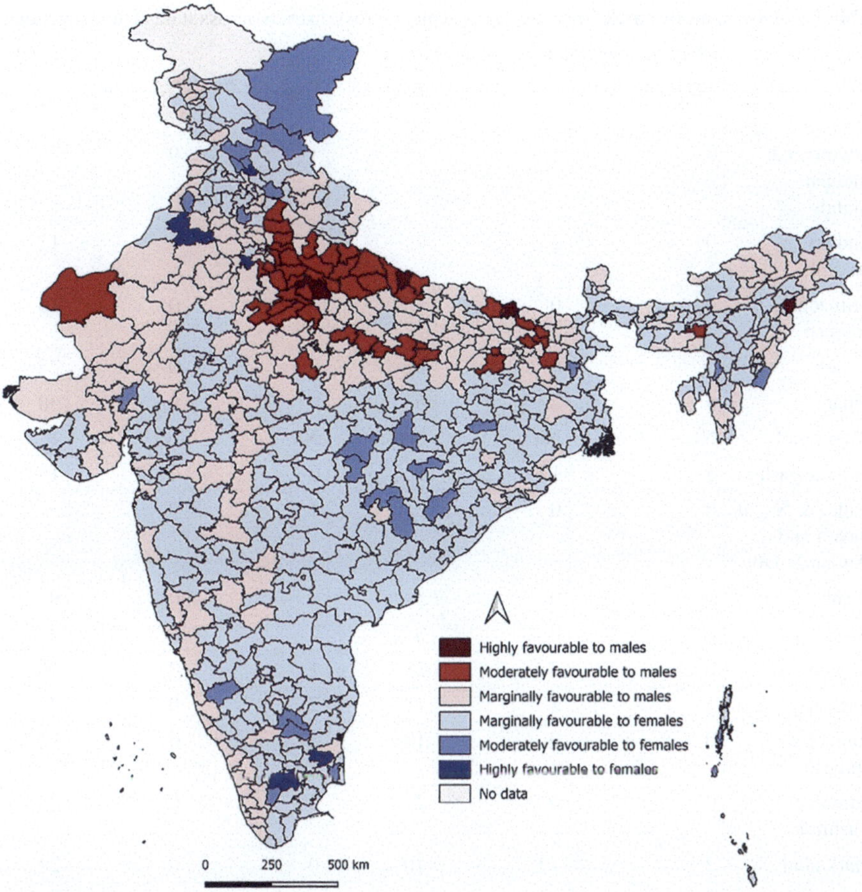

Fig 5.6 Female-male difference in the life expectancy at birth which is attributed to district-specific factors. *Source* Author

the impact of health policies and programme in meeting the health needs of the people. Monitoring population health at the district level is important as the National Health Mission emphasises district-based decentralised planning for the delivery of health care services to meet the health needs of the people (Government of India 2013). District level life tables is also necessary to highlight inter-district disparities in population health and in identifying hotspot districts which may be given focused attention in planning for the delivery of health care services. The inter-district disparity in population health in India is quite pervasive and appears to have persisted over time. It has also been repeatedly argued that reducing inter-district disparities in population health may contribute, substantially, towards accelerating the pace of improvement in population health in the country which remains low by international standards. According to the estimates prepared by the United Nations, male life expectancy at birth in India was 70.5 years in the year 2023 while female

Table 5.1 Distribution of male life expectancy at birth across districts across states/Union territories

	Male life expectancy at birth (Years)					Total
	Very low <64	Low 64–66	Average 66–68	High 68–70	Very high ≥70	
Andaman & Nicobar Islands	0	1	1	1	0	3
Andhra Pradesh	0	2	9	2	0	13
Arunachal Pradesh	0	0	4	12	0	16
Assam	0	6	21	0	0	27
Bihar	5	27	6	0	0	38
Chandigarh	Na	Na	Na	Na	Na	1
Chhattisgarh	5	12	1	0	0	18
Dadra & Nagar Haveli and Daman & Diu	0	0	2	1	0	3
Delhi	0	0	5	4	0	9
Goa	0	0	0	2	0	2
Gujarat	0	2	23	1	0	26
Haryana	1	3	15	2	0	21
Himachal Pradesh	0	0	10	2	0	12
Jammu & Kashmir	0	0	2	20	0	22
Jharkhand	2	12	10	0	0	24
Karnataka	0	0	18	12	0	30
Kerala	0	0	0	0	14	14
Lakshadweep	0	0	0	1	0	1
Madhya Pradesh	7	27	16	0	0	50
Maharashtra	0	1	18	16	0	35
Manipur	0	0	8	1	0	9
Meghalaya	0	1	6	0	0	7
Mizoram	0	0	3	5	0	8
Nagaland	0	0	7	4	0	11
Odisha	2	7	21	0	0	30
Puducherry	0	0	0	1	3	4
Punjab	0	2	16	2	0	20
Rajasthan	1	4	24	4	0	33

(continued)

Table 5.1 (continued)

	Male life expectancy at birth (Years)					Total
	Very low <64	Low 64–66	Average 66–68	High 68–70	Very high ≥70	
Sikkim	0	0	0	4	0	4
Tamil Nadu	1	2	5	24	0	32
Telangana	0	0	7	3	0	10
Tripura	0	2	2	0	0	4
Uttar Pradesh	12	51	8	0	0	71
Uttarakhand	0	4	9	0	0	13
West Bengal	0	0	5	14	0	19
India	36	166	282	138	17	639

Source Author

life expectancy at birth was 73.6 years. India ranked 153 among 237 countries and areas of the world in terms of the life expectancy at birth (United Nations 2024).

The life expectancy at birth, however, is not the appropriate measure for analysing the variation in longevity as it does not take into account the variation in the distribution of the age at death or the variation in the length of life. The life expectancy at birth is a composite measure of the person-years lived in different ages with the simple arithmetic mean as the aggregation function. The limitations of the simple arithmetic mean as the aggregation function in the construction of the composite measure are well known (OECD 2008). The alternative that has been suggested is to use the generalised mean or the power mean (Bullen 2003). When the exponent of the power mean is equal to 1, it is equal to the simple arithmetic mean. When the exponent of the power mean is equal to 0, it is equal to the geometric mean. The geometric mean is always less than or equal to the simple arithmetic mean. The difference between the simple arithmetic mean and the geometric mean of the person-years lived in different ages reflects the inequality in the distribution of the age at death. When there is no difference between the simple arithmetic and the geometric mean of the person-years lived in different ages, there is no inequality in the age at death and the higher the difference the higher the inequality. The inequality in the age at death, therefore, can be measured in terms of the difference between the simple arithmetic mean and the geometric mean of the person-years lived in different ages. We have also presented estimates of the inequality adjusted life expectancy at birth for the districts of the country. The inequality adjusted life expectancy at birth is always less than the life expectancy at birth and the difference reflects the loss of life due to the unequal distribution of the age at death.

The present chapter first analyses the variation in the life expectancy at birth in 639 districts of the country as they existed at the 2011 population census and then examines inter-district variation in the inequality in the distribution of the age at death across districts. The chapter also analyses the inter-district variation in the sex differentials in the life expectancy at birth and in the inequality adjusted life expectancy at

Table 5.2 Inter-district distribution of female life expectancy at birth across states/Union Territories

India/State/ Union Territory	Female life expectancy at birth (Years)					Total
	Very low <66	Low 66–68	Average 68–70	High 70–72	Very high ≥72	
Andaman & Nicobar Islands	0	0	0	1	2	3
Andhra Pradesh	0	0	2	11	0	13
Arunachal Pradesh	0	0	0	5	11	16
Assam	0	0	6	21	0	27
Bihar	1	20	17	0	0	38
Chandigarh	Na	Na	Na	Na	Na	1
Chhattisgarh	0	6	7	5	0	18
Dadra & Nagar Haveli and Daman & Diu	0	0	0	2	1	3
Delhi	0	1	1	5	2	9
Goa	0	0	0	0	2	2
Gujarat	0	0	4	22	0	26
Haryana	0	1	7	13	0	21
Himachal Pradesh	0	0	0	4	8	12
Jammu & Kashmir	0	0	1	1	20	22
Jharkhand	0	4	9	11	0	24
Karnataka	0	0	0	26	4	30
Kerala	0	0	0	0	14	14
Lakshadweep	0	0	0	1	0	1
Madhya Pradesh	0	11	28	11	0	50
Maharashtra	0	0	0	24	11	35
Manipur	0	0	0	7	2	9
Meghalaya	0	0	3	4	0	7
Mizoram	0	0	1	3	4	8
Nagaland	0	1	1	8	1	11
Odisha	0	4	9	17	0	30
Puducherry	0	0	0	0	4	4
Punjab	0	0	1	18	1	20
Rajasthan	0	1	14	18	0	33

(continued)

Table 5.2 (continued)

India/State/ Union Territory	Female life expectancy at birth (Years)					Total
	Very low <66	Low 66–68	Average 68–70	High 70–72	Very high ≥72	
Sikkim	0	0	0	0	4	4
Tamil Nadu	0	0	0	6	26	32
Telangana	0	0	0	9	1	10
Tripura	0	0	3	1	0	4
Uttar Pradesh	13	34	24	0	0	71
Uttarakhand	0	1	4	8	0	13
West Bengal	0	0	0	9	10	19
India	14	84	142	271	128	639

Source Author

birth. The analysis is limited to 639 of the 640 districts of the country as they existed at the time of 2011 population census. District Chandigarh has been excluded from the analysis as life tables for the district are not available. The number of districts in the country has, however, increased substantially since the 2011 population census because of the creation of new districts on political and administrative considerations. In 2024, there were 783 districts in the country (Government of India 2024) and this number is continuously increasing. Life tables for the newly created districts are not available to obtain estimates of life expectancy at birth and inequality adjusted life expectancy at birth. There has been no population census in the country after 2011 so that construction of life tables and estimation of life expectancy at birth and inequality adjusted life expectancy at birth is not possible for the districts that have been created after the 2011 population census. In any case, the analysis reflects the diversity in both longevity and population health across the districts of the country.

Inter-district Variation in Life Expectancy at Birth

Estimates of male and female life expectancy at birth, derived from the life tables constructed by Chaurasia (2024) have been found to vary widely across 639 of the 640 districts of the country as they existed at the time of the 2011 population census. Estimates of the life expectancy at birth for district Chandigarh are not available. Among the 639 districts, male life expectancy at birth is estimated to be the lowest in district Rewari of Haryana (62 years) but the highest in district Mahe of the Union Territory of Puducherry (71 years). The female life expectancy at birth, on the other hand, has been found to vary from the lowest in district Sitapur of Uttar Pradesh (65 years) to the highest in district Karaikal of Puducherry (74 years). The inter-district variation in male life expectancy at birth is depicted in Fig 5.1 while inter-district variation in female life expectancy at birth is depicted in Fig 5.2. There are 36 districts

Table 5.3 The distribution of districts in different states and Union Territories by the district-specific component of the female-male difference in the life expectancy at birth

India/State/Union Territory	Female-male difference (years) in observed life expectancy at birth attributed to factors specific to the district						Total
	Favourable to females			Favourable to males			
	≥ 2	1–2	0–1	0–1	1–2	≥ 2	
Andaman & Nicobar Islands	0	1	2	0	0	0	3
Andhra Pradesh	0	0	13	0	0	0	13
Arunachal Pradesh	0	0	6	10	0	0	16
Assam	0	1	20	6	0	0	27
Bihar	0	0	2	26	9	1	38
Chandigarh	Na	Na	Na	Na	Na	Na	1
Chhattisgarh	0	4	13	1	0	0	18
Dadra & Nagar Haveli and Daman & Diu	0	0	3	0	0	0	3
Delhi	0	0	1	7	0	1	9
Goa	0	0	1	1	0	0	2
Gujarat	0	1	15	10	0	0	26
Haryana	1	1	4	13	2	0	21
Himachal Pradesh	1	5	6	0	0	0	12
Jammu & Kashmir	0	1	13	8	0	0	22
Jharkhand	0	2	17	5	0	0	24
Karnataka	0	1	20	9	0	0	30
Kerala	0	0	1	13	0	0	14
Lakshadweep	0	0	0	1	0	0	1
Madhya Pradesh	0	1	26	21	2	0	50
Maharashtra	0	1	27	7	0	0	35
Manipur	0	1	6	2	0	0	9
Meghalaya	0	0	3	3	1	0	7
Mizoram	0	0	4	4	0	0	8
Nagaland	0	0	6	4	0	1	11
Odisha	0	1	23	6	0	0	30
Puducherry	0	0	1	3	0	0	4
Punjab	0	1	13	6	0	0	20
Rajasthan	1	0	8	21	3	0	33
Sikkim	0	0	2	2	0	0	4
Tamil Nadu	3	4	24	1	0	0	32
Telangana	0	0	10	0	0	0	10
Tripura	0	0	3	1	0	0	4

(continued)

Table 5.3 (continued)

India/State/Union Territory	Female-male difference (years) in observed life expectancy at birth attributed to factors specific to the district						Total
	Favourable to females			Favourable to males			
	≥ 2	1–2	0–1	0–1	1–2	≥ 2	
Uttar Pradesh	0	0	2	35	31	3	71
Uttarakhand	0	0	7	6	0	0	13
West Bengal	0	0	18	1	0	0	19
India	6	26	320	233	48	6	640

Source Author

in the country in which male life expectancy at birth is estimated to be less than 64 years. Twelve of these 36 districts are in Uttar Pradesh, 7 in Madhya Pradesh and 5 each in Bihar and Chhattisgarh. By comparison, there are only 17 districts in the country which male life expectancy at birth is estimated to be either equal to or more than 70 years and 14 of these 17 districts are in Kerala alone (Table 5.1). Kerala is the only state/Union Territory in the country in which male life expectancy at birth is either equal to or more than 70 years in all districts. In Puducherry, male life expectancy at birth is found to be either equal to or more than 70 years in 3 of the 4 districts. There is no other state/Union Territory in the country in which male life expectancy at birth is estimated to be either equal to or more than 70 years in any district. In 166 districts, male life expectancy at birth has been found to range between 64 and 66 years.

The female life expectancy at birth, on the other hand, is estimated to be less than 66 years in 14 districts of the country. There is no district in the country in which the female life expectancy at birth is found to be less than 64 years. Thirteen of the 14 districts in which female life expectancy at birth is estimated to be less than 66 years are in Uttar Pradesh alone while 1 district is in Bihar In other states/Union Territories of the country, there is no district in which female life expectancy at birth is estimated to be less than 66 years. By comparison, female life expectancy at birth is estimated to be at least 72 years in 128 districts (Table 5.2). There are, however, only six districts in which female life expectancy at birth is estimated to be either 74 years or more than 74 years. Three of these six districts are in Kerala while the remaining three are in the Puducherry. Kerala, Puducherry, Sikkim and Goa are the only four states/Union Territories in which female life expectancy at birth is estimated to be at least 72 years in all districts. In 271 districts of the country, the female life expectancy at birth is estimated to range between 70 and 72 years whereas in 142 districts, it ranges between 68 and 70 years.

Regional pattern in male and female life expectancy at birth in the country is also evident from Figs. 5.1 and 5.2. There are 202 districts in which male life expectancy at birth is estimated to be either low or very low (<66 years) and 150 of these districts are in the central region of the country, in Bihar, Chhattisgarh, Jharkhand, Madhya Pradesh and Uttar Pradesh. On the other hand, male life expectancy at birth is estimated to be either high or very high (≥68 years) in 155 districts and 102 of these districts are in the southern region of the country, in Kerala, Puducherry, Tamil Nadu, Maharashtra and Karnataka, and in West Bengal and Jammu & Kashmir. In 10 states/Union Territories, there is no district in which male life expectancy at birth is either high or very high (≥68 years) whereas there are 15 states/Union Territories where there is no district in which male life expectancy at birth is either low or very low (<66 years). Similarly, there are 98 districts in which female life expectancy at birth is estimated to be either low or very low (<68 years) and 79 of these districts are in the central region of the country, in Uttar Pradesh, Madhya Pradesh and Bihar. On the other hand, there are 128 districts in which female life expectancy at birth is very high (≥72 years) and 82 of these districts are in Kerala, Tamil Nadu, Maharashtra, Jammu & Kashmir and Arunachal Pradesh. There is no district in Kerala, Goa, Puducherry and Sikkim where female life expectancy at birth is estimated to be less than 72 years. The inter-district disparity in female life expectancy at birth is more marked than that in male life expectancy at birth.

The inter-district disparity in male and female life expectancy at birth depicted in Figs. 5.1 and 5.2 indicate substantial variation in longevity and population health across the districts of the country and may be a contributing factor to the low life expectancy at birth in the country as a whole. It may be argued that there are district-specific factors that have a strong impact on longevity and population health in the district. Identification of these factors and incorporating then in district level planning and programming for the delivery of health care services is necessary for reducing inter-district disparity in longevity and population health in India. The National Health Mission of the Government of India calls for decentralised district-based planning and programming for the delivery of health care services to meet the health care needs of the people. Estimating life expectancy at birth in the district may provide the much needed empirical evidence to support decentralised district-based planning and programming and measuring the impact of health care services on longevity and population health.

Inequality in Age at Death Distribution

Let μ denotes the simple arithmetic mean while g denotes the geometric mean of the person-years lived in different ages of the life span. Then the index of inequality in the age at death is defined as (Gisbert 2020)

$$A = 1 - \frac{g}{\mu} \tag{5.1}$$

The inequality index A ranges between 0 and 1 when there is no inequality in the age at death distribution or when g=μ, A=0. On the other hand, the larger the difference between g and μ in a district, the larger the inequality in the age at death distribution in that district. Figure 5.3 depicts the variation in the index A in the male population across the districts whereas Fig. 5.4 depicts the variation across districts in the inequality in the age at death distribution in the female population. The inequality in the age at death distribution in the male population has been found to be the highest in district Rewari of Haryana but the lowest in district Mahe of Puducherry among the 639 districts of the country. Table 5.3 gives the distribution of districts by the inequality in the age at death distribution in the male population by states/Union Territories of the country. There is only one district—district Mahe in Puducherry—where the inequality in the age at distribution in the male population may be termed as very low (A<0.15). In 113 districts, on the other hand, the inequality in the age at distribution in the male population may be termed as low (0.15≤A<0.20). On the other hand, there are 14 districts where the inequality in the age at death distribution in the male population is found to be very high (A>0.30). The inequality in age at death distribution in the male population is found to be very high in four districts of Uttar Pradesh; three districts of Chhattisgarh; and two districts of Madhya Pradesh and one district each in Haryana, Bihar, Jharkhand, Odisha and Tamil Nadu.

In the female population, the inequality in the age at death distribution is found to be the highest in district Sitapur of Uttar Pradesh but the lowest in district Mahe of Puducherry. Table 5.4 gives the distribution of districts by the inequality in the age at death distribution in the female population. Contrary to the male population, the inequality in the age at death distribution in the female population is found to be very low in 244 districts and low in 201 districts. There are only four districts where the inequality in the age at death distribution in the female population is found to be very high. These districts are districts Budaun, Sitapur, Hardoi and Shrawasti. All the four districts are in Uttar Pradesh. District Budaun is the only district in the country in which the inequality in the age at death distribution is found to be very high in both male and female populations. There are only 67 districts in which the inequality in the age at death distribution in the female population is found to be high (0.25≤A<0.30).

The inequality in the age at death distribution is found to be higher in the male population relative to the inequality in the age at death distribution in the female population in most of the districts of the country. There are, however, 26 districts in which the inequality in the age at death distribution is found to be higher in the female population relative to the male population. Twenty-one of these 26 districts ae in Uttar Pradesh alone. The inequality in the age at death distribution is found to be higher in the female population relative to the male population in 3 districts of Bihar and one district each in Delhi and Nagaland.

Sex Differentials in Longevity

In all districts of the country, females live longer than males as the female life expectancy at birth is estimated to be higher than the male life expectancy at birth. However, female-male difference in the life expectancy at birth varies widely (Fig. 5.5). There are 11 districts in which female life expectancy at birth exceeds male life expectancy at birth by at least 5 years. The female-male difference in the life expectancy at birth is found to be the highest in district Ariyalur of Tamil Nadu where female life expectancy at birth exceeds male life expectancy at birth by more than 9 years. On the other hand, there is virtually no difference between female and male in life expectancy at birth in district Shrawasti of Uttar Pradesh. District Mon of Nagaland and district Etah of Uttar Pradesh are the only other districts in the country in which the difference between female and male life expectancy at birth is found to be less than 1 year. In 440 or almost 60 per cent districts, female life expectancy at birth exceeds male life expectancy at birth by 3-4 years whereas in 188 or in about one-fourth districts of the country, female life expectancy at birth exceeds male life expectancy at birth by 4–5 years. There are only 123 or about 15 per cent districts in which female life expectancy at birth exceeds male life expectancy at birth by less than 2 years.

The observed difference between female and male life expectancy at birth in a district may be thought of comprising of two components: 1) female-male difference in the life expectancy at birth that is common to all districts and 2) female-male difference in the life expectancy at birth that is specific to the district. Understanding sex difference in longevity requires decomposing the observed female-male difference in the life expectancy at birth in a district into the common component and district-specific component . Let e_{md} denotes the male life expectancy at birth in district d; e_{fd} denotes the female life expectancy at birth in district d; $e_{d.}$ denotes the geometric mean of male and female life expectancy at birth in district d, $e_{.m}$ and $e_{.f}$ denote, respectively, the geometric mean of male and female life expectancy at birth across all districts. Then, we can write

$$e_{md} = e_{m.} \times e_{.d} \times \frac{e_{md}}{e_{m.} \times e_{.d}} = e_{m.} \times e_{.d} \times r_{md} \tag{5.2}$$

$$e_{fd} = e_{f.} \times e_{.d} \times \frac{e_{fd}}{e_{f.} \times e_{.d}} = e_{f.} \times e_{.d} \times r_{fd} \tag{5.3}$$

Now

$$\nabla e_d = e_{fd} - e_{md} = \frac{e_{fd} - e_{md}}{\ln\left(\frac{e_{fd}}{e_{md}}\right)} \times \ln\left(\frac{e_{fd}}{e_{md}}\right) = \left\{LM_d \times \ln\left(\frac{e_{f.}}{e_{m.}}\right)\right\} + \left\{LM_d \times \ln\left(\frac{r_{fd}}{r_{md}}\right)\right\} \tag{5.4}$$

$$\nabla e_d = e_{fd} - e_{md} = C + S \tag{5.5}$$

where

$$LM_d = \frac{e_{fd} - e_{md}}{\ln\left(\frac{e_{fd}}{e_{md}}\right)} \tag{5.6}$$

is the logarithmic mean of female and male life expectancy at birth in district *d* and

$$C = \left\{LM_d \times \ln\left(\frac{e_{f.}}{e_{m.}}\right)\right\} \tag{5.7}$$

$$S = \left\{LM_d \times \ln\left(\frac{r_{fd}}{r_{md}}\right)\right\} \tag{5.8}$$

Equation (5.5) decomposes the observed female-male difference in life expectancy at birth into female-male difference common to all districts and female-male difference specific to the district. When female-male difference in life expectancy at birth common to all districts is separated, female-male difference in the life expectancy at birth specific to the district is favourable to females in only 352 districts. In 287 districts, female-male difference in the life expectancy at birth specific to the district is found to be favourable to males (Fig. 5.6). The geometric mean of life expectancy at birth across districts is estimated to be 66.8 years for males and 70.3 years for females.

Table 5.4 gives the distribution of districts in different states and Union Territories by the female-male difference in the life expectancy at birth that is attributed to female-male difference in the life expectancy at birth specific to the district. In Kerala, there is only one district where female-male difference in life expectancy at birth specific to the district is favourable to females. In rest of the districts, it is favourable to males. This is in quite contrast to the observed female-male difference in the life expectancy at birth which is favourable to females in all districts of the state. In Delhi also, female-male difference in life expectancy at birth specific to the district is found to be favourable to males in 8 of the 9 districts. Similarly, in Uttar Pradesh, female-male difference in the life expectancy at birth specific to the district is found to be favourable to males in 69 of the 71 districts. In Bihar female-male difference in the life expectancy at birth specific to the district is found to be favourable to males in 36 of the 38 districts. By contrast, in Andhra Pradesh, Himachal Pradesh and Telangana, female-male difference in the life expectancy at birth specific to the district is found to be favourable to females in all districts. In West Bengal, female-male difference in life expectancy at birth specific to the district is found to be favourable to females in 18 of the 19 districts. There is, however, no state in which the female-male difference in the life expectancy at birth specific to the district is found to be favourable to males in all districts. It is clear from Table 5.4 that in a large proportion of districts, longer female life expectancy at birth relative to the male life expectancy at birth is due to longer average female life expectancy at birth relative to the average male life expectancy at birth across districts. If the average of female-male difference in life expectancy at birth across districts is separated, then inter-district distribution of female-male life expectancy at birth that is specific to the district is markedly different.

Conclusions

This chapter reveals very wide variation in both male and female longevity in the country as reflected by the life expectancy at birth and inter-district variation in female life expectancy at birth is more marked than inter-district variation in male life expectancy at birth. Reducing inter-district variation in the life expectancy at birth may go a long way in improving the longevity in the country as the life expectancy at birth in India remains low by international standards. The analysis also suggests that a district-based approach should be adopted for improving longevity of both males and females in the country. The first requirement to this direction is the estimation of life expectancy at birth at the district level at regular intervals. At present, such a system does not exists in India, primarily, because the civil registration system in the country is deficient, although registration of births and deaths is mandatory under the Registration of Births and Deaths Act, 1969 of the Government of India (Government of India 1969) which has been amended recently (Government of India 2023).

The analysis presented in this chapter also indicates that the male-female difference in the life expectancy at birth is influenced by both male-female difference in the life expectancy at birth that is common to all districts of the country and male-female difference in the life expectancy at birth that is specific to the district. When the male-female difference in the life expectancy at birth that is common to all districts is excluded, female-male difference in the life expectancy at birth is found to be unfavourable to females in a large number of districts. However, the observed female-male difference in the life expectancy at birth suggests that female life expectancy at birth is higher than the male life expectancy at birth in all districts.

Separating the female-male difference in life expectancy at birth that is common to all districts is important to understand the inter-district variation in the female-male difference in life expectancy at birth.

Inter-district variation in the life expectancy at birth also implies that population health varies markedly across the districts of the country. There are districts in which males have better health status than females and there are districts in which females health status is better than males. This means that there are district-specific factors that influence, differentially, the health status of males and females. Identification of these factors is important from the perspective of improving population health. An analysis of deaths by age and cause of death in each district is needed to understand the mortality profile of the district which determines the longevity of the district population. Currently, there is little information about the causes of death in the districts of the country. The Registration of Births and Deaths Act of the Government of India has provision of recording the medically certified cause of death of all deaths registered under the civil registration system but the information is grossly deficient as the cause of death of only a very small proportion of deaths registered under the system is medically certified. There is also a medical certification of causes of death scheme (Government of India 2022d), but the coverage of the scheme is limited and the scheme does not provide data on causes of death for the districts of the country. Understanding the pattern of causes of death in the district is necessary to evolve a district-specific approach to improving the life expectancy at birth.

Appendix

See Table 5.4.

Table 5.4 Male and female life expectancy at birth in districts of India and decomposition of the female-male difference in the life expectancy at birth

State/Union Territory/ District	Life expectancy at birth (years)		Female-male difference in life expectancy at birth (years)		
	Male	Female	Common to all district	Specific to the district	Total
		Jammu & Kashmir			
Kupwara	68.56	72.03	3.58	–0.11	3.47
Badgam	69.62	72.81	3.63	–0.44	3.19
Leh(Ladakh)	66.89	71.77	3.53	1.35	4.87
Kargil	66.24	69.94	3.47	0.23	3.70
Punch	68.55	72.26	3.58	0.12	3.71

(continued)

Table 5.4 (continued)

State/Union Territory/ District	Life expectancy at birth (years)		Female-male difference in life expectancy at birth (years)		
	Male	Female	Common to all district	Specific to the district	Total
Rajouri	69.31	72.63	3.61	–0.29	3.32
Kathua	69.49	73.04	3.63	–0.08	3.54
Baramula	69.05	72.64	3.61	–0.02	3.59
Bandipore	68.49	72.15	3.58	0.08	3.66
Srinagar	69.24	72.67	3.61	–0.18	3.43
Ganderbal	69.00	72.32	3.60	–0.28	3.32
Pulwama	69.36	73.13	3.63	0.14	3.77
Shupiyan	69.11	72.71	3.61	–0.01	3.60
Anantnag	69.06	72.81	3.61	0.14	3.74
Kulgam	68.58	72.43	3.59	0.26	3.85
Doda	68.89	72.72	3.61	0.22	3.82
Ramban	68.72	72.61	3.60	0.29	3.89
Kishtwar	68.57	72.30	3.59	0.14	3.73
Udhampur	69.00	72.71	3.61	0.10	3.70
Reasi	68.65	72.40	3.59	0.16	3.75
Jammu	69.49	73.26	3.63	0.14	3.77
Samba	69.73	73.56	3.65	0.18	3.83
Himachal Pradesh					
Chamba	67.84	71.76	3.55	0.37	3.92
Kangra	66.69	72.11	3.53	1.89	5.42
Lahul & Spiti	67.49	72.12	3.55	1.08	4.64
Kullu	67.97	72.06	3.56	0.53	4.09
Mandi	68.67	72.38	3.59	0.12	3.71
Hamirpur	67.42	72.63	3.56	1.65	5.21
Una	66.52	71.78	3.52	1.74	5.26
Bilaspur	66.40	72.07	3.52	2.15	5.66
Solan	68.01	72.00	3.56	0.43	3.98
Sirmaur	66.84	71.55	3.52	1.19	4.70
Shimla	67.96	72.01	3.56	0.49	4.06
Kinnaur	67.94	72.18	3.57	0.67	4.24

(continued)

Table 5.4 (continued)

State/Union Territory/ District	Life expectancy at birth (years)		Female-male difference in life expectancy at birth (years)		
	Male	Female	Common to all district	Specific to the district	Total
			Punjab		
Gurdaspur	67.84	71.49	3.55	0.10	3.65
Kapurthala	67.60	71.44	3.54	0.30	3.83
Jalandhar	67.74	71.60	3.55	0.31	3.85
Hoshiarpur	67.78	71.65	3.55	0.32	3.88
Shahid Bhagat Singh Nagar	67.06	71.13	3.52	0.55	4.07
Fatehgarh Sahib	67.80	71.38	3.54	0.04	3.58
Ludhiana	67.90	71.63	3.55	0.18	3.73
Moga	65.78	70.13	3.46	0.89	4.35
Firozpur	67.37	70.89	3.52	0.00	3.53
Muktsar	65.76	70.49	3.47	1.26	4.73
Faridkot	67.40	70.37	3.51	–0.54	2.96
Bathinda	67.33	71.01	3.52	0.16	3.68
Mansa	66.64	69.72	3.47	–0.39	3.08
Patiala	67.67	71.28	3.54	0.07	3.61
Amritsar	68.08	71.32	3.55	–0.31	3.24
Tarn Taran	67.91	71.00	3.54	–0.45	3.09
Rupnagar	67.61	71.09	3.53	–0.05	3.49
Sahibzada Ajit Singh Nagar	68.32	72.04	3.57	0.15	3.72
Sangrur	66.68	70.21	3.49	0.04	3.53
Barnala	67.01	71.38	3.52	0.85	4.37
			Chandigarh		
Chandigarh					
			Uttarakhand		
Uttarkashi	65.25	68.43	3.40	–0.22	3.18
Chamoli	67.20	71.05	3.52	0.33	3.85
Rudraprayag	66.80	70.95	3.51	0.64	4.16
Tehri Garhwal	66.40	69.70	3.47	–0.17	3.30
Dehradun	66.71	70.46	3.49	0.26	3.74
Garhwal	67.42	70.92	3.52	–0.02	3.51
Pithoragarh	67.40	70.76	3.52	–0.16	3.36
Bageshwar	67.07	71.03	3.52	0.44	3.96
Almora	66.74	70.74	3.50	0.50	4.00

(continued)

Table 5.4 (continued)

State/Union Territory/ District	Life expectancy at birth (years)		Female-male difference in life expectancy at birth (years)		
	Male	Female	Common to all district	Specific to the district	Total
Champawat	65.86	68.84	3.43	–0.45	2.98
Nainital	66.58	70.10	3.48	0.04	3.51
Udham Singh Nagar	65.33	68.91	3.42	0.16	3.59
Hardwar	64.56	67.24	3.36	–0.68	2.68
Haryana					
Panchkula	67.53	71.25	3.53	0.19	3.72
Ambala	68.00	71.43	3.55	–0.12	3.42
Yamunanagar	67.30	70.87	3.52	0.05	3.58
Kurukshetra	67.38	70.61	3.51	–0.28	3.22
Kaithal	65.33	69.95	3.44	1.18	4.62
Karnal	65.62	69.54	3.44	0.48	3.93
Panipat	67.42	70.54	3.51	–0.39	3.12
Sonipat	67.88	71.06	3.54	–0.36	3.18
Jind	66.93	69.76	3.48	–0.65	2.83
Fatehabad	66.69	69.97	3.48	–0.20	3.28
Sirsa	67.39	70.42	3.51	–0.48	3.03
Hisar	67.29	70.28	3.50	–0.51	2.99
Bhiwani	67.44	70.66	3.52	–0.30	3.23
Rohtak	67.61	71.08	3.53	–0.06	3.47
Jhajjar	67.58	71.20	3.53	0.09	3.62
Mahendragarh	66.96	69.96	3.49	–0.49	3.00
Rewari	61.96	69.57	3.35	4.26	7.61
Gurgaon	68.17	71.52	3.56	–0.21	3.35
Mewat	64.65	66.72	3.34	–1.28	2.07
Faridabad	67.95	71.21	3.54	–0.28	3.26
Palwal	66.91	69.28	3.47	–1.10	2.37
National Capital Territory of Delhi					
North West	68.48	71.67	3.57	–0.38	3.18
North	68.67	72.08	3.58	–0.17	3.41
North East	67.99	71.60	3.55	0.06	3.62
East	68.16	71.69	3.56	–0.03	3.53
New Delhi	66.64	67.81	3.42	–2.25	1.16
Central	68.66	72.22	3.59	–0.03	3.56
West	67.96	71.30	3.55	–0.21	3.34

(continued)

Table 5.4 (continued)

State/Union Territory/ District	Life expectancy at birth (years)		Female-male difference in life expectancy at birth (years)		
	Male	Female	Common to all district	Specific to the district	Total
South West	67.23	69.93	3.49	–0.79	2.70
South	67.51	70.45	3.51	–0.57	2.94
Rajasthan					
Ganganagar	67.58	71.18	3.53	0.07	3.60
Hanumangarh	63.44	68.99	3.37	2.18	5.55
Bikaner	67.91	71.39	3.55	–0.07	3.48
Churu	67.73	71.12	3.54	–0.15	3.38
Jhunjhunun	68.04	71.48	3.55	–0.11	3.44
Alwar	67.40	70.77	3.52	–0.15	3.37
Bharatpur	67.53	70.20	3.51	–0.84	2.67
Dhaulpur	67.31	69.24	3.48	–1.55	1.93
Karauli	67.46	69.55	3.49	–1.40	2.10
Sawai Madhopur	67.34	69.89	3.49	–0.94	2.55
Dausa	66.89	69.53	3.47	–0.83	2.64
Jaipur	68.31	71.79	3.57	–0.09	3.49
Sikar	68.21	71.45	3.56	–0.32	3.24
Nagaur	67.17	70.08	3.49	–0.58	2.91
Jodhpur	67.85	70.93	3.53	–0.45	3.07
Jaisalmer	68.15	70.62	3.53	–1.06	2.47
Barmer	67.69	70.50	3.52	–0.71	2.81
Jalor	67.32	70.37	3.51	–0.46	3.04
Sirohi	66.53	69.82	3.47	–0.18	3.29
Pali	66.34	69.75	3.46	–0.05	3.40
Ajmer	66.18	69.70	3.46	0.06	3.52
Tonk	66.59	70.10	3.48	0.03	3.51
Bundi	67.03	70.52	3.50	–0.01	3.49
Bhilwara	65.44	69.33	3.43	0.46	3.89
Rajsamand	65.93	69.25	3.44	–0.12	3.32
Dungarpur	66.81	70.11	3.49	–0.19	3.29
Banswara	64.75	67.80	3.37	–0.32	3.05
Chittaurgarh	66.00	69.67	3.45	0.22	3.67
Kota	67.87	71.62	3.55	0.20	3.75
Baran	66.49	69.90	3.47	–0.06	3.41
Jhalawar	67.03	70.62	3.50	0.09	3.59

(continued)

Table 5.4 (continued)

State/Union Territory/ District	Life expectancy at birth (years)		Female-male difference in life expectancy at birth (years)		
	Male	Female	Common to all district	Specific to the district	Total
Udaipur	66.14	69.07	3.44	–0.51	2.93
Pratapgarh	65.86	69.45	3.44	0.15	3.59
			Uttar Pradesh		
Saharanpur	65.31	66.81	3.36	–1.86	1.51
Muzaffarnagar	65.09	67.21	3.37	–1.25	2.12
Bijnor	64.41	67.07	3.35	–0.69	2.66
Moradabad	63.92	66.14	3.31	–1.09	2.21
Rampur	64.84	67.43	3.37	–0.78	2.59
Jyotiba Phule Nagar	64.39	66.89	3.34	–0.84	2.49
Meerut	65.64	68.03	3.40	–1.01	2.39
Baghpat	65.93	68.74	3.43	–0.62	2.81
Ghaziabad	65.40	67.71	3.39	–1.08	2.31
Gautam Buddha Nagar	65.98	68.27	3.42	–1.13	2.29
Bulandshahr	64.60	66.68	3.34	–1.26	2.08
Aligarh	64.85	66.65	3.35	–1.55	1.80
Mahamaya Nagar	66.03	68.03	3.41	–1.41	2.00
Mathura	64.48	66.14	3.33	–1.67	1.66
Agra	66.02	67.65	3.40	–1.77	1.63
Firozabad	65.34	66.88	3.37	–1.83	1.54
Mainpuri	64.69	65.77	3.32	–2.24	1.08
Budaun	63.29	64.67	3.26	–1.88	1.38
Bareilly	64.19	65.92	3.31	–1.58	1.73
Pilibhit	64.58	66.07	3.33	–1.84	1.48
Shahjahanpur	64.07	65.84	3.31	–1.54	1.77
Kheri	63.84	65.56	3.29	–1.58	1.72
Sitapur	63.29	64.59	3.26	–1.96	1.30
Hardoi	63.42	64.87	3.27	–1.82	1.45
Unnao	64.13	67.06	3.34	–0.41	2.93
Lucknow	65.55	68.85	3.42	–0.12	3.30
Rae Bareli	63.94	67.52	3.35	0.23	3.57
Farrukhabad	65.09	66.69	3.36	–1.76	1.60
Kannauj	65.13	67.54	3.38	–0.97	2.41
Etawah	65.84	68.30	3.42	–0.96	2.46
Auraiya	65.44	68.44	3.41	–0.41	3.00

(continued)

Table 5.4 (continued)

State/Union Territory/ District	Life expectancy at birth (years)		Female-male difference in life expectancy at birth (years)		
	Male	Female	Common to all district	Specific to the district	Total
Kanpur Dehat	65.33	67.97	3.39	–0.75	2.65
Kanpur Nagar	65.98	69.06	3.44	–0.36	3.07
Jalaun	66.46	69.08	3.45	–0.83	2.62
Jhansi	65.70	68.98	3.43	–0.15	3.28
Lalitpur	63.92	66.02	3.31	–1.21	2.09
Hamirpur	65.62	67.82	3.40	–1.20	2.20
Mahoba	65.16	67.59	3.38	–0.95	2.43
Banda	65.10	67.02	3.36	–1.44	1.92
Chitrakoot	65.14	66.99	3.36	–1.51	1.85
Fatehpur	64.23	66.70	3.33	–0.86	2.47
Pratapgarh	65.25	68.34	3.40	–0.31	3.09
Kaushambi	62.53	65.39	3.26	–0.40	2.86
Allahabad	63.58	65.39	3.28	–1.47	1.81
Bara Banki	62.86	65.82	3.28	–0.32	2.96
Faizabad	64.53	67.69	3.37	–0.21	3.16
Ambedkar Nagar	65.16	68.45	3.40	–0.11	3.29
Sultanpur	65.34	68.53	3.41	–0.22	3.19
Bahraich	64.14	66.14	3.32	–1.32	2.01
Shrawasti	64.45	64.79	3.29	–2.95	0.34
Balrampur	64.47	66.41	3.33	–1.39	1.93
Gonda	65.42	67.74	3.39	–1.07	2.32
Siddharthnagar	64.43	67.25	3.35	–0.53	2.82
Basti	65.45	68.39	3.41	–0.47	2.94
Sant Kabir Nagar	65.65	68.69	3.42	–0.38	3.03
Mahrajganj	63.99	67.06	3.34	–0.27	3.08
Gorakhpur	66.16	69.40	3.45	–0.21	3.23
Kushinagar	64.04	67.64	3.35	0.25	3.60
Deoria	66.42	69.73	3.47	–0.16	3.30
Azamgarh	66.39	69.81	3.47	–0.05	3.42
Mau	65.39	68.71	3.41	–0.09	3.31
Ballia	66.17	69.43	3.45	–0.19	3.26
Jaunpur	65.21	68.22	3.40	–0.39	3.01
Ghazipur	65.03	68.09	3.39	–0.33	3.06
Chandauli	66.48	69.50	3.46	–0.44	3.02

(continued)

Table 5.4 (continued)

State/Union Territory/ District	Life expectancy at birth (years)		Female-male difference in life expectancy at birth (years)		
	Male	Female	Common to all district	Specific to the district	Total
Varanasi	65.35	68.19	3.40	–0.56	2.85
Sant Ravidas Nagar (Bhadohi)	64.22	66.09	3.32	–1.45	1.87
Mirzapur	64.18	66.19	3.32	–1.31	2.01
Sonbhadra	64.87	67.56	3.37	–0.68	2.69
Etah	65.09	65.76	3.33	–2.66	0.67
Kanshiram Nagar	63.44	65.55	3.28	–1.17	2.12
Bihar					
Pashchim Champaran	64.67	67.38	3.36	–0.65	2.71
Purba Champaran	64.95	66.90	3.36	–1.41	1.95
Sheohar	64.17	65.60	3.30	–1.87	1.43
Sitamarhi	64.76	66.04	3.33	–2.05	1.28
Madhubani	65.89	68.37	3.42	–0.94	2.48
Supaul	65.60	68.56	3.42	–0.46	2.96
Araria	63.93	66.32	3.32	–0.93	2.39
Kishanganj	62.85	66.17	3.28	0.04	3.32
Purnia	63.69	66.49	3.31	–0.51	2.80
Katihar	64.07	67.30	3.34	–0.11	3.23
Madhepura	65.74	68.39	3.42	–0.77	2.65
Saharsa	66.02	68.36	3.42	–1.08	2.34
Darbhanga	64.96	67.13	3.36	–1.19	2.17
Muzaffarpur	65.30	67.93	3.39	–0.76	2.63
Gopalganj	64.91	68.33	3.39	0.03	3.43
Siwan	65.83	69.07	3.43	–0.19	3.24
Saran	65.99	68.94	3.44	–0.49	2.95
Vaishali	65.69	68.31	3.41	–0.79	2.62
Samastipur	66.18	68.66	3.43	–0.95	2.47
Begusarai	65.97	68.33	3.42	–1.06	2.36
Khagaria	66.43	68.65	3.44	–1.22	2.23
Bhagalpur	66.59	69.39	3.46	–0.66	2.80
Banka	65.91	68.31	3.42	–1.02	2.40
Munger	66.36	69.05	3.45	–0.76	2.69
Lakhisarai	66.29	69.03	3.45	–0.71	2.74
Sheikhpura	65.37	67.93	3.39	–0.83	2.56

(continued)

Table 5.4 (continued)

State/Union Territory/ District	Life expectancy at birth (years)		Female-male difference in life expectancy at birth (years)		
	Male	Female	Common to all district	Specific to the district	Total
Nalanda	65.34	67.95	3.39	–0.78	2.61
Patna	64.96	67.39	3.37	–0.94	2.42
Bhojpur	65.48	67.91	3.40	–0.97	2.42
Buxar	64.71	67.30	3.36	–0.77	2.59
Kaimur (Bhabua)	63.61	66.24	3.31	–0.68	2.63
Rohtas	65.13	67.96	3.39	–0.56	2.83
Aurangabad	65.20	67.94	3.39	–0.65	2.74
Gaya	64.55	66.86	3.35	–1.04	2.30
Nawada	65.82	68.60	3.42	–0.64	2.78
Jamui	65.64	68.23	3.41	–0.82	2.59
Jehanabad	64.89	67.25	3.36	–1.00	2.36
Arwal	63.82	66.22	3.31	–0.91	2.40
Sikkim					
North District	69.70	73.16	3.64	–0.18	3.45
West District	69.72	73.45	3.64	0.08	3.73
South District	69.72	73.44	3.64	0.08	3.72
East District	69.99	73.45	3.65	–0.19	3.47
Arunachal Pradesh					
Tawang	68.84	72.23	3.59	–0.20	3.39
West Kameng	68.70	72.61	3.60	0.31	3.92
East Kameng	67.12	70.64	3.51	0.01	3.52
Papum Pare	69.38	72.97	3.62	–0.03	3.59
Upper Subansiri	68.12	71.55	3.56	–0.13	3.43
West Siang	69.83	72.81	3.63	–0.65	2.98
East Siang	69.89	73.04	3.64	–0.49	3.15
Upper Siang	69.38	73.05	3.63	0.04	3.67
Changlang	69.34	73.02	3.62	0.06	3.68
Tirap	68.93	72.09	3.59	–0.43	3.16
Lower Subansiri	69.43	73.01	3.63	–0.05	3.58
Kurung Kumey	67.28	70.54	3.51	–0.25	3.26
Dibang Valley	67.42	71.77	3.54	0.81	4.34
Lower Dibang Valley	69.12	72.64	3.61	–0.09	3.52
Lohit	69.31	72.71	3.62	–0.22	3.40
Anjaw	67.46	71.48	3.54	0.48	4.02

(continued)

Table 5.4 (continued)

State/Union Territory/ District	Life expectancy at birth (years)		Female-male difference in life expectancy at birth (years)		
	Male	Female	Common to all district	Specific to the district	Total
			Nagaland		
Mon	67.03	67.40	3.42	–3.05	0.37
Mokokchung	68.15	71.31	3.55	–0.39	3.16
Zunheboto	68.29	71.97	3.57	0.11	3.68
Wokha	67.92	71.59	3.55	0.12	3.66
Dimapur	67.78	71.67	3.55	0.34	3.89
Phek	67.73	71.09	3.53	–0.17	3.36
Tuensang	67.08	70.54	3.50	–0.04	3.46
Longleng	68.01	71.71	3.56	0.14	3.70
Kiphire	66.53	69.79	3.47	–0.21	3.26
Kohima	68.57	72.35	3.59	0.19	3.77
Peren	67.02	70.75	3.51	0.22	3.73
			Manipur		
Senapati	68.00	71.07	3.54	–0.47	3.07
Tamenglong	67.34	71.51	3.53	0.64	4.17
Churachandpur	67.79	71.56	3.55	0.22	3.77
Bishnupur	68.13	71.75	3.56	0.06	3.62
Thoubal	67.57	72.12	3.56	0.99	4.54
Imphal West	67.61	72.00	3.55	0.84	4.38
Imphal East	67.56	72.10	3.56	0.98	4.54
Ukhrul	67.19	70.48	3.51	–0.22	3.29
Chandel	66.44	71.07	3.50	1.13	4.64
			Mizoram		
Mamit	67.63	70.71	3.52	–0.44	3.08
Kolasib	68.38	72.05	3.58	0.09	3.67
Aizawl	69.13	72.97	3.62	0.22	3.83
Champhai	68.98	72.47	3.60	–0.11	3.49
Serchhip	68.91	72.76	3.61	0.24	3.85
Lunglei	68.07	71.51	3.55	–0.11	3.44
Lawngtlai	66.15	69.58	3.46	–0.03	3.43
			Tripura		
Saiha	67.45	71.81	3.55	0.81	4.36
West Tripura	66.99	71.04	3.51	0.54	4.05

(continued)

Table 5.4 (continued)

State/Union Territory/ District	Life expectancy at birth (years)		Female-male difference in life expectancy at birth (years)		
	Male	Female	Common to all district	Specific to the district	Total
South Tripura	66.81	69.44	3.47	–0.84	2.62
Dhalai	65.03	68.73	3.41	0.29	3.70
North Tripura	65.04	68.97	3.41	0.52	3.93
Meghalaya					
West Garo Hills	65.27	69.38	3.43	0.68	4.11
East Garo Hills	66.73	70.59	3.50	0.36	3.85
South Garo Hills	66.06	69.36	3.45	–0.15	3.30
West Khasi Hills	67.30	70.43	3.51	–0.38	3.13
Ribhoi	67.51	69.49	3.49	–1.51	1.99
East Khasi Hills	67.69	71.01	3.53	–0.21	3.32
Jaintia Hills	66.28	70.05	3.47	0.30	3.76
Assam					
Kokrajhar	65.88	69.06	3.44	–0.26	3.18
Dhubri	65.42	69.09	3.42	0.25	3.68
Goalpara	66.61	70.05	3.48	–0.04	3.44
Barpeta	66.97	69.98	3.49	–0.48	3.01
Morigaon	65.98	69.72	3.45	0.29	3.74
Nagaon	66.55	70.29	3.48	0.26	3.74
Sonitpur	66.95	70.20	3.49	–0.24	3.24
Lakhimpur	67.26	70.79	3.51	0.02	3.53
Dhemaji	67.78	71.16	3.54	–0.16	3.38
Tinsukia	67.84	71.38	3.54	0.00	3.53
Dibrugarh	67.79	71.68	3.55	0.34	3.89
Sivasagar	67.73	71.74	3.55	0.46	4.01
Jorhat	67.65	71.65	3.55	0.45	4.00
Golaghat	67.13	71.44	3.53	0.78	4.31
Karbi Anglong	65.87	69.64	3.45	0.32	3.76
Dima Hasao	67.30	71.08	3.52	0.26	3.78
Cachar	67.20	71.10	3.52	0.38	3.90
Karimganj	66.12	70.06	3.47	0.47	3.94
Hailakandi	65.67	70.13	3.46	1.00	4.45
Bongaigaon	66.55	70.60	3.49	0.56	4.05
Chirang	66.63	70.30	3.49	0.18	3.67
Kamrup	67.03	71.07	3.52	0.52	4.04

(continued)

Table 5.4 (continued)

State/Union Territory/ District	Life expectancy at birth (years)		Female-male difference in life expectancy at birth (years)		
	Male	Female	Common to all district	Specific to the district	Total
Kamrup Metropolitan	67.83	71.82	3.56	0.43	3.99
Nalbari	67.83	71.63	3.55	0.25	3.80
Baksa	66.80	70.62	3.50	0.32	3.83
Darrang	65.60	69.68	3.44	0.64	4.08
Udalguri	66.43	70.59	3.49	0.67	4.16
West Bengal					
Darjiling	68.54	72.22	3.58	0.10	3.68
Jalpaiguri	68.07	71.74	3.56	0.11	3.67
Koch Bihar	68.23	71.93	3.57	0.13	3.70
Uttar Dinajpur	67.49	71.34	3.53	0.32	3.84
Dakshin Dinajpur	67.75	71.57	3.55	0.27	3.82
Maldah	67.37	71.17	3.53	0.27	3.80
Murshidabad	67.75	71.42	3.54	0.13	3.67
Birbhum	68.05	71.80	3.56	0.19	3.75
Barddhaman	68.49	72.18	3.58	0.11	3.69
Nadia	68.53	72.40	3.59	0.28	3.87
North Twenty Four Parganas	68.46	72.12	3.58	0.08	3.66
Hugli	68.86	72.63	3.60	0.17	3.77
Bankura	68.92	72.45	3.60	–0.07	3.54
Puruliya	68.68	72.40	3.59	0.13	3.72
Haora	68.69	72.60	3.60	0.31	3.91
Kolkata	67.88	71.72	3.55	0.29	3.85
South Twenty Four Parganas	68.13	71.90	3.56	0.20	3.78
Paschim Medinipur	68.69	72.67	3.60	0.38	3.98
Purba Medinipur	68.55	72.16	3.58	0.03	3.61
Jharkhand					
Garhwa	65.05	68.36	3.40	–0.09	3.32
Chatra	65.15	68.84	3.41	0.28	3.69
Kodarma	67.05	70.80	3.51	0.24	3.75
Giridih	66.87	70.18	3.49	–0.18	3.31
Deoghar	66.97	70.14	3.49	–0.32	3.17
Godda	65.89	69.11	3.44	–0.22	3.23
Sahibganj	64.96	68.96	3.41	0.59	4.00

(continued)

Table 5.4 (continued)

State/Union Territory/ District	Life expectancy at birth (years)		Female-male difference in life expectancy at birth (years)		
	Male	Female	Common to all district	Specific to the district	Total
Pakur	63.40	68.37	3.35	1.62	4.97
Dhanbad	66.41	70.39	3.48	0.50	3.98
Bokaro	66.57	70.53	3.49	0.47	3.96
Lohardaga	65.76	70.22	3.46	1.00	4.46
Purbi Singhbhum	67.63	71.88	3.55	0.70	4.25
Palamu	65.76	69.47	3.44	0.27	3.71
Latehar	65.52	69.14	3.43	0.19	3.62
Hazaribagh	66.36	70.22	3.48	0.38	3.87
Ramgarh	66.82	70.49	3.50	0.17	3.67
Dumka	65.95	69.62	3.45	0.22	3.67
Jamtara	64.99	67.65	3.38	–0.72	2.66
Ranchi	66.72	70.53	3.49	0.32	3.81
Khunti	64.16	67.65	3.36	0.13	3.49
Gumla	65.52	69.41	3.43	0.45	3.90
Simdega	63.21	67.64	3.33	1.10	4.43
Pashchimi Singhbhum	64.05	67.44	3.35	0.04	3.39
Saraikela–Kharsawan	66.27	70.35	3.48	0.60	4.08
Odisha					
Bargarh	67.65	71.63	3.55	0.43	3.98
Jharsuguda	67.17	71.19	3.52	0.50	4.02
Sambalpur	67.14	70.72	3.51	0.07	3.58
Debagarh	66.22	69.58	3.46	–0.10	3.36
Sundargarh	66.50	70.57	3.49	0.58	4.07
Kendujhar	66.96	70.65	3.50	0.19	3.69
Mayurbhanj	67.46	71.14	3.53	0.15	3.69
Baleshwar	67.40	70.88	3.52	–0.04	3.48
Bhadrak	67.38	70.76	3.52	–0.14	3.37
Kendrapara	67.06	70.73	3.51	0.16	3.67
Jagatsinghapur	67.71	71.22	3.54	–0.03	3.51
Cuttack	67.30	70.68	3.51	–0.13	3.38
Jajapur	67.40	71.05	3.52	0.13	3.65
Dhenkanal	67.06	70.60	3.50	0.04	3.54
Anugul	66.36	69.93	3.47	0.10	3.58
Nayagarh	66.47	69.58	3.46	–0.35	3.11

(continued)

Table 5.4 (continued)

State/Union Territory/ District	Life expectancy at birth (years)		Female-male difference in life expectancy at birth (years)		
	Male	Female	Common to all district	Specific to the district	Total
Khordha	67.13	71.01	3.52	0.36	3.88
Puri	67.04	70.86	3.51	0.31	3.82
Ganjam	66.35	69.98	3.47	0.16	3.64
Gajapati	63.89	67.69	3.35	0.45	3.80
Kandhamal	62.57	66.77	3.29	0.91	4.20
Baudh	65.63	69.82	3.45	0.74	4.19
Subarnapur	67.40	70.99	3.52	0.07	3.60
Balangir	66.22	70.21	3.47	0.52	3.99
Nuapada	65.52	69.94	3.45	0.97	4.42
Kalahandi	64.61	69.26	3.41	1.24	4.65
Rayagada	64.28	67.88	3.36	0.24	3.60
Nabarangapur	64.59	68.74	3.39	0.76	4.15
Koraput	64.18	68.34	3.37	0.79	4.16
Malkangiri	64.12	67.94	3.36	0.46	3.82
Chhattisgarh					
Koriya	63.87	67.91	3.35	0.69	4.04
Surguja	65.37	69.13	3.42	0.34	3.77
Jashpur	65.04	68.71	3.41	0.26	3.68
Raigarh	65.77	70.04	3.46	0.81	4.27
Korba	65.18	69.57	3.43	0.96	4.39
Janjgir–Champa	66.07	70.20	3.47	0.66	4.13
Bilaspur	64.96	69.42	3.42	1.04	4.47
Kabeerdham	64.70	68.78	3.40	0.68	4.08
Rajnandgaon	64.26	68.34	3.38	0.70	4.07
Durg	65.89	70.25	3.47	0.89	4.36
Raipur	65.96	70.24	3.47	0.81	4.27
Mahasamund	63.48	67.97	3.35	1.14	4.48
Dhamtari	65.80	70.06	3.46	0.80	4.26
Uttar Bastar Kanker	65.31	69.78	3.44	1.03	4.47
Bastar	62.99	67.57	3.32	1.26	4.58
Narayanpur	64.09	67.32	3.35	–0.12	3.23
Dakshin Bastar Dantewada	62.85	66.60	3.30	0.45	3.74
Bijapur	63.20	67.14	3.32	0.62	3.94

(continued)

Table 5.4 (continued)

State/Union Territory/ District	Life expectancy at birth (years)		Female-male difference in life expectancy at birth (years)		
	Male	Female	Common to all district	Specific to the district	Total
		Madhya Pradesh			
Sheopur	64.06	66.62	3.33	–0.77	2.56
Morena	67.12	69.16	3.47	–1.43	2.04
Bhind	67.31	69.68	3.49	–1.12	2.38
Gwalior	66.11	69.44	3.45	–0.12	3.33
Datia	65.11	68.32	3.40	–0.19	3.21
Shivpuri	64.78	67.36	3.36	–0.78	2.58
Tikamgarh	65.71	68.53	3.42	–0.60	2.82
Chhatarpur	65.06	67.97	3.39	–0.48	2.91
Panna	63.63	66.73	3.32	–0.22	3.10
Sagar	65.29	68.78	3.41	0.08	3.49
Damoh	65.50	68.38	3.41	–0.53	2.88
Satna	64.43	67.47	3.36	–0.32	3.04
Rewa	65.85	69.16	3.44	–0.13	3.31
Umaria	63.39	66.74	3.31	0.04	3.35
Neemuch	66.39	70.29	3.48	0.42	3.90
Mandsaur	66.52	70.40	3.49	0.39	3.89
Ratlam	65.91	69.75	3.45	0.39	3.83
Ujjain	67.01	70.51	3.50	0.00	3.50
Shajapur	66.62	70.05	3.48	–0.05	3.43
Dewas	67.05	70.48	3.50	–0.07	3.43
Dhar	67.11	70.61	3.51	–0.01	3.49
Indore	67.32	71.15	3.53	0.30	3.84
Khargone (West Nimar)	66.64	70.59	3.49	0.46	3.95
Barwani	65.11	69.30	3.42	0.77	4.20
Rajgarh	65.79	69.23	3.44	0.00	3.43
Vidisha	65.36	68.47	3.41	–0.30	3.11
Bhopal	67.06	70.82	3.51	0.25	3.75
Sehore	65.51	69.26	3.43	0.32	3.75
Raisen	65.63	69.15	3.43	0.09	3.53
Betul	64.47	68.62	3.39	0.76	4.15
Harda	65.09	68.09	3.39	–0.39	2.99
Hoshangabad	65.80	69.45	3.44	0.21	3.65
Katni	63.08	67.02	3.31	0.63	3.93

(continued)

Table 5.4 (continued)

State/Union Territory/ District	Life expectancy at birth (years)		Female-male difference in life expectancy at birth (years)		
	Male	Female	Common to all district	Specific to the district	Total
Jabalpur	65.04	69.39	3.42	0.93	4.35
Narsimhapur	65.27	69.50	3.43	0.80	4.23
Dindori	64.36	68.26	3.38	0.52	3.90
Mandla	65.05	69.42	3.42	0.95	4.36
Chhindwara	65.28	69.43	3.43	0.72	4.15
Seoni	66.10	70.00	3.46	0.44	3.90
Balaghat	65.08	69.60	3.43	1.09	4.52
Guna	66.13	69.16	3.44	–0.41	3.03
Ashoknagar	65.12	68.03	3.39	–0.48	2.91
Shahdol	63.32	67.21	3.32	0.57	3.89
Anuppur	63.87	68.18	3.36	0.95	4.31
Sidhi	64.08	67.12	3.34	–0.30	3.04
Singrauli	62.81	66.27	3.29	0.17	3.45
Jhabua	64.71	68.38	3.39	0.28	3.66
Alirajpur	63.94	67.91	3.36	0.61	3.97
Khandwa (East Nimar)	66.05	69.45	3.45	–0.05	3.40
Burhanpur	67.21	70.51	3.51	–0.21	3.29
Gujarat					
Kachchh	67.27	70.77	3.51	–0.01	3.51
Banas Kantha	67.26	70.60	3.51	–0.17	3.34
Patan	66.85	70.08	3.49	–0.26	3.23
Mahesana	66.21	70.79	3.49	1.09	4.58
Sabar Kantha	66.37	70.30	3.48	0.45	3.93
Gandhinagar	66.83	69.96	3.48	–0.35	3.13
Ahmedabad	67.67	70.45	3.52	–0.74	2.78
Surendranagar	67.98	71.44	3.55	–0.09	3.46
Rajkot	67.15	70.89	3.51	0.23	3.75
Jamnagar	67.40	71.12	3.53	0.19	3.72
Porbandar	67.21	71.16	3.52	0.43	3.95
Junagadh	67.20	71.02	3.52	0.30	3.82
Amreli	67.49	71.01	3.53	–0.01	3.51
Bhavnagar	68.02	71.22	3.55	–0.35	3.20
Anand	66.04	69.69	3.46	0.19	3.65
Kheda	65.93	69.65	3.45	0.27	3.71

(continued)

Table 5.4 (continued)

State/Union Territory/ District	Life expectancy at birth (years)		Female-male difference in life expectancy at birth (years)		
	Male	Female	Common to all district	Specific to the district	Total
Panch Mahals	66.36	70.25	3.48	0.41	3.89
Dahod	65.81	69.31	3.44	0.06	3.50
Vadodara	66.59	70.28	3.48	0.21	3.69
Narmada	66.14	70.30	3.47	0.69	4.16
Bharuch	66.33	70.52	3.48	0.71	4.19
The Dangs	66.50	70.49	3.49	0.50	4.00
Navsari	67.10	71.11	3.52	0.49	4.01
Valsad	67.74	71.20	3.54	–0.08	3.46
Surat	67.91	71.14	3.54	–0.31	3.23
Tapi	66.84	70.49	3.50	0.15	3.64
Daman & Diu					
Diu	67.72	71.39	3.54	0.13	3.67
Daman	68.05	72.31	3.57	0.69	4.26
Dadra & Nagar Haveli					
Dadra & Nagar Haveli	66.13	70.11	3.47	0.51	3.98
Maharashtra					
Nandurbar	67.27	70.85	3.52	0.06	3.58
Dhule	67.56	71.01	3.53	–0.08	3.45
Jalgaon	67.73	71.22	3.54	–0.05	3.49
Buldana	67.81	71.34	3.54	–0.01	3.54
Akola	67.56	71.53	3.54	0.43	3.97
Washim	68.02	71.74	3.56	0.16	3.72
Amravati	68.05	72.04	3.57	0.42	3.99
Wardha	68.27	72.51	3.58	0.66	4.24
Nagpur	67.82	71.56	3.55	0.19	3.74
Bhandara	67.47	71.21	3.53	0.21	3.74
Gondiya	65.98	70.55	3.48	1.09	4.57
Gadchiroli	66.50	70.53	3.49	0.54	4.03
Chandrapur	67.07	71.23	3.52	0.64	4.16
Yavatmal	67.19	71.05	3.52	0.34	3.86
Nanded	67.82	71.47	3.55	0.10	3.65
Hingoli	67.82	71.45	3.55	0.08	3.63
Parbhani	68.01	71.64	3.56	0.07	3.62
Jalna	67.85	71.29	3.54	–0.10	3.44

(continued)

Table 5.4 (continued)

State/Union Territory/ District	Life expectancy at birth (years)		Female-male difference in life expectancy at birth (years)		
	Male	Female	Common to all district	Specific to the district	Total
Aurangabad	67.93	71.70	3.55	0.22	3.77
Nashik	67.88	71.68	3.55	0.25	3.80
Thane	67.82	71.50	3.55	0.13	3.68
Mumbai Suburban	68.34	72.09	3.58	0.17	3.75
Mumbai	67.90	71.98	3.56	0.52	4.08
Raigarh	68.16	71.50	3.56	–0.22	3.33
Pune	68.64	72.39	3.59	0.16	3.75
Ahmadnagar	68.32	72.16	3.58	0.26	3.84
Bid	68.66	72.05	3.58	–0.19	3.39
Latur	67.52	71.59	3.54	0.53	4.07
Osmanabad	68.23	71.87	3.57	0.07	3.64
Solapur	68.33	71.96	3.57	0.06	3.63
Satara	68.56	72.21	3.58	0.07	3.65
Ratnagiri	69.38	73.15	3.63	0.14	3.78
Sindhudurg	68.08	72.08	3.57	0.43	4.00
Kolhapur	68.72	72.25	3.59	–0.06	3.53
Sangli	68.45	72.21	3.58	0.18	3.77
		Andhra Pradesh			
Adilabad	67.56	71.36	3.54	0.26	3.80
Nizamabad	67.61	71.58	3.54	0.43	3.97
Karimnagar	68.79	72.48	3.60	0.09	3.69
Medak	68.00	71.89	3.56	0.33	3.89
Hyderabad	68.05	71.64	3.56	0.03	3.59
Rangareddy	67.92	71.74	3.56	0.26	3.81
Mahbubnagar	66.87	70.78	3.50	0.41	3.91
Nalgonda	67.85	71.57	3.55	0.17	3.72
Warangal	68.10	71.89	3.56	0.23	3.80
Khammam	67.41	71.63	3.54	0.68	4.22
Srikakulam	66.52	70.29	3.48	0.29	3.77
Vizianagaram	64.94	69.24	3.42	0.88	4.31
Visakhapatnam	66.23	70.46	3.48	0.75	4.23
East Godavari	67.65	71.61	3.55	0.41	3.96
West Godavari	67.79	71.65	3.55	0.31	3.86
Krishna	66.20	70.67	3.48	0.99	4.48

(continued)

Table 5.4 (continued)

State/Union Territory/ District	Life expectancy at birth (years)		Female-male difference in life expectancy at birth (years)		
	Male	Female	Common to all district	Specific to the district	Total
Guntur	68.09	71.90	3.56	0.25	3.81
Prakasam	67.94	71.79	3.56	0.29	3.85
Sri Potti Sriramulu Nellore	68.18	71.95	3.57	0.20	3.76
Y.S.R.	67.84	71.68	3.55	0.29	3.85
Kurnool	67.04	70.75	3.51	0.20	3.71
Anantapur	65.46	69.27	3.43	0.38	3.81
Chittoor	66.67	70.98	3.50	0.81	4.31
		Karnataka			
Belgaum	68.12	71.64	3.56	–0.04	3.52
Bagalkot	67.29	71.16	3.52	0.35	3.87
Bijapur	68.21	71.60	3.56	–0.17	3.39
Bidar	68.67	72.25	3.59	–0.01	3.58
Raichur	67.50	71.02	3.53	–0.01	3.52
Koppal	66.67	70.69	3.50	0.52	4.02
Gadag	67.19	70.97	3.52	0.26	3.78
Dharwad	68.05	71.93	3.56	0.32	3.88
Uttara Kannada	68.55	71.87	3.58	–0.26	3.32
Haveri	68.03	71.67	3.56	0.08	3.64
Bellary	66.85	70.75	3.50	0.40	3.90
Chitradurga	67.48	71.47	3.54	0.45	3.99
Davanagere	67.56	71.65	3.54	0.55	4.09
Shimoga	67.63	71.59	3.54	0.42	3.95
Udupi	68.15	71.53	3.56	–0.18	3.38
Chikmagalur	66.52	71.28	3.51	1.25	4.76
Tumkur	67.51	71.27	3.53	0.23	3.76
Bangalore	68.11	71.79	3.56	0.12	3.68
Mandya	67.90	71.59	3.55	0.1A.4	3.69
Hassan	67.77	71.62	3.55	0.30	3.85
Dakshina Kannada	68.90	72.30	3.59	–0.20	3.39
Kodagu	68.31	72.24	3.58	0.35	3.94
Mysore	67.80	71.57	3.55	0.22	3.77
Chamarajanagar	67.85	71.26	3.54	–0.13	3.41
Gulbarga	68.15	71.55	3.56	–0.16	3.40
Yadgir	67.45	71.28	3.53	0.30	3.83

(continued)

Table 5.4 (continued)

State/Union Territory/ District	Life expectancy at birth (years)		Female-male difference in life expectancy at birth (years)		
	Male	Female	Common to all district	Specific to the district	Total
Kolar	67.92	71.83	3.56	0.35	3.90
Chikkaballapura	67.58	71.50	3.54	0.38	3.93
Bangalore Rural	67.68	71.94	3.55	0.71	4.26
Ramanagara	68.42	72.29	3.58	0.29	3.88
Goa					
North Goa	69.91	73.47	3.65	–0.09	3.56
South Goa	69.69	73.56	3.65	0.22	3.86
Kerala					
Kasaragod	70.53	73.85	3.68	–0.36	3.31
Kannur	70.56	73.80	3.68	–0.44	3.24
Wayanad	70.36	73.97	3.67	–0.06	3.61
Kozhikode	70.16	73.87	3.67	0.04	3.71
Malappuram	70.51	74.01	3.68	–0.18	3.51
Palakkad	70.48	73.81	3.67	–0.34	3.33
Thrissur	70.47	73.88	3.68	–0.27	3.42
Ernakulam	70.52	74.00	3.68	–0.20	3.48
Idukki	70.40	74.03	3.68	–0.05	3.62
Kottayam	70.46	73.91	3.68	–0.23	3.45
Alappuzha	70.40	73.84	3.67	–0.23	3.45
Pathanamthitta	70.44	73.93	3.68	–0.19	3.49
Kollam	70.40	73.97	3.68	–0.11	3.57
Thiruvananthapuram	70.35	73.85	3.67	–0.17	3.50
Lakshadweep					
Lakshadweep	68.26	71.12	3.55	–0.69	2.87
Tamil Nadu					
Thiruvallur	68.85	72.58	3.60	0.13	3.73
Chennai	69.04	73.02	3.62	0.36	3.99
Kancheepuram	68.73	72.62	3.60	0.29	3.89
Vellore	68.43	72.28	3.58	0.27	3.84
Tiruvannamalai	68.34	72.24	3.58	0.32	3.89
Viluppuram	68.47	72.47	3.59	0.41	4.00
Salem	68.34	72.21	3.58	0.29	3.86
Namakkal	68.53	72.23	3.58	0.12	3.70
Erode	68.51	72.49	3.59	0.39	3.98

(continued)

Table 5.4 (continued)

State/Union Territory/ District	Life expectancy at birth (years)		Female-male difference in life expectancy at birth (years)		
	Male	Female	Common to all district	Specific to the district	Total
The Nilgiris	68.77	72.87	3.61	0.49	4.10
Dindigul	65.56	71.84	3.50	2.78	6.29
Karur	68.77	72.40	3.59	0.04	3.63
Tiruchirappalli	68.66	72.65	3.60	0.39	3.99
Perambalur	65.19	71.57	3.48	2.90	6.38
Ariyalur	62.27	71.35	3.40	5.68	9.09
Cuddalore	68.93	72.52	3.60	–0.01	3.59
Nagapattinam	68.86	72.75	3.61	0.28	3.90
Thiruvarur	67.84	72.66	3.58	1.24	4.82
Thanjavur	68.72	72.68	3.60	0.36	3.96
Pudukkottai	68.91	72.74	3.61	0.22	3.83
Sivaganga	68.58	72.66	3.60	0.48	4.07
Madurai	68.70	72.38	3.59	0.09	3.68
Theni	67.15	71.79	3.54	1.10	4.65
Virudhunagar	67.91	72.28	3.57	0.80	4.37
Ramanathapuram	68.88	72.58	3.60	0.10	3.69
Thoothukkudi	68.87	72.55	3.60	0.08	3.68
Tirunelveli	68.56	72.49	3.59	0.34	3.93
Kanniyakumari	69.47	73.23	3.63	0.13	3.77
Dharmapuri	66.41	71.64	3.51	1.72	5.22
Krishnagiri	66.93	71.90	3.53	1.44	4.96
Coimbatore	69.11	72.96	3.62	0.23	3.85
Tiruppur	68.78	72.62	3.60	0.24	3.83
Puducherry					
Yanam	69.88	73.87	3.66	0.33	3.99
Puducherry	70.38	74.02	3.68	–0.04	3.63
Mahe	70.79	74.19	3.69	–0.29	3.40
Karaikal	70.71	74.20	3.69	–0.20	3.49
Andaman & Nicobar Islands					
Nicobar	65.85	70.01	3.46	0.70	4.16
North & Middle Andaman	68.48	72.41	3.59	0.34	3.93
South Andaman	67.88	72.88	3.58	1.42	5.00

Source Author

References

Bhatia R (2008) The logarithmic mean. Resonance 13(6):583–594

Carlson BC (1972) The logarithmic mean. Am Math Mon 79(6):615–618

Chaurasia AR (2021) Child Mortality in District of India 2019–2021. MLC Foundation and Shyam Institute, Bhopal, Madhya Pradesh, India

Chaurasia AR (2023) A non-parametric approach to the construction of life tables at the local level with application to Andhra Pradesh, India. https://doi.org/10.2139/ssrn.4377361

Chaurasia AR (2024) Life Tables for 640 Districts of India. Bhopal, Mewalal Chaurasia Foundation and 'Shyam' Institute.

Government of India (1969) The Registration of Birth and Death Act 1969, New Delhi

Government of India (2013) National Health Mission. Ministry of Health and Family Welfare, New Delhi

Government of India (2022a) SRS Based Abridged Life Tables 2016–2020. Office of the Registrar General and Census Commissioner, India, New Delhi

Government of India (2022b) SRS Bulletin 55(1). New Delhi, Office of the Registrar General, India. Vital Statistics Division

Government of India (2022c) National Family Health Survey (NFHS-5), 2019–21. India Report. New Delhi, Ministry of Health and Family Welfare.

Government of India (2022d) Report on Medical Certification of Cause of Death 2020. New Delhi, Ministry of Home Affairs. Registrar General of India, Vital Statistics Division.

Government of India (2023) The Registration of Birth and Death (Amendment) Act 2023. New Delhi, Ministry of Home Affairs

Kesarwani R (2015) Estimation of life expectancy from infant mortality rate at districts level. Int Res J Soc Sci 4(6):52–63

Maultrie T, Doringtin R, Hill A, Hill K, Timaeus I, Zaba B (2013) Tools for Demographic Estimation. International Union for the Scientific Study of Population, Paris

Namboodiri K, Suchindran CM (1987) Life Table Techniques and Their Applications. Academic Press, New York

OECD (2008) Handbook on Constructing Composite Indicators: Methodology and User Guide. Organization of Economic Cooperation and Development, Paris

Ponnapalli KM (2011) Construction of abridged life tables for districts of India: Using UNCOMBIN of MORTPAK-LITE. Paper presented in the 2011 Annual Meeting of the Population Association of America, Washington DC.

Saikia P, Borah M (2016) Construction of complete life tables for 23 districts of Assam for both male and female. Int J Sci Res 5(12):1–11

Singh R (2024) Review of civil registration system in Madhya Pradesh, India. Indian J Popul Dev 3(2):297–318

United Nations (1983) Manual X: Indirect Techniques of Demographic Estimation. Department of Economic and Social Affairs, New York

United Nations (2013) MORTPAK. The *United Nations Software Package for Mortality Measurement*. New York, United Nations, Population Division.

United Nations (2022) Technical Notes. United Nations, Department of Economic and Social Affairs. Population Division. https://hdr.undp.org/sites/default/files/2021-22_HDR/hdr2021-22_technical_notes.pdf

Chapter 6
Variation in Human Development

Abstract This chapter measures the level of human development across 707 districts of India as they existed at the time of National Family Health Survey 2019–2022. The chapter uses an alternative approach to measure progress in the three core dimensions of human development that is different from the conventional approach. The alternative approach adopted in this chapter addresses some of the problems associated with the conventional human development index developed and used by the United Nations. The chapter highlights stark variation in the progress in human development across the districts. The analysis also reveals that there are 78 districts in the country where progress in different dimensions of human development is starkly different. The analysis also reveals that in 231 districts of the country, progress is below the average in all the three core dimensions of human development. The chapter emphasises the need of measuring and monitoring human development in the districts on a regular basis to imbibe human development sensitiveness in district development planning and programming process. This chapter is a modified version of an earlier paper (Chaurasia 2023).

Keywords India · Districts · Human development · Surface measure

Introduction

Human development movement in India has never been strong. The lopsided attention accorded to human development in the development discourse of the country is reflected from the fact that since 1990, India could produce only two national level human development reports (Government of India 2002, 2011). India launched its first national human development report in 2001, more than a decade after the United Nations launched the first human development report in 1990 (United Nations 1990). The second national level human development report in India could be launched in 2011, ten years after the launch of the first report. It is now more than 15 years since the second report was launched but there is little indication that India will be launching any report to highlight the human development progress in the country.

A. R. Chaurasia, *Population and Human Development in Districts of India*,
https://doi.org/10.1007/978-981-95-8169-6_6

This is so when India ranks 132 out of 192 countries and territories in 2021–2022 in terms of the progress in human development, down from 130 in 2000 according to the assessment made by the United Nations (2022).

An integral feature of the human development reports prepared annually by the United Nations and human development reports prepared by different countries is the human development index (HDI), a simple, yet straightforward and widely appealing, indicator to measure human progress. HDI is now used universally to monitor human progress. Although HDI has gone through its evolutionary ups and downs since its introduction in 1990, yet the core dimensions of the index have remained unchanged (Kovacevic 2010). United Nations ranks human progress in its member-countries based on HDI—the higher the HDI in a country, the more advanced the human progress and vice versa. A high rank in HDI is now globally used as a means of national aggrandisement whereas a low rank in HDI is commonly used to highlight national insufficiencies (Chaurasia 2013). Estimates prepared by the United Nations suggest that HDI in India increased from 0.434 in 1990 to 0.633 in 2022 but the rank of the country vis-à-vis other countries of the world has decreased in terms of HDI.

Within India, all evidence suggests that human progress varies widely across constituent states/Union Territories and districts. However, estimates of HDI for states and Union Territories are not available on an annual basis to chart human progress. One reason is that estimation of HDI for the country and for its constituent states and Union Territories on an annual basis, as is done by the United Nations Development Programme, has not been institutionalised in the public and development administration system of the country. The second reason is that there has been little attempt to develop a data system that can generate the data required to calculate HDI annually for the country, for its states and Union Territories and possibly districts. The Government of India had, in the past, launched two projects—Strengthening State Plans for Human Development (Government of India 2010) and Human Development towards Bridging Inequalities (Government of India 2015)—to give human development orientation to development planning and programming in the country. Under these projects, most of the states and Union Territories prepared state/Union Territory-specific human development reports and estimated HDI for the constituent districts of the state/Union Territory. However, the exercise remained ad-hoc in nature. There is no state/Union Territory which prepared its human development report annually or at regular interval and estimated HDI to chart the variation in human progress across districts. A major limitation of this exercise has been that a standardised approach has not been adopted for the estimation of HDI by different states/Union Territories so that HDI estimated by different states/Union Territories is not strictly comparable. Different States and Union Territories have used different set of indicators for the construction of HDI. Even today, estimates of HDI based on a common set of indicators and common methodology are not available for all states and Union Territories and districts of the country to provide a pan India scenario of human progress that can provide human development orientation to development planning and programming in the country. Estimation of HDI at the district level is particularly relevant in this context as it helps in identifying hotspot districts in terms of human progress.

This chapter analyses progress in human development in the districts of India using an alternative approach which is different from the approach adopted by United Nations, although the core elements of the two approaches remain the same—living standard, education and health. The approach adopted here attempts to address some of the weaknesses of the approach adopted by the United Nations which has been subject to criticism from different perspectives. The progress in the three core dimensions of human development has been measured in terms of proportionate achievement rather than the average achievement as is followed by the United Nations and the concept of human development surface has been used to combine the progress in the three core dimensions of human development into a single composite index of human development.

Surface Measure of Human Development

The Human Development Index (HDI) adopted and used by the United Nations is a summary measure of average achievement in the three core dimensions of human development: health, education and standard of living. The HDI is the geometric mean of normalised indices of these three core dimensions of human development. The health dimension of human development is measured in terms of the health index which is based on the life expectancy at birth, the education dimension is measured in terms of the education index which is based on the mean number of years of schooling for adults aged 25 years and above, and expected years of schooling for children of school going age, and the standard of living dimension is measured in terms of income index which is based on the gross national income per capita adjusted for purchasing power parity. The three indexes range between 0 and 1 so that the HDI ranges between 0 and 1—the higher the HDI the better the human development. Technical and computational details of the construction of HDI are given elsewhere (United Nations 2022).

The HDI has been criticised on the grounds of both its conceptual foundations and its method of construction. A detailed review of the limitations of HDI is given elsewhere and is not repeated here (Kovacevic 2010; Klugman et al. 2011). The first concern about HDI is the selection of the aggregation function used to combine the three indexes reflecting the progress in the three core dimensions of human development. In its original formulation, United Nations had used the simple arithmetic mean as the aggregation function. However, since 2010, United Nations uses geometric mean as the aggregation function which embodies imperfect substitutability across the three dimensions of human development (United Nations 2010). Anand (2018) has recently raised concerns about the appropriateness of geometric mean as the aggregation function to construct the HDI and has recommended that simple arithmetic mean should be used in place of geometric mean for constructing HDI.

The second limitation of HDI relates to the implied trade-offs across the three core dimensions of human development. The magnitude of this trade-offs depends upon the aggregation function used. It is well known that the three indexes reflecting the progress in the three core dimensions of human development are highly correlated and, therefore, it is argued that HDI may not reveal more than what is revealed by its individual dimensions (Ghislandi et al. 2019). Finally, the third key limitation of HDI is that it gives equal weight to the progress in the three core dimensions of human development. It may be argued from the perspective of progress in human development that relatively more weight should be given to that dimension of human development in the construction of HDI in which the progress is lagging compared to that dimension in which the progress is advanced.

The indicators used to construct HDI by the United Nations are, by definition, average indicators. One requirement for the statistical validity of the average or the simple arithmetic mean to describe the underlying distribution, along with the standard deviation, is that the underlying distribution must be normally distributed. If the underlying distribution is not statistically normally distributed, then the average or the simple arithmetic mean and standard deviation do not truly describe the underlying distribution. In situations where the underlying distribution is not statistically normally distributed, interpretation of simple arithmetic mean and standard deviation is difficult and robust measures of central tendency and dispersion are commonly used to describe the distribution which is not statistically normal. It is well known that the distribution of the population by standard of living, level of education and life expectancy at birth is not normally distributed. As such construction of HDI using the average indicators is statistically imperfect. The use of averages indicators also requires goalposts to be fixed to normalise indicators so that they range between 0 (minimum) and 1 (maximum). The setting of the goalposts is, however, arbitrary. The United Nations fixes goalposts for per capita income as 100 (minimum) and 75,000 (maximum) at 2011 PPP$. For the expected years of schooling, the goalposts are 0 years (minimum) and 18 years (maximum) and for the mean years of schooling, the goalposts are 0 years (minimum) and 15 years (maximum). Finally for the life expectancy at birth, the goalposts are 20 years (minimum) and 85 years (maximum) (United Nations 2016). However, there are countries where prevailing values of the three indicators exceed the upper goalpost. In such a situation, the value of the indicator is set to be equal to the upper goalpost which is again an arbitrary procedure. Fixing the goalposts means that the index of progress is measured relative to the goalposts. If the goalposts are changed the index of progress will change. This means that the HDI measures human progress in relative terms and not in absolute terms.

The problems associated with average indicators can be addressed by using proportion indicators. The advantage of using proportion indicators is that a proportion always ranges between 0 and 1, so that there is no need to fix goal posts to construct the indexes required for the construction of HDI. Another advantage of using proportion indicators is that the index based on proportion indicators can measure progress in both absolute and relative terms whereas the progress based on average indicators can be measured in relative terms, in relation to the goalposts fixed.

The present chapter uses the following three proportion indicators which measure, respectively, the standard of living, the level of education and the state of health to construct an alternative index of human development:

1. The standard of living is measured in terms of the proportion of households having the wealth index equal to or more than the second quintile of the inter-household distribution of wealth index (*SL*). The wealth index is a composite measure of the cumulative living standard of a household. It is calculated using data on household ownership of selected assets; materials used for house construction; and types of water access and sanitation facilities (Rutstein and Johnson 2004).
2. The level of education is measured in terms of the secondary school net attendance ratio (*ED*) which is defined as the proportion of secondary school age children attending a secondary school (Croft et al. 2018).
3. The state of health is measured in terms of the probability of survival in the first five years of life (*HE*).

On the other hand, in view of the limitations of the simple arithmetic mean and the geometric mean as the aggregation function in the construction of HDI, Chaurasia (2022) has followed the human development surface approach to construct a surface measure of human development. The human development surface approach considers progress in the three dimensions of human development simultaneously, not independent of each other. In other words, the human development surface approach takes into consideration the correlation in progress that exists among the different dimensions of human development.

Following the human development surface approach proposed by Chaurasia (2022), the surface measure of human development, HDS, is defined as

$$\text{HDS} = \frac{\sqrt{SL * ED} + \sqrt{ED * HE} + \sqrt{HE * SL}}{3}$$

It is obvious that HDS does not treat the three dimensions of human development independent of each other. It considers the correlation between different dimensions of human development in measuring the human progress. The rationale behind the construction of HDS has been discussed in detail by Chaurasia (2022).

It can be shown that HDS is the maximum when $SL = ED = HE$. In this case HDS is nothing but the simple arithmetic mean of *SL*, *ED* and *HE* or

$$\text{HDS}_{\text{m}} = \frac{SL + ED + HE}{3}$$

When $SL \neq ED \neq HE$, HDS is always less than HDS_{m} so that the difference between HDS_{m} and HDS measures the inequality in progress in the three dimensions of human development. When the progress in the three dimensions of human development is the same, $\text{HDS} = \text{HDS}_{\text{m}}$ irrespective of the level of progress and the larger the difference the larger the inequality in progress in different dimensions of human development.

Data Source

Data for the construction of HDS for the districts of India are derived from the fifth round of the National Family Health Survey (NFHS) 2019–2021 (Government of India 2022). The NFHS programme was instituted by the Government of India, Ministry of Health and Family Welfare in the year 1992 and five rounds of NFHS have so far been carried out. The first round of the survey provided data for the constituent states of the country but not for the Union Territories and districts. The second and third rounds of the survey, provided data for all states and Union Territories but not for districts. The fourth and the fifth rounds of the survey, have provided data for all states, Union Territories and districts as they existed at the time of the survey which permit measurement of human progress at the district level. The NFHS is the only source of data in India which allows estimation of HDS at the district level.

The NFHS is a household survey of a statistically representative sample of households selected through a multistage sampling procedure. It covers the entire country and is designed in such a way that it provides reliable estimates of key indicators of population, health, family planning and development for all the districts of the country as they existed at the time of the survey. The fifth round of NFHS, conducted during the period 2019–2021, covered 636,699 households from 707 districts of the country as they existed at the time of the survey. Technical details of the survey and its organisation including the sampling design are given elsewhere (Government of India 2022).

Estimates of the proportion of households with wealth index equal to or more than the second quintile of the inter-household distribution of wealth index (*SL*) and the secondary school net attendance ratio (*ED*) have been derived for each district of the country using the SPSS software package. Estimates of the probability of survival in the first five years of life (*HE*) have been derived from the full birth history data available through the NFHS. The CMRJack software package (Pedersen and Liu 2012) has been used for the purpose. Estimates of the probability of death or the probability of survival during the first five years of life, estimated from the NFHS data, reflect the average survival experience in the first five years of life during the three years prior to the survey or approximately during the period 2016–2018.

Summary measures of the inter-district variation in the three indicators used to construct HDS are presented in Table 6.1. The inter-district variation is the largest in *SL* but the narrowest in *HE*. The indicator *HE* ranges from just around 8 per cent in district Bahraich of Uttar Pradesh which is the lowest in the country to 100 per cent in district Mahe of Puducherry. Similarly, the secondary school net attendance ratio (*ED*) is estimated to be less than 40 per cent in district Bahraich of Uttar Pradesh which is also the lowest in the country whereas this ratio is 100 per cent in 19 districts. Finally, the probability of survival in the first five years of life (*HE*) is estimated to be only around 0.883 in district Koriya of Chhattisgarh, the lowest in the country, whereas there are 36 districts where the probability of survival in the first five years of life is estimated to be 100 per cent. District Mahe in Puducherry is the only district in the country where all the three indicators that constitute the

HDS are equal to 1. The standard of living, measured in terms of the indicator *SL*, is estimated to be very low ($SL < 0.300$) in 308 districts of the country but very high ($SL \geq 0.950$) in only 36 districts. The level of education measured in terms of the indicator *ED* is estimated to be very low ($ED < 0750$) in 146 districts but very high ($ED \geq 0.950$) in only 78 districts. Finally, the state of health, measured in terms of the indicator *HE* is estimated to be very low ($HE < 0.950$) in 177 districts but very high ($HE \geq 0.990$) in only 76 districts of the country. The shape of the inter-district distribution of the three indicators has also been found to be different. The indicator *SL* is nearly statistically normally distributed across the districts whereas the inter-district distribution of indicators *ED* and *HE* is found to be negatively skewed. The inter-distribution of *SL* is platykurtic whereas that of *ED* and *HE* are leptokurtic.

Estimates of HDS for each of the 707 districts of the country are presented in the appendix Table 6.7 to this chapter along with estimates of the indexes *SL*, *ED* and *HE*. Human development in a district may be termed as very low if HDS < 0.500; low if 0.500 ≤ HDS < 0.700; average if 0.700 ≤ HDS < 0.850; high if 0.850 ≤ HDS < 0.950; and very high if HDS ≥ 0.950. If the HDS is an indication, then

Table 6.1 Inter-district variation in the three indicators used for the construction of HDS, 2019–2021

Summary measure	Proportion of households having wealth index equal to or more than second wealth index quintiles (*SL*)		Secondary school net attendance ratio (*ED*)		Probability of survival in the first five years of life (*HE*)	
Frequencies						
Very low	<0.500	293	<0.700	146	<0.950	177
Low	$0.500 \leq SL < 0.750$	160	$0.700 \leq ED < 0.800$	210	$0.950 \leq HE < 0.970$	238
Medium	$0.750 \leq SL < 0.850$	122	$0.800 \leq ED < 0.900$	142	$0.970 \leq HE < 0.980$	126
High	$0.850 \leq SL < 0.950$	86	$0.900 \leq ED < 0.950$	131	$0.980 \leq HE < 0.990$	90
Very high	≥0.950	46	≥0.950	78	≥0.990	76
Summary measures of distribution						
Minimum	0.081		0.395		0.883	
First quartile	0.359		0.773		0.950	
Median	0.570		0.849		0.965	
Third quartile	0.812		0.908		0.979	
Maximum	1.000		1.000		1.000	
IQR	0.452		0.135		0.029	
Skewness	−0.047		−0.817		−0.582	
Excess kurtosis	−1.211		0.577		0.420	
N	707		707		707	

Source Author

human development is relatively the poorest in district Supaul in Bihar (HDS = 0.430) against the minimum possible value of 0 but the highest in district Mahe in Puducherry (HDS = 1.000). In district Supaul of Bihar, household wealth index was either equal to or more than the second quintile of the inter-household distribution of the wealth index in only about 8 per cent households according to NFHS and this proportion is the lowest in the country. Similarly, the secondary school net attendance ratio in this district was less than 64 per cent while the probability of survival in the first five years of life was less than 0.957. By contrast, there is no household in district Mahe of Puducherry which has the wealth index less than the second quintile of the inter-household distribution of wealth index; the secondary school net attendance ratio was 100 per cent and there was no death in children below five years of age in the district during the three years before the survey so that the probability of survival in the first five years of life in the districts was 100 per cent. District Mahe is the only district in the country in which HDS = 1.

It may also be seen from Table 6.2 that in 20 (2.8 per cent) districts of the country, HDS < 0.500 which suggests that human development is very low in these districts. Seven of these 20 districts are in Bihar while five are in Jharkhand. Other states where there is at least one district where human development is very low as reflected through HDS are Assam, Chhattisgarh, Madhya Pradesh, Meghalaya, Odisha and Uttar Pradesh. In Bihar, Jharkhand, Chhattisgarh and Uttar Pradesh, these districts constitute a geographical continuity as may be seen from Fig. 6.1. Human development has also been found to be low ($0.500 \leq$ HDS < 0.700) in 219 (31 per cent) districts. This means that there are 239 districts in the country where human development is either low or very low according to HDS. These districts may be termed as hotspot districts of the country in terms of human progress. In Assam, human development is either low or very low in 29 of the 33 districts. Similarly, in Bihar, human development is either low or very low in 31 of the 38 districts. Other states/ Union Territories where human development is either low or very low in at least half of the districts are Chhattisgarh, Jharkhand, Madhya Pradesh, Meghalaya, Nagaland and Odisha.

On the other hand, there are only 36 (5 per cent) districts in the country where human development may be termed as very high as HDS $\geq$ 0.950 in these districts. Twelve of these 36 districts are in Kerala alone while five are in Delhi. Other states/ Union Territories where there is at least one district where human development is very high are Goa, Haryana, Jammu and Kashmir, Karnataka, Lakshadweep, Maharashtra, Puducherry, Punjab, Tamil Nadu and Telangana. In other states/Union Territories, there is no district where human development may be termed as very high in terms of HDS. There are, however, 198 (28 per cent) districts where human development may be termed as high as $0.850 \leq$ HDS < 0.950 in these districts. This means that human development is above the average in only 234 of the 707 or around one-third districts of the country that existed at the time of the National Family Health Survey, 2019–2021. In rest of the districts, human development may be termed as either average or below average as the HDS < 0.850 in these districts. Table 6.2 and Fig. 6.1 reflect the wide variation in human progress across the districts.

Table 6.2 Inter-district variation in HDS in states and Union Territories, 2019–2021

State/Union Territory	Surface measure of human development (HDS)					
	Very low	Low	Medium	High	Very high	Total
Andaman & Nicobar Islands	0	0	2	1	0	3
Andhra Pradesh	0	0	5	8	0	13
Arunachal Pradesh	0	8	9	3	0	20
Assam	1	28	3	1	0	33
Bihar	7	24	7	0	0	38
Chandigarh	0	0	0	1	0	1
Chhattisgarh	2	14	9	2	0	27
Delhi	0	0	0	6	5	11
Dadra & Nagar Haveli and Daman & Diu	0	0	1	2	0	3
Goa	0	0	0	0	2	2
Gujarat	0	10	17	6	0	33
Haryana	0	1	0	19	2	22
Himachal Pradesh	0	0	1	11	0	12
Jammu & Kashmir	0	1	12	7	2	22
Jharkhand	5	12	7	0	0	24
Karnataka	0	1	11	17	1	30
Kerala	0	0	0	2	12	14
Lakshadweep	0	0	0	0	1	1
Madhya Pradesh	1	26	21	3	0	51
Maharashtra	0	1	19	14	2	36
Manipur	0	3	5	1	0	9
Meghalaya	1	8	2	0	0	11
Mizoram	0	0	2	6	0	8
Nagaland	0	7	3	1	0	11
Odisha	1	20	8	1	0	30
Puducherry	0	0	0	3	1	4
Punjab	0	0	0	18	4	22
Rajasthan	0	3	21	9	0	33
Sikkim	0	0	0	4	0	4
Tamil Nadu	0	0	7	23	2	32
Telangana	0	1	11	17	2	31
Tripura	0	7	1	0	0	8
Uttar Pradesh	2	32	36	5	0	75
Uttarakhand	0	0	7	6	0	13

(continued)

Table 6.2 (continued)

State/Union Territory	Surface measure of human development (HDS)					
	Very low	Low	Medium	High	Very high	Total
West Bengal	0	12	7	1	0	20
India	20	219	234	198	36	707

Remarks:
Human development is very low if HDS < 0.500
Human development is low if 0.500 ≤ HDS < 0.700
Human development is medium if 0.700 ≤ HDS < 0.850
Human development is high if 0.850 ≤ HDS < 0.950
Human development is very high if HDS ≥ 0.950
Source Author, based on the data available from NFHS, 2019–2021

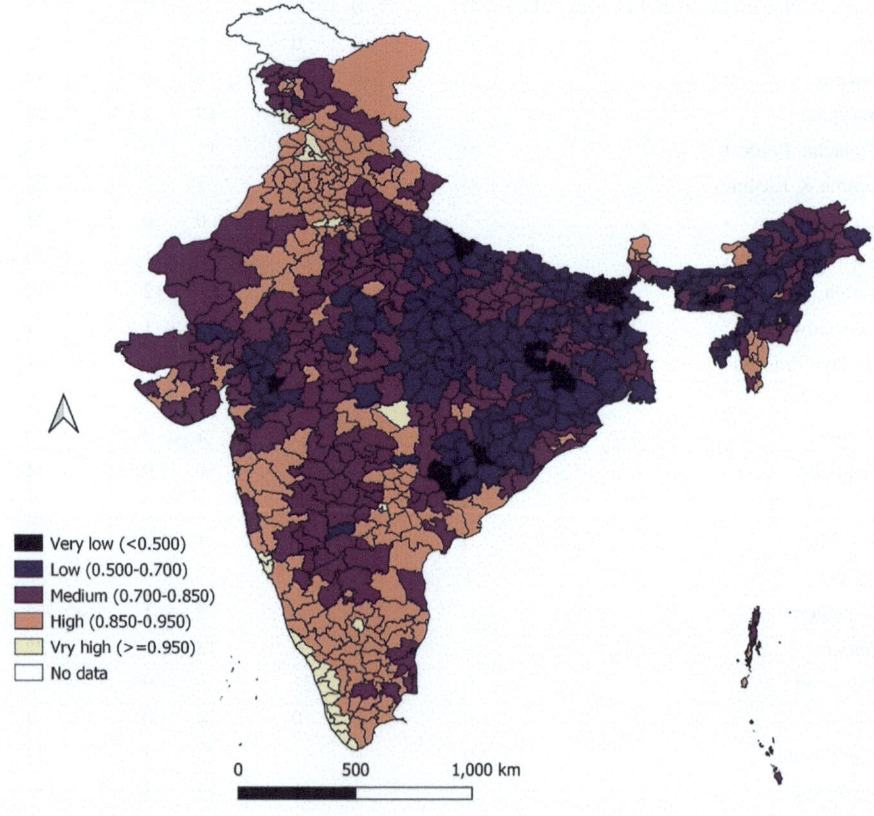

Fig. 6.1 Inter-district variation in HDS in India, 2019-2021. *Source* Author

An important aspect of human progress in the districts of the country is that, in general, the progress in the three core dimensions of human development is not the same. In some districts, progress in the health dimension is the most advanced while in other districts, progress in either the education or the standard of living dimension is the most advanced. Similarly, there are districts in which progress in the health dimension is the slowest while in other districts progress is the slowest either in the education dimension or in the standard of living dimension of human development. The inequality resulting from the differential progress in the three dimensions of human development has been found to be the highest in district Latehar of Jharkhand whereas district Mahe in Puducherry is the only district in the country where there is no inequality in progress in the three dimensions of human development. There are 303 districts where the inequality in progress in the three dimensions of human development, as measured by the difference between HDS_m and HDS, is found to be small (Fig. 6.2). There are, however, 78 districts where there is marked difference in the progress in the three core dimensions of human development so that the inequality in human development in these districts is very high. The progress in human development in these districts may be termed as highly imbalanced as advanced progress in one of the three core dimensions of human development in these districts is found to be associated with either slow to very slow progress in the remaining two dimensions human development. The inequality in progress in different dimensions appears to be one of the reasons behind low level of human development in these districts. It is also obvious that reducing this imbalance in human progress may significantly contribute to accelerating human progress in these districts. The regional pattern in the inequality in human development in the district is also evident from Fig. 6.2. The inequality in human development is generally found to be either high or very high in the eastern region of the country. In the western and the southern regions of the country progress in the three core dimensions of human development is found to be nearly the same so that the inequality in human development in districts of these regions is generally low. In the central region of the country, the inequality in progress in different dimensions of human development is found to be high in majority of the districts.

The within state/Union Territory variation in the progress in the three dimensions of human development across districts is presented in Tables 6.3, 6.4 and 6.5. There are 59 districts where the progress in all the three dimensions of human development is very low and 40 of these 59 districts are in Bihar, Madhya Pradesh and Uttar Pradesh. Other states/Union Territories where there is at least one district in which the progress is very poor in all the three dimensions of human development are Assam, Chhattisgarh, Gujarat, Jharkhand, Karnataka, Maharashtra, Meghalaya and Odisha. On the other hand, there are only 11 districts where the progress is very high in all the three dimensions of human development and six of these 11 districts are in Kerala alone. Other states/Union Territories where progress in all the three dimensions of human development is very high in at least one district of the state/ Union Territory are Goa, Haryana, Lakshadweep, Puducherry and Tamil Nadu. In 231 or almost one third districts, the progress in all the three dimensions of human development is found to be below the average (Table 6.6) whereas there are only 39

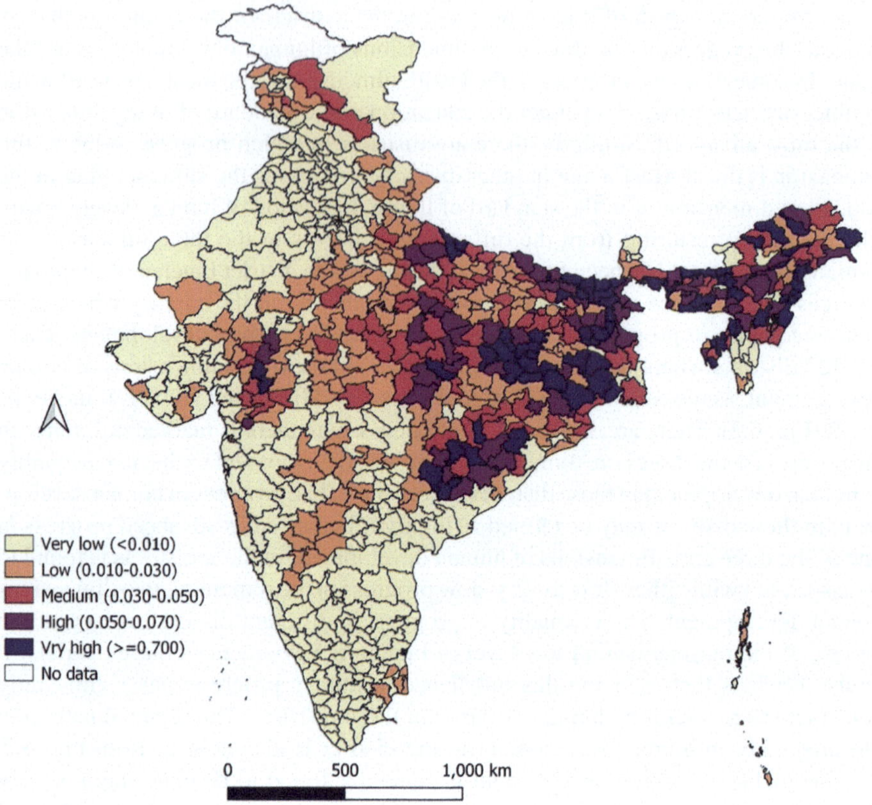

Fig. 6.2 Inter-district variation in the inequality in progress in different dimensions of human development. *Source* Author

or just around 5 per cent districts where progress in all the three dimensions of human development is found to be above the average. Similarly, there are only 10 districts where progress in all the three dimensions of human development may be termed as average. In 50 districts, the progress in the standard of living dimension and in the health dimension of human development is found to be below the average but above the average in the dimension of education. In majority of the districts. Table 6.6 highlights the importance of adopting a decentralised district-based approach to address the challenge of human development in the country. Since the progress in different dimensions of human development is different in different districts, priorities for human progress are different in different districts. Table 6.6 may serve as the guide for prioritising investment in human development in districts.

Table 6.3 Inter-district variation in *SL* 2019–2021

State/Union Territory	Proportion of households having wealth index equal to or more than second quintile of inter-household distribution of wealth index (*SL*)					
	Very low	Low	Medium	High	Very high	Total
Andaman & Nicobar Islands	0	2	0	1	0	3
Andhra Pradesh	0	3	8	2	0	13
Arunachal Pradesh	12	6	2	0	0	20
Assam	32	0	1	0	0	33
Bihar	35	3	0	0	0	38
Chandigarh	0	0	0	0	1	1
Chhattisgarh	20	5	1	1	0	27
Delhi	0	0	0	1	10	11
Dadra & Nagar Haveli and Daman & Diu	0	1	0	1	1	3
Goa	0	0	0	0	2	2
Gujarat	9	5	13	6	0	33
Haryana	0	1	1	18	2	22
Himachal Pradesh	1	3	3	5	0	12
Jammu & Kashmir	2	11	5	3	1	22
Jharkhand	20	4	0	0	0	24
Karnataka	3	8	11	7	1	30
Kerala	0	0	2	5	7	14
Lakshadweep	0	0	0	0	1	1
Madhya Pradesh	33	15	1	2	0	51
Maharashtra	1	19	11	3	2	36
Manipur	7	2	0	0	0	9
Meghalaya	11	0	0	0	0	11
Mizoram	1	1	5	1	0	8
Nagaland	8	2	1	0	0	11
Odisha	24	6	0	0	0	30
Puducherry	0	0	1	1	2	4
Punjab	0	0	1	10	11	22
Rajasthan	6	16	11	0	0	33
Sikkim	0	2	1	1	0	4
Tamil Nadu	0	8	13	10	1	32
Telangana	1	9	15	4	2	31
Tripura	7	1	0	0	0	8
Uttar Pradesh	45	14	12	3	1	75
Uttarakhand	1	8	3	0	1	13

(continued)

Table 6.3 (continued)

State/Union Territory	Proportion of households having wealth index equal to or more than second quintile of inter-household distribution of wealth index (*SL*)					
	Very low	Low	Medium	High	Very high	Total
West Bengal	14	5	0	1	0	20
India	293	160	122	86	46	707

Remarks:

Standard of living is very low if $SL < 0.500$

Standard of living is low if $0.500 \leq SL < 0.700$

Standard of living is medium if $0.700 \leq SL < 0.850$

Standard of living is high if $0.850 \leq SL < 0.950$

Standard of living is very high if $SL \geq 0.950$

Source Author, based on the data available from NFHS, 2019–2021

Discussions and Conclusions

This chapter has used an alternative approach to measure the progress in human development in districts of India which addresses many of the oft repeated limitations of the conventional human development index (HDI) developed and used by the United Nations. This alternative combines the progress in the three core dimensions of human development into a single composite index following the concept of human development surface which does not treat the progress in the three dimensions independent of each other and in a linear manner but recognises that the progress in different dimensions of human development is correlated, although the correlation is not perfect. The index is constructed using proportion indicators rather than average indicators. Average indicators are not appropriate to measure the progress, especially, when the underlying distribution is not statistically normally distributed. The use of proportion indicators automatically obliviates the need of setting up of goalposts to construct normalised index of progress as the proportion always ranges between 0 and 1. Setting up of goalposts is always an arbitrary process and setting up different goalposts leads to different index of progress. Any index of progress based on proportion indicators permits measuring the progress in both absolute and relative terms.

The analysis presented in this chapter highlights the diversity in the progress in human development across the districts of the country. The very marked degree of inter-district variation in the progress in human development suggests that reducing the inter-district inequality in the progress in human development may contribute substantially towards accelerating human progress in the country which remains low by international standards. The inter-district inequality in the progress in human development also emphasises the need of a decentralised, district-based approach for planning and programming for accelerated progress in human development in the country. The inter-district disparity in human development also suggests that progress in human development in a district is conditioned by the factors that are

Table 6.4 Inter-district variation in *ED*, 2019–2021

State/Union Territory	Secondary school net attendance ratio (*ED*)					
	Very low	Low	Medium	High	Very high	Total
Andaman & Nicobar Islands	0	0	0	1	2	3
Andhra Pradesh	0	4	7	2	0	13
Arunachal Pradesh	0	6	6	1	7	20
Assam	5	18	9	1	0	33
Bihar	14	19	5	0	0	38
Chandigarh	0	0	1	0	0	1
Chhattisgarh	7	10	5	4	1	27
Delhi	0	1	4	6	0	11
Dadra & Nagar Haveli and Daman & Diu	1	2	0	0	0	3
Goa	0	0	0	1	1	2
Gujarat	30	3	0	0	0	33
Haryana	1	1	6	11	3	22
Himachal Pradesh	0	0	3	6	3	12
Jammu & Kashmir	0	3	4	8	7	22
Jharkhand	5	12	6	1	0	24
Karnataka	2	4	9	11	4	30
Kerala	0	0	0	0	14	14
Lakshadweep	0	0	0	0	1	1
Madhya Pradesh	22	26	3	0	0	51
Maharashtra	1	9	11	6	9	36
Manipur	0	0	3	6	0	9
Meghalaya	4	3	1	3	0	11
Mizoram	0	2	3	2	1	8
Nagaland	1	6	0	3	1	11
Odisha	11	11	4	4	0	30
Puducherry	0	1	1	1	1	4
Punjab	0	7	5	8	2	22
Rajasthan	1	16	10	6	0	33
Sikkim	0	0	0	2	2	4
Tamil Nadu	0	2	4	21	5	32
Telangana	3	3	9	11	5	31
Tripura	0	2	5	1	0	8
Uttar Pradesh	37	30	7	1	0	75
Uttarakhand	1	1	2	0	9	13

(continued)

Table 6.4 (continued)

State/Union Territory	Secondary school net attendance ratio (*ED*)					
	Very low	Low	Medium	High	Very high	Total
West Bengal	0	8	9	3	0	20
India	146	210	142	131	78	707

Remarks:
Education development is very low if $ED < 0.750$
Education development is low if $0.750 \le ED < 0.850$
Education development is medium if $0.850 \le ED < 0.900$
Education development is high if $0.900 \le ED < 0.950$
Education development is very high if $ED \ge 0.950$
Source Author, based on data available from NFHS, 2019–2021

specific to the district. Identification of these factors and incorporating them in the local level human development planning and programming, therefore, may go a long way towards accelerating the pace of human progress in the country which has implications for the economy of the country also. The diversity across districts in terms of such exogenous factors as the degree of urbanisation and variation in the religious and social class composition of the population is well known and need to be explored further. Human development is a multidimensional concept. There is a possibility that the same exogenous factor may have different impact on different dimensions of human development in different districts. There is, therefore, a need to explore how these and many other factors exogenous to the human development system influence the progress in the three core dimensions of human development as human progress is a composite of the progress, at least in its three core dimensions. One impediment to such investigation is the lack of necessary data at the district level. The only source of data related to the social and cultural composition of the population of a district in India is the population census. The last population census was conducted in the country in 2011 and so the data available at the district level are dated. The problem is further compounded by frequent changes in the administrative boundaries of the constituent districts because of the formation of new districts.

The analysis carried out in this chapter also emphasises the need of imbibing human development sensitiveness in the development planning and programming process in the country with a particular focus on district development planning and programming. Imbibing human development sensitiveness requires regular monitoring of the human development progress at the district level. Efforts made in this direction have been fraught primarily because of the lack of the necessary data. The present chapter uses an alternative surface measure of human development (HDS) that can be constructed at the district level from the data available through the National Family Health Survey (NFHS) or similar other survey. The NFHS is now being organised at regular intervals in the country. It is, therefore, possible to measure and monitor the progress in human development in the districts of the country regularly if not annually. What is required is to institutionalise a system of measuring

Table 6.5 Inter-district variation in *HE*, 2019–2021

State/Union Territory	Probability of survival in the first five years of life (*HE*)					
	Very low	Low	Medium	High	Very high	Total
Andaman & Nicobar Islands	0	1	1	0	1	3
Andhra Pradesh	1	7	4	1	0	13
Arunachal Pradesh	0	2	3	5	10	20
Assam	7	10	12	3	1	33
Bihar	22	15	1	0	0	38
Chandigarh	0	0	0	1	0	1
Chhattisgarh	11	10	4	1	1	27
Delhi	1	5	1	3	1	11
Dadra & Nagar Haveli and Daman & Diu	0	2	0	0	1	3
Goa	0	0	0	1	1	2
Gujarat	8	9	11	5	0	33
Haryana	4	13	3	1	1	22
Himachal Pradesh	1	3	3	3	2	12
Jammu & Kashmir	0	3	4	10	5	22
Jharkhand	5	17	2	0	0	24
Karnataka	4	10	7	7	2	30
Kerala	0	0	0	3	11	14
Lakshadweep	0	0	0	0	1	1
Madhya Pradesh	23	18	7	2	1	51
Maharashtra	4	13	7	9	3	36
Manipur	1	3	3	1	1	9
Meghalaya	3	2	2	0	4	11
Mizoram	0	1	3	1	3	8
Nagaland	2	3	2	4	0	11
Odisha	10	11	5	3	1	30
Puducherry	0	1	0	1	2	4
Punjab	3	13	2	3	1	22
Rajasthan	5	15	10	2	1	33
Sikkim	0	1	0	1	2	4
Tamil Nadu	0	10	7	5	10	32
Telangana	6	6	9	4	6	31
Tripura	3	4	1	0	0	8
Uttar Pradesh	49	19	3	4	0	75
Uttarakhand	3	4	3	2	1	13

(continued)

Table 6.5 (continued)

State/Union Territory	Probability of survival in the first five years of life (*HE*)					
	Very low	Low	Medium	High	Very high	Total
West Bengal	1	7	6	4	2	20
India	177	238	126	90	76	707

Remarks:
State of health is very low if $HE < 0.950$
State of health is low if $0.950 \leq HE < 0.970$
State of health is medium if $0.970 \leq HE < 0.980$
State of health is high if $0.980 \leq HE < 0.990$
State of health is very high if $HE \geq 0.990$
Source Author, based on data available from NFHS, 2019–2021

and monitoring human development within the district development administration system.

The surface measure of human development used in the present chapter and the conventional human development index used by United Nations to measure human development or any other similar index to measure human progress are composite indexes. The main limitation of any composite index is that it presents a broad, rounded picture of progress. A composite index can be used to rank countries or administrative units within countries in terms of the overall progress in human development but it tells little about the progress in different dimensions that constitute human development. The rank of a district in terms of the progress in different dimensions of human development, as a rule, is different from the rank in of the district in terms of the composite index of human development. Because of this limitation, any composite index of human development contributes little to prioritising investment in human development. Since the progress in different dimensions that constitute human development in a district is rarely the same, prioritising investment in human development requires going beyond simply constructing a composite index of human development.

Table 6.6 Classification of districts according to progress in different dimensions of human development

Progress in			Districts	
Standard of living	Education	Health	Number	Per cent
Below average	Below average	Below average	231	32.7
Below average	Below average	Average	34	4.8
Below average	Below average	Above average	24	3.4
Below average	Average	Below average	50	7.1
Below average	Average	Average	18	2.5
Below average	Average	Above average	15	2.1
Below average	Above average	Below average	33	4.7
Below average	Above average	Average	17	2.4
Below average	Above average	Above average	31	4.4
Average	Below average	Below average	21	3.0
Average	Below average	Average	11	1.6
Average	Below average	Above average	11	1.6
Average	Average	Below average	12	1.7
Average	Average	Average	10	1.4
Average	Average	Above average	7	1.0
Average	Above average	Below average	14	2.0
Average	Above average	Average	15	2.1
Average	Above average	Above average	21	3.0
Above average	Below average	Below average	12	1.7
Above average	Below average	Average	5	0.7
Above average	Below average	Above average	7	1.0
Above average	Average	Below average	15	2.1
Above average	Average	Average	4	0.6
Above average	Average	Above average	11	1.6
Above average	Above average	Below average	27	3.8
Above average	Above average	Average	12	1.7
Above average	Above average	Above average	39	5.5
Total			707	100.0

Source Author

Appendix

See Table 6.7.

Table 6.7 Indexes *SL*, *ED*, *HE* and HDS in districts of India, 2019–2021 and inequality in progress in humand development

State/Union Territory/District		*SL*	*ED*	*HE*	HDS	Rank in HDS	Inequality in progress
Andaman & Nicobar Islands							
	South Andaman	0.919	0.952	0.979	0.950	37	0.000
	Nicobar	0.523	1.000	0.959	0.803	307	0.024
	North & Middle Andaman	0.526	0.909	0.993	0.788	331	0.021
Andhra Pradesh							
	West Godavari	0.867	0.871	0.981	0.905	132	0.001
	Y.S.R	0.864	0.872	0.928	0.888	162	0.000
	Chittoor	0.768	0.897	0.970	0.875	186	0.003
	East Godavari	0.762	0.876	0.977	0.869	198	0.003
	Prakasam	0.788	0.869	0.953	0.868	200	0.002
	Guntur	0.811	0.817	0.976	0.866	203	0.002
	Srikakulam	0.676	0.950	0.979	0.860	213	0.009
	Visakhapatnam	0.735	0.870	0.961	0.852	231	0.004
	Anantapur	0.711	0.878	0.969	0.847	239	0.005
	Krishna	0.755	0.825	0.958	0.843	242	0.003
	Sri Potti Sriramulu Nellore	0.785	0.772	0.950	0.833	252	0.003
	Vizianagaram	0.583	0.941	0.960	0.813	288	0.015
	Kurnool	0.649	0.804	0.954	0.795	318	0.007
Arunachal Pradesh							
	Tawang	0.824	1.000	0.999	0.938	58	0.003
	West Kameng	0.850	0.857	0.973	0.892	155	0.001
	Papum Pare	0.686	0.889	0.998	0.850	234	0.008
	Lower Dibang Valley	0.550	1.000	1.000	0.828	267	0.022
	East Siang	0.584	0.875	1.000	0.805	303	0.015
	West Siang	0.540	0.875	0.977	0.779	347	0.018
	Lower Subansiri	0.534	0.857	0.988	0.774	361	0.019
	Kra Daadi	0.443	1.000	0.974	0.770	367	0.036
	Lohit	0.553	0.800	1.000	0.768	375	0.017
	Dibang Valley	0.412	1.000	0.982	0.756	391	0.042
	Upper Siang	0.367	1.000	1.000	0.737	414	0.052
	Changlang	0.350	0.929	1.000	0.709	454	0.051
	Tirap	0.394	0.800	0.964	0.685	489	0.034
	Siang	0.279	1.000	0.987	0.682	500	0.073
	East Kameng	0.358	0.857	0.956	0.681	501	0.042

(continued)

Table 6.7 (continued)

State/Union Territory/District		*SL*	*ED*	*HE*	HDS	Rank in HDS	Inequality in progress
	Kurung Kumey	0.239	0.833	0.991	0.614	583	0.074
	Anjaw	0.179	1.000	0.982	0.611	584	0.109
	Namsai	0.233	0.750	0.985	0.586	622	0.070
	Upper Subansiri	0.192	0.800	0.995	0.574	637	0.088
	Longding	0.162	0.750	1.000	0.539	668	0.098
Assam							
	Kamrup Metropolitan	0.776	0.884	0.972	0.874	188	0.003
	Jorhat	0.466	0.867	0.943	0.735	418	0.024
	Sivasagar	0.463	0.800	0.979	0.723	434	0.025
	Kamrup	0.411	0.831	0.982	0.707	457	0.034
	Dima Hasao	0.392	0.826	0.991	0.699	473	0.037
	Hojai	0.361	0.852	0.971	0.685	490	0.043
	Bongaigaon	0.349	0.865	0.976	0.684	494	0.046
	Dibrugarh	0.413	0.736	0.977	0.678	504	0.031
	Nalbari	0.329	0.847	0.987	0.671	516	0.050
	Kokrajhar	0.308	0.866	0.961	0.658	534	0.054
	Tinsukia	0.354	0.781	0.937	0.653	540	0.038
	Karbi Anglong	0.294	0.867	0.968	0.652	542	0.058
	Lakhimpur	0.283	0.840	0.976	0.640	557	0.060
	Sonitpur	0.320	0.743	0.959	0.629	568	0.045
	Barpeta	0.252	0.840	0.972	0.619	579	0.068
	Golaghat	0.261	0.785	0.971	0.609	590	0.063
	Goalpara	0.251	0.800	0.969	0.607	592	0.066
	Chirang	0.230	0.852	0.963	0.607	593	0.075
	Cachar	0.259	0.754	0.975	0.601	605	0.062
	Biswanath	0.251	0.808	0.929	0.600	606	0.063
	Darrang	0.232	0.788	0.974	0.593	613	0.072
	Dhemaji	0.181	0.913	0.976	0.591	617	0.100
	Majuli	0.183	0.895	0.958	0.583	624	0.096
	Udalguri	0.218	0.804	0.949	0.583	625	0.075
	Dhubri	0.214	0.802	0.962	0.582	626	0.077
	Baksa	0.190	0.840	0.971	0.578	632	0.090
	Nagaon	0.243	0.735	0.936	0.576	633	0.062
	Charaideo	0.246	0.709	0.951	0.574	635	0.061
	Morigaon	0.214	0.784	0.948	0.574	636	0.075

(continued)

Table 6.7 (continued)

State/Union Territory/District		*SL*	*ED*	*HE*	HDS	Rank in HDS	Inequality in progress
	Karimganj	0.216	0.741	0.954	0.565	646	0.072
	Hailakandi	0.203	0.768	0.898	0.551	660	0.072
	West Karbi Anglong	0.121	0.865	0.969	0.527	674	0.124
	South Salmara Mancachar	0.113	0.750	0.981	0.494	692	0.121
Bihar							
	Patna	0.608	0.753	0.955	0.762	387	0.010
	Rohtas	0.518	0.885	0.906	0.752	396	0.017
	Munger	0.516	0.813	0.968	0.747	400	0.018
	Bhojpur	0.456	0.842	0.963	0.728	425	0.026
	Buxar	0.466	0.823	0.940	0.720	440	0.023
	Bhagalpur	0.462	0.825	0.936	0.718	443	0.023
	Siwan	0.442	0.852	0.940	0.717	444	0.027
	Nawada	0.404	0.815	0.971	0.696	476	0.034
	Aurangabad	0.390	0.842	0.953	0.693	479	0.035
	Nalanda	0.446	0.745	0.950	0.690	480	0.024
	Jehanabad	0.358	0.882	0.952	0.688	485	0.043
	Sheikhpura	0.392	0.829	0.927	0.683	496	0.033
	Lakhisarai	0.377	0.783	0.962	0.671	515	0.036
	Gopalganj	0.367	0.781	0.959	0.665	525	0.038
	Saran	0.358	0.800	0.940	0.661	531	0.039
	Muzaffarpur	0.315	0.851	0.961	0.657	536	0.052
	Vaishali	0.329	0.811	0.928	0.646	549	0.044
	Arwal	0.301	0.855	0.925	0.641	553	0.052
	Gaya	0.328	0.766	0.962	0.641	554	0.045
	Begusarai	0.304	0.797	0.950	0.633	563	0.050
	Kaimur (Bhabua)	0.300	0.814	0.923	0.629	567	0.050
	Banka	0.239	0.787	0.960	0.594	611	0.068
	Darbhanga	0.272	0.714	0.949	0.591	616	0.054
	Purba Champaran	0.277	0.706	0.941	0.590	618	0.052
	Jamui	0.258	0.751	0.926	0.588	619	0.057
	Khagaria	0.247	0.726	0.950	0.579	629	0.062
	Madhubani	0.220	0.740	0.955	0.567	644	0.071
	Sheohar	0.207	0.757	0.936	0.559	650	0.074
	Sitamarhi	0.212	0.702	0.918	0.543	664	0.067
	Samastipur	0.174	0.768	0.942	0.540	665	0.088

(continued)

Table 6.7 (continued)

State/Union Territory/District		*SL*	*ED*	*HE*	HDS	Rank in HDS	Inequality in progress
	Pashchim Champaran	0.220	0.658	0.906	0.533	670	0.062
	Katihar	0.155	0.661	0.954	0.500	688	0.090
	Purnia	0.179	0.612	0.922	0.496	690	0.075
	Saharsa	0.153	0.673	0.926	0.495	691	0.088
	Kishanganj	0.172	0.573	0.945	0.485	693	0.079
	Araria	0.135	0.639	0.944	0.476	697	0.097
	Madhepura	0.101	0.643	0.947	0.448	703	0.116
	Supaul	0.081	0.639	0.957	0.429	707	0.130
Chandigarh							
	Chandigarh	0.967	0.873	0.983	0.940	54	0.001
Chhattisgarh							
	Durg	0.867	0.868	0.979	0.903	137	0.001
	Raipur	0.757	0.843	0.978	0.856	225	0.004
	Dhamtari	0.618	0.908	0.986	0.825	272	0.012
	Balod	0.573	0.964	0.992	0.825	273	0.018
	Rajnandgaon	0.550	0.916	0.947	0.788	332	0.017
	Bemetara	0.527	0.868	0.965	0.768	374	0.018
	Kabeerdham	0.504	0.886	0.975	0.766	378	0.022
	Janjgir - Champa	0.460	0.909	0.970	0.751	398	0.028
	Bilaspur	0.474	0.813	0.954	0.725	431	0.022
	Baloda Bazar	0.467	0.812	0.952	0.721	437	0.023
	Raigarh	0.425	0.854	0.908	0.701	468	0.028
	Uttar Bastar Kanker	0.342	0.932	0.978	0.699	471	0.052
	Korba	0.463	0.767	0.923	0.697	475	0.021
	Mahasamund	0.404	0.791	0.964	0.687	487	0.032
	Mungeli	0.370	0.830	0.944	0.677	509	0.038
	Koriya	0.352	0.890	0.883	0.668	521	0.040
	Gariyaband	0.348	0.824	0.952	0.666	523	0.042
	Surguja	0.254	0.807	0.935	0.603	599	0.062
	Surajpur	0.228	0.839	0.966	0.602	600	0.075
	Dantewada	0.252	0.721	0.952	0.581	627	0.060
	Bastar	0.269	0.621	0.925	0.555	654	0.050
	Narayanpur	0.217	0.647	0.962	0.540	666	0.068
	Jashpur	0.146	0.795	0.951	0.528	673	0.103
	Kodagaon	0.169	0.737	0.937	0.527	675	0.087

(continued)

Table 6.7 (continued)

State/Union Territory/District		*SL*	*ED*	*HE*	HDS	Rank in HDS	Inequality in progress
	Balrampur	0.146	0.748	0.937	0.513	679	0.098
	Bijapur	0.148	0.552	0.917	0.455	702	0.084
	Sukma	0.092	0.625	0.929	0.432	706	0.117
Delhi							
	South West	0.981	0.934	0.990	0.968	22	0.000
	North West	0.987	0.931	0.984	0.967	23	0.000
	South East	0.972	0.927	0.985	0.961	28	0.000
	South	0.989	0.932	0.948	0.956	32	0.000
	North East	0.990	0.897	0.970	0.951	35	0.001
	New Delhi	0.960	0.902	0.979	0.946	44	0.000
	East	0.961	0.914	0.959	0.945	46	0.000
	Central	0.963	0.885	0.985	0.944	48	0.001
	West	0.969	0.886	0.963	0.939	56	0.001
	Shahdara	0.987	0.869	0.951	0.935	65	0.001
	North	0.928	0.827	0.969	0.907	130	0.001
Dadra & Nagar Haveli and Daman & Diu							
	Diu	0.955	0.750	1.000	0.897	147	0.005
	Daman	0.867	0.800	0.958	0.873	191	0.002
	Dadra & Nagar Haveli	0.520	0.694	0.950	0.705	463	0.016
Goa							
	North Goa	0.966	0.965	1.000	0.977	11	0.000
	South Goa	0.971	0.940	0.984	0.965	25	0.000
Gujarat							
	Rajkot	0.934	0.827	0.980	0.912	116	0.002
	Ahmadabad	0.934	0.733	0.961	0.872	194	0.005
	Jamnagar	0.915	0.713	0.979	0.863	207	0.006
	Porbandar	0.896	0.719	0.982	0.860	210	0.005
	Surat	0.870	0.739	0.974	0.857	219	0.004
	Gandhinagar	0.816	0.785	0.957	0.850	233	0.002
	Valsad	0.725	0.805	0.977	0.831	261	0.005
	Vadodara	0.831	0.699	0.977	0.830	262	0.006
	Junagadh	0.817	0.724	0.961	0.830	263	0.004
	Morbi	0.868	0.653	0.989	0.828	268	0.009
	Gir Somnath	0.751	0.712	0.976	0.807	296	0.006
	Navsari	0.729	0.727	0.982	0.806	300	0.006

(continued)

Table 6.7 (continued)

State/Union Territory/District		*SL*	*ED*	*HE*	HDS	Rank in HDS	Inequality in progress
	Kachchh	0.821	0.613	0.962	0.789	330	0.010
	Botad	0.810	0.596	0.980	0.783	338	0.012
	Amreli	0.781	0.616	0.980	0.782	341	0.011
	Bhavnagar	0.717	0.656	0.980	0.775	360	0.009
	Bharuch	0.711	0.673	0.947	0.770	366	0.007
	Surendranagar	0.721	0.646	0.967	0.769	370	0.009
	Anand	0.697	0.676	0.935	0.763	383	0.006
	Mahesana	0.665	0.692	0.937	0.757	390	0.007
	Patan	0.648	0.641	0.910	0.725	428	0.008
	Sabar Kantha	0.587	0.626	0.968	0.713	447	0.014
	Devbhumi Dwarka	0.776	0.450	0.971	0.707	459	0.026
	Aravali	0.458	0.687	0.961	0.679	502	0.023
	Kheda	0.555	0.577	0.934	0.673	513	0.015
	Panch Mahals	0.474	0.616	0.939	0.656	537	0.020
	Banas Kantha	0.479	0.581	0.949	0.648	543	0.022
	Mahisagar	0.394	0.664	0.970	0.644	550	0.032
	Tapi	0.404	0.639	0.959	0.638	558	0.030
	Narmada	0.305	0.667	0.954	0.596	608	0.046
	Chhota Udaipur	0.329	0.549	0.926	0.563	647	0.038
	The Dangs	0.227	0.571	0.975	0.526	676	0.066
	Dohad	0.184	0.594	0.973	0.505	684	0.079
Haryana							
	Jhajjar	0.954	0.971	0.991	0.972	17	0.000
	Charkhi Dadri	0.935	0.984	0.967	0.962	27	0.000
	Panchkula	0.968	0.915	0.960	0.948	42	0.000
	Sonipat	0.940	0.937	0.965	0.947	43	0.000
	Rewari	0.926	0.936	0.969	0.944	47	0.000
	Rohtak	0.927	0.922	0.975	0.941	50	0.000
	Gurgaon	0.940	0.911	0.966	0.939	57	0.000
	Kaithal	0.914	0.935	0.954	0.934	67	0.000
	Jind	0.906	0.946	0.944	0.932	70	0.000
	Hisar	0.910	0.910	0.971	0.930	74	0.000
	Faridabad	0.949	0.873	0.967	0.929	78	0.001
	Panipat	0.897	0.926	0.959	0.927	80	0.000
	Mahendragarh	0.851	0.970	0.965	0.927	82	0.001

(continued)

Table 6.7 (continued)

State/Union Territory/District		*SL*	*ED*	*HE*	HDS	Rank in HDS	Inequality in progress
	Fatehabad	0.885	0.931	0.961	0.925	87	0.000
	Karnal	0.912	0.892	0.972	0.925	88	0.000
	Yamunanagar	0.918	0.880	0.967	0.921	91	0.001
	Ambala	0.939	0.889	0.932	0.920	94	0.000
	Bhiwani	0.891	0.928	0.929	0.916	104	0.000
	Kurukshetra	0.914	0.865	0.958	0.912	117	0.001
	Sirsa	0.879	0.870	0.958	0.902	141	0.001
	Palwal	0.805	0.841	0.981	0.873	190	0.002
	Mewat	0.561	0.526	0.941	0.658	533	0.018
Himachal Pradesh							
	Bilaspur	0.899	0.886	0.992	0.925	89	0.001
	Una	0.900	0.882	0.980	0.920	93	0.001
	Solan	0.902	0.900	0.941	0.914	109	0.000
	Hamirpur	0.872	0.900	0.972	0.914	110	0.001
	Shimla	0.837	0.942	0.965	0.913	111	0.001
	Mandi	0.820	0.947	0.979	0.913	112	0.002
	Sirmaur	0.824	0.909	0.980	0.903	139	0.002
	Kangra	0.855	0.889	0.955	0.899	145	0.001
	Kinnaur	0.695	1.000	0.968	0.879	177	0.008
	Kullu	0.661	0.957	0.984	0.857	220	0.010
	Chamba	0.673	0.942	0.970	0.854	226	0.008
	Lahul & Spiti	0.481	1.000	0.996	0.794	319	0.031
Jammu & Kashmir							
	Jammu	0.952	0.981	0.985	0.973	15	0.000
	Srinagar	0.947	0.979	0.988	0.971	18	0.000
	Samba	0.876	0.946	0.991	0.937	61	0.001
	Pulwama	0.869	0.950	0.974	0.930	75	0.001
	Kathua	0.847	0.961	0.981	0.928	79	0.001
	Anantnag	0.830	0.953	1.000	0.926	86	0.002
	Shupiyan	0.803	0.909	1.000	0.901	143	0.003
	Badgam	0.773	0.925	0.995	0.894	153	0.004
	Leh (Ladakh)	0.742	0.900	0.969	0.866	202	0.004
	Udhampur	0.606	0.950	0.980	0.832	258	0.014
	Ganderbal	0.687	0.839	0.987	0.831	260	0.007
	Punch	0.581	0.949	0.972	0.818	281	0.016

(continued)

Table 6.7 (continued)

State/Union Territory/District		*SL*	*ED*	*HE*	HDS	Rank in HDS	Inequality in progress
	Rajouri	0.621	0.875	0.989	0.817	283	0.011
	Kupwara	0.581	0.914	0.961	0.804	304	0.014
	Baramula	0.648	0.805	0.976	0.801	309	0.008
	Kulgam	0.542	0.926	0.983	0.797	315	0.019
	Bandipore	0.584	0.848	0.969	0.787	333	0.013
	Kishtwar	0.522	0.906	0.990	0.785	335	0.021
	Doda	0.514	0.897	0.981	0.775	359	0.022
	Reasi	0.506	0.897	0.986	0.773	363	0.023
	Kargil	0.388	1.000	0.975	0.742	406	0.046
	Ramban	0.370	0.865	0.993	0.700	469	0.043
Jharkhand							
	Ranchi	0.546	0.891	0.958	0.782	343	0.017
	Dhanbad	0.555	0.847	0.976	0.777	353	0.016
	Purbi Singhbhum	0.551	0.871	0.952	0.776	356	0.015
	Bokaro	0.541	0.873	0.957	0.774	362	0.017
	Ramgarh	0.484	0.861	0.959	0.745	402	0.023
	Hazaribagh	0.434	0.883	0.954	0.727	427	0.030
	Kodarma	0.399	0.912	0.963	0.720	439	0.038
	Giridih	0.331	0.806	0.956	0.653	541	0.045
	Saraikela-Kharsawan	0.298	0.844	0.929	0.638	559	0.053
	Lohardaga	0.234	0.863	0.959	0.611	586	0.074
	Palamu	0.235	0.814	0.944	0.595	609	0.069
	Jamtara	0.197	0.794	0.972	0.571	640	0.084
	Chatra	0.196	0.772	0.952	0.560	648	0.080
	Garhwa	0.173	0.827	0.952	0.557	653	0.094
	Godda	0.213	0.723	0.930	0.552	658	0.069
	Deoghar	0.194	0.723	0.967	0.548	661	0.080
	Dumka	0.173	0.785	0.941	0.544	663	0.089
	Simdega	0.119	0.813	0.964	0.511	681	0.120
	Sahibganj	0.161	0.648	0.969	0.504	685	0.089
	Gumla	0.108	0.822	0.933	0.497	689	0.124
	Khunti	0.112	0.725	0.961	0.482	694	0.117
	Latehar	0.094	0.776	0.963	0.479	695	0.132
	Pashchimi Singhbhum	0.091	0.760	0.955	0.470	698	0.132
	Pakur	0.117	0.623	0.959	0.460	700	0.107

(continued)

Table 6.7 (continued)

State/Union Territory/District		*SL*	*ED*	*HE*	HDS	Rank in HDS	Inequality in progress
Karnataka							
	Bangalore	0.969	0.949	0.982	0.966	24	0.000
	Dakshina Kannada	0.903	0.950	0.992	0.948	41	0.001
	Kolar	0.892	0.942	0.988	0.940	53	0.001
	Mandya	0.864	0.952	1.000	0.937	59	0.001
	Bangalore Rural	0.891	0.931	0.969	0.930	76	0.000
	Udupi	0.837	0.963	0.985	0.926	83	0.002
	Kodagu	0.868	0.927	0.974	0.922	90	0.001
	Ramanagara	0.901	0.901	0.941	0.914	108	0.000
	Hassan	0.812	0.945	0.990	0.913	113	0.002
	Tumkur	0.823	0.991	0.926	0.911	121	0.002
	Chikmagalur	0.826	0.940	0.969	0.910	123	0.002
	Mysore	0.858	0.889	0.987	0.910	124	0.001
	Shimoga	0.833	0.922	0.979	0.910	125	0.001
	Chikkaballapura	0.849	0.856	0.978	0.893	154	0.001
	Dharwad	0.762	0.927	0.960	0.880	176	0.003
	Uttara Kannada	0.786	0.892	0.963	0.878	181	0.002
	Chamarajanagar	0.748	0.887	0.966	0.864	205	0.004
	Davanagere	0.743	0.861	0.977	0.856	224	0.004
	Chitradurga	0.700	0.887	0.957	0.843	244	0.005
	Belgaum	0.670	0.907	0.975	0.843	245	0.008
	Bellary	0.643	0.862	0.932	0.805	302	0.007
	Gadag	0.567	0.901	0.984	0.801	311	0.016
	Bidar	0.539	0.882	0.969	0.779	348	0.018
	Haveri	0.582	0.827	0.965	0.779	350	0.013
	Bagalkot	0.533	0.857	0.972	0.769	369	0.018
	Gulbarga	0.554	0.777	0.978	0.755	392	0.015
	Bijapur	0.520	0.814	0.981	0.753	395	0.019
	Raichur	0.497	0.739	0.946	0.709	453	0.018
	Koppal	0.448	0.792	0.957	0.707	458	0.025
	Yadgir	0.458	0.735	0.958	0.694	478	0.023
Kerala							
	Kozhikode	0.980	1.000	1.000	0.993	3	0.000
	Ernakulam	0.982	0.990	0.989	0.987	4	0.000
	Malappuram	0.966	1.000	0.995	0.987	5	0.000

(continued)

Table 6.7 (continued)

State/Union Territory/District		*SL*	*ED*	*HE*	HDS	Rank in HDS	Inequality in progress
	Alappuzha	0.968	0.987	1.000	0.985	6	0.000
	Thrissur	0.971	0.984	1.000	0.985	7	0.000
	Kannur	0.968	0.967	1.000	0.978	9	0.000
	Kollam	0.943	0.993	1.000	0.978	10	0.000
	Kottayam	0.956	0.970	1.000	0.975	12	0.000
	Pathanamthitta	0.934	0.989	1.000	0.974	14	0.000
	Thiruvananthapuram	0.925	0.993	1.000	0.972	16	0.000
	Kasaragod	0.916	0.974	0.995	0.961	29	0.000
	Palakkad	0.888	0.988	1.000	0.958	30	0.001
	Idukki	0.832	0.967	0.989	0.927	81	0.002
	Wayanad	0.815	0.954	0.987	0.916	100	0.002
Lakshadweep							
	Lakshadweep	0.981	1.000	1.000	0.993	2	0.000
Madhya Pradesh							
	Indore	0.895	0.849	0.971	0.904	135	0.001
	Gwalior	0.828	0.851	0.972	0.882	170	0.002
	Bhopal	0.850	0.799	0.984	0.875	185	0.003
	Ujjain	0.667	0.818	0.976	0.813	289	0.007
	Neemuch	0.657	0.823	0.963	0.807	297	0.007
	Bhind	0.573	0.854	0.957	0.781	344	0.013
	Harda	0.597	0.779	0.963	0.769	372	0.011
	Mandsaur	0.576	0.824	0.940	0.768	373	0.012
	Dewas	0.575	0.802	0.952	0.764	381	0.012
	Khargone (West Nimar)	0.647	0.708	0.959	0.763	384	0.008
	Khandwa (East Nimar)	0.533	0.803	0.973	0.753	394	0.017
	Agar Malwa	0.522	0.787	0.981	0.745	403	0.018
	Ratlam	0.592	0.714	0.953	0.742	407	0.011
	Morena	0.545	0.782	0.936	0.741	409	0.014
	Shajapur	0.526	0.802	0.940	0.740	410	0.016
	Burhanpur	0.612	0.687	0.946	0.739	412	0.010
	Sehore	0.495	0.835	0.940	0.737	415	0.020
	Datia	0.516	0.802	0.942	0.736	416	0.017
	Hoshangabad	0.531	0.765	0.952	0.734	419	0.015
	Tikamgarh	0.476	0.798	0.960	0.722	435	0.022

(continued)

Table 6.7 (continued)

State/Union Territory/District		*SL*	*ED*	*HE*	HDS	Rank in HDS	Inequality in progress
	Raisen	0.466	0.809	0.955	0.720	442	0.023
	Sagar	0.421	0.830	0.961	0.707	461	0.031
	Dhar	0.478	0.731	0.973	0.706	462	0.022
	Chhindwara	0.462	0.766	0.951	0.704	466	0.023
	Narsimhapur	0.438	0.735	0.976	0.689	481	0.027
	Betul	0.391	0.797	0.965	0.683	497	0.035
	Jabalpur	0.395	0.744	1.000	0.678	505	0.035
	Vidisha	0.460	0.684	0.942	0.674	511	0.021
	Guna	0.428	0.695	0.950	0.665	524	0.026
	Chhatarpur	0.413	0.723	0.934	0.663	526	0.027
	Satna	0.391	0.741	0.946	0.661	530	0.031
	Rajgarh	0.350	0.794	0.957	0.659	532	0.041
	Shivpuri	0.394	0.718	0.935	0.653	539	0.030
	Balaghat	0.285	0.863	0.955	0.642	552	0.059
	Mandla	0.297	0.818	0.956	0.637	560	0.054
	Ashoknagar	0.331	0.728	0.957	0.629	566	0.043
	Singrauli	0.320	0.748	0.938	0.625	572	0.044
	Katni	0.348	0.730	0.896	0.624	573	0.034
	Anuppur	0.286	0.798	0.947	0.622	575	0.055
	Seoni	0.305	0.755	0.950	0.622	576	0.048
	Damoh	0.307	0.758	0.941	0.621	578	0.047
	Umaria	0.310	0.723	0.929	0.610	587	0.044
	Shahdol	0.280	0.778	0.909	0.604	596	0.052
	Barwani	0.371	0.572	0.968	0.601	603	0.036
	Sheopur	0.323	0.658	0.934	0.598	607	0.040
	Sidhi	0.232	0.723	0.931	0.565	645	0.064
	Panna	0.234	0.725	0.901	0.559	649	0.060
	Rewa	0.246	0.674	0.928	0.559	652	0.057
	Dindori	0.146	0.779	0.945	0.523	677	0.101
	Jhabua	0.181	0.657	0.933	0.513	680	0.077
	Alirajpur	0.171	0.540	0.970	0.479	696	0.082
Maharashtra							
	Mumbai	0.968	0.956	0.983	0.969	21	0.000
	Nagpur	0.932	0.951	0.990	0.957	31	0.000
	Mumbai Suburban	0.979	0.966	0.886	0.943	49	0.001

(continued)

Table 6.7 (continued)

State/Union Territory/District		*SL*	*ED*	*HE*	HDS	Rank in HDS	Inequality in progress
	Thane	0.896	0.923	0.979	0.932	69	0.000
	Kolhapur	0.817	0.958	0.967	0.912	119	0.002
	Sangli	0.813	0.955	0.971	0.911	122	0.002
	Pune	0.882	0.850	1.000	0.909	126	0.002
	Wardha	0.836	0.894	0.984	0.903	138	0.002
	Chandrapur	0.760	0.953	0.953	0.885	166	0.004
	Amravati	0.754	0.915	0.981	0.879	178	0.004
	Raigarh	0.814	0.848	0.980	0.879	179	0.002
	Satara	0.755	0.883	1.000	0.875	187	0.004
	Akola	0.776	0.883	0.969	0.873	192	0.003
	Sindhudurg	0.687	0.975	0.977	0.871	195	0.008
	Aurangabad	0.740	0.902	0.977	0.869	199	0.004
	Ahmadnagar	0.724	0.895	0.974	0.859	214	0.005
	Buldana	0.667	0.937	0.969	0.849	236	0.009
	Osmanabad	0.667	0.910	0.995	0.848	237	0.009
	Bhandara	0.677	0.919	0.957	0.844	241	0.007
	Latur	0.686	0.875	0.972	0.838	248	0.006
	Jalgaon	0.711	0.823	0.980	0.833	254	0.005
	Solapur	0.661	0.866	0.981	0.828	266	0.008
	Gondiya	0.607	0.958	0.946	0.824	274	0.013
	Ratnagiri	0.586	0.953	0.961	0.818	280	0.015
	Yavatmal	0.644	0.882	0.950	0.817	282	0.008
	Palghar	0.675	0.793	0.988	0.811	292	0.008
	Hingoli	0.608	0.873	0.965	0.804	305	0.011
	Bid	0.577	0.881	0.980	0.798	313	0.015
	Washim	0.590	0.878	0.959	0.796	317	0.012
	Nashik	0.640	0.815	0.948	0.793	321	0.008
	Jalna	0.577	0.849	0.960	0.782	340	0.013
	Nanded	0.597	0.817	0.964	0.782	342	0.011
	Parbhani	0.559	0.848	0.970	0.777	352	0.015
	Dhule	0.594	0.764	0.984	0.769	371	0.012
	Gadchiroli	0.507	0.878	0.964	0.762	386	0.021
	Nandurbar	0.362	0.703	0.942	0.634	562	0.035
Manipur							
	Imphal West	0.697	0.933	0.991	0.867	201	0.007

(continued)

Table 6.7 (continued)

State/Union Territory/District		*SL*	*ED*	*HE*	HDS	Rank in HDS	Inequality in progress
	Imphal East	0.597	0.907	0.957	0.808	294	0.013
	Bishnupur	0.430	0.917	0.972	0.740	411	0.033
	Thoubal	0.422	0.894	0.971	0.728	424	0.034
	Churachandpur	0.422	0.870	0.976	0.723	433	0.033
	Chandel	0.410	0.857	0.962	0.710	452	0.033
	Senapati	0.302	0.917	0.986	0.674	512	0.061
	Tamenglong	0.189	0.900	0.950	0.587	620	0.093
	Ukhrul	0.164	0.900	0.966	0.571	639	0.105
Meghalaya							
	West Garo Hills	0.480	0.940	0.993	0.776	355	0.028
	East Khasi Hills	0.472	0.780	1.000	0.725	429	0.025
	East Garo Hills	0.330	0.867	0.991	0.678	506	0.051
	South Garo Hills	0.268	0.941	0.994	0.662	528	0.073
	North Garo Hills	0.277	0.923	0.974	0.658	535	0.067
	South West Garo Hills	0.301	0.826	0.973	0.646	548	0.055
	Ribhoi	0.261	0.739	0.964	0.595	610	0.060
	East Jantia Hills	0.274	0.696	0.947	0.586	621	0.053
	South West Khasi Hills	0.203	0.813	0.943	0.573	638	0.080
	West Jaintia Hills	0.236	0.587	0.951	0.531	672	0.060
	West Khasi Hills	0.112	0.652	0.929	0.457	701	0.107
Mizoram							
	Aizawl	0.935	0.896	0.970	0.933	68	0.000
	Kolasib	0.820	0.900	0.991	0.902	142	0.002
	Champhai	0.813	0.900	0.990	0.899	146	0.002
	Serchhip	0.840	0.857	0.981	0.891	157	0.002
	Saiha	0.707	1.000	0.979	0.887	163	0.008
	Lunglei	0.754	0.833	1.000	0.858	217	0.004
	Mamit	0.655	0.857	0.980	0.822	276	0.008
	Lawngtlai	0.472	0.750	0.956	0.704	464	0.022
Nagaland							
	Dimapur	0.794	0.939	0.986	0.904	136	0.003
	Kohima	0.655	0.933	0.984	0.848	238	0.010
	Mokokchung	0.534	0.909	0.984	0.789	327	0.020
	Wokha	0.467	1.000	0.958	0.777	354	0.031
	Peren	0.345	0.800	0.900	0.644	551	0.038

(continued)

Table 6.7 (continued)

State/Union Territory/District		*SL*	*ED*	*HE*	HDS	Rank in HDS	Inequality in progress
	Zunheboto	0.240	0.800	0.976	0.602	602	0.070
	Phek	0.225	0.833	0.975	0.601	604	0.077
	Tuensang	0.187	0.739	0.967	0.548	662	0.084
	Kiphire	0.159	0.800	0.917	0.532	671	0.094
	Mon	0.124	0.800	0.982	0.517	678	0.118
	Longleng	0.130	0.750	0.953	0.503	686	0.108
Odisha							
	Khordha	0.699	0.891	0.989	0.853	227	0.007
	Puri	0.570	0.926	0.979	0.809	293	0.016
	Ganjam	0.639	0.802	0.975	0.796	316	0.009
	Cuttack	0.583	0.786	0.983	0.771	364	0.013
	Jagatsinghapur	0.515	0.884	0.972	0.770	368	0.021
	Jharsuguda	0.511	0.911	0.940	0.767	377	0.020
	Nayagarh	0.452	0.887	0.993	0.747	399	0.030
	Kendrapara	0.430	0.906	0.939	0.727	426	0.031
	Subarnapur	0.433	0.814	0.941	0.703	467	0.027
	Baleshwar	0.362	0.881	0.971	0.694	477	0.044
	Sambalpur	0.380	0.833	0.966	0.689	483	0.038
	Jajapur	0.412	0.798	0.941	0.687	486	0.029
	Bhadrak	0.343	0.903	0.957	0.686	488	0.048
	Dhenkanal	0.383	0.807	0.965	0.682	498	0.036
	Bargarh	0.370	0.825	0.972	0.682	499	0.040
	Sundargarh	0.411	0.780	0.925	0.677	507	0.028
	Anugul	0.418	0.697	0.985	0.670	519	0.030
	Balangir	0.362	0.783	0.959	0.663	527	0.039
	Baudh	0.329	0.717	0.949	0.623	574	0.042
	Kendujhar	0.290	0.747	0.945	0.610	588	0.051
	Debagarh	0.247	0.808	0.958	0.604	597	0.067
	Nuapada	0.220	0.779	0.952	0.578	631	0.073
	Gajapati	0.283	0.636	0.936	0.570	641	0.048
	Kandhamal	0.206	0.745	0.952	0.559	651	0.075
	Kalahandi	0.247	0.645	0.938	0.553	656	0.057
	Rayagada	0.260	0.604	0.962	0.553	657	0.056
	Koraput	0.220	0.635	0.965	0.539	667	0.067
	Mayurbhanj	0.154	0.698	0.953	0.509	682	0.093

(continued)

Table 6.7 (continued)

State/Union Territory/District		*SL*	*ED*	*HE*	HDS	Rank in HDS	Inequality in progress
	Malkangiri	0.189	0.633	0.889	0.502	687	0.068
	Nabarangapur	0.132	0.538	0.957	0.447	704	0.096
Puducherry							
	Mahe	1.000	1.000	1.000	1.000	1	0.000
	Puducherry	0.914	0.911	1.000	0.941	51	0.001
	Yanam	0.972	0.800	0.959	0.908	128	0.003
	Karaikal	0.839	0.893	0.986	0.904	133	0.002
Punjab							
	Hoshiarpur	0.964	0.971	0.989	0.975	13	0.000
	Pathankot	0.962	0.982	0.967	0.970	19	0.000
	Shahid Bhagat Singh Nagar	0.965	0.949	0.997	0.970	20	0.000
	Jalandhar	0.985	0.911	0.968	0.954	33	0.000
	Fatehgarh Sahib	0.970	0.895	0.985	0.949	38	0.001
	Sahibzada Ajit Singh Nagar	0.967	0.944	0.937	0.949	39	0.000
	Rupnagar	0.957	0.916	0.975	0.949	40	0.000
	Gurdaspur	0.937	0.940	0.959	0.945	45	0.000
	Kapurthala	0.949	0.905	0.969	0.941	52	0.000
	Barnala	0.947	0.914	0.950	0.937	60	0.000
	Patiala	0.963	0.879	0.967	0.936	62	0.001
	Amritsar	0.956	0.867	0.975	0.932	71	0.001
	Sangrur	0.960	0.915	0.920	0.932	72	0.000
	Ludhiana	0.955	0.880	0.960	0.931	73	0.001
	Moga	0.929	0.843	0.966	0.911	120	0.001
	Mansa	0.878	0.881	0.960	0.906	131	0.001
	Tarn Taran	0.873	0.820	0.987	0.891	156	0.002
	Muktsar	0.894	0.840	0.935	0.889	160	0.001
	Firozpur	0.901	0.805	0.960	0.887	164	0.002
	Bathinda	0.907	0.787	0.964	0.884	167	0.002
	Faridkot	0.901	0.783	0.956	0.877	183	0.002
	Fazilka	0.820	0.809	0.951	0.858	216	0.002
Rajasthan							
	Jaipur	0.836	0.925	0.966	0.908	127	0.001
	Pali	0.839	0.900	0.976	0.904	134	0.001
	Jhunjhunun	0.835	0.905	0.967	0.901	144	0.001

(continued)

Table 6.7 (continued)

State/Union Territory/District		*SL*	*ED*	*HE*	HDS	Rank in HDS	Inequality in progress
	Kota	0.835	0.902	0.956	0.897	148	0.001
	Sikar	0.804	0.880	0.964	0.881	173	0.002
	Ajmer	0.821	0.886	0.939	0.881	174	0.001
	Hanumangarh	0.786	0.858	0.975	0.871	196	0.003
	Ganganagar	0.778	0.849	0.971	0.863	206	0.003
	Nagaur	0.780	0.852	0.959	0.861	208	0.002
	Churu	0.700	0.875	0.971	0.843	243	0.006
	Bikaner	0.754	0.793	0.977	0.837	249	0.004
	Jodhpur	0.743	0.793	0.974	0.832	255	0.004
	Rajsamand	0.652	0.862	0.969	0.820	278	0.008
	Alwar	0.675	0.803	0.960	0.806	299	0.006
	Jalor	0.652	0.806	0.977	0.803	306	0.008
	Dausa	0.567	0.915	0.968	0.801	312	0.016
	Bhilwara	0.605	0.862	0.960	0.798	314	0.011
	Barmer	0.588	0.799	0.992	0.780	346	0.013
	Udaipur	0.507	0.908	0.980	0.776	357	0.023
	Chittaurgarh	0.592	0.802	0.939	0.767	376	0.010
	Jaisalmer	0.637	0.715	0.972	0.765	379	0.010
	Baran	0.544	0.836	0.959	0.764	382	0.016
	Tonk	0.532	0.851	0.953	0.762	388	0.017
	Bharatpur	0.568	0.777	0.975	0.760	389	0.014
	Bundi	0.513	0.854	0.945	0.752	397	0.019
	Sirohi	0.560	0.759	0.960	0.746	401	0.013
	Sawai Madhopur	0.525	0.826	0.928	0.744	405	0.016
	Dhaulpur	0.479	0.833	0.947	0.731	421	0.022
	Karauli	0.437	0.838	0.973	0.720	441	0.029
	Jhalawar	0.441	0.811	0.952	0.708	455	0.026
	Dungarpur	0.361	0.886	0.988	0.699	470	0.046
	Pratapgarh	0.286	0.810	0.953	0.628	569	0.056
	Banswara	0.282	0.774	0.952	0.615	582	0.055
Sikkim							
	East District	0.869	0.900	0.992	0.919	95	0.001
	South District	0.739	0.917	1.000	0.880	175	0.005
	North District	0.667	1.000	0.955	0.864	204	0.010
	West District	0.635	1.000	0.985	0.860	212	0.013

(continued)

Table 6.7 (continued)

State/Union Territory/District		*SL*	*ED*	*HE*	HDS	Rank in HDS	Inequality in progress
Tamil Nadu							
	Chennai	0.976	0.974	1.000	0.983	8	0.000
	Kanniyakumari	0.938	0.953	1.000	0.963	26	0.000
	Thiruvallur	0.908	0.931	0.981	0.940	55	0.000
	Vellore	0.854	0.976	0.980	0.935	64	0.001
	The Nilgiris	0.849	0.974	0.985	0.935	66	0.002
	Tiruppur	0.880	0.917	0.994	0.929	77	0.001
	Coimbatore	0.896	0.927	0.957	0.926	84	0.000
	Namakkal	0.846	0.938	0.971	0.917	97	0.001
	Theni	0.853	0.940	0.960	0.917	98	0.001
	Thoothukkudi	0.871	0.907	0.973	0.916	101	0.001
	Madurai	0.876	0.877	1.000	0.916	103	0.001
	Tirunelveli	0.831	0.920	0.993	0.913	114	0.002
	Kancheepuram	0.862	0.889	0.988	0.912	118	0.001
	Salem	0.778	0.944	0.970	0.894	150	0.003
	Dharmapuri	0.759	0.965	0.972	0.894	151	0.004
	Erode	0.852	0.842	0.994	0.894	152	0.002
	Ramanathapuram	0.751	0.943	0.991	0.890	158	0.005
	Krishnagiri	0.808	0.894	0.968	0.888	161	0.002
	Tiruchirappalli	0.739	0.944	0.991	0.886	165	0.005
	Thanjavur	0.741	0.946	0.971	0.882	171	0.005
	Perambalur	0.734	0.939	0.978	0.879	180	0.005
	Sivaganga	0.756	0.923	0.966	0.878	182	0.004
	Karur	0.713	0.934	0.978	0.869	197	0.006
	Virudhunagar	0.763	0.820	1.000	0.857	221	0.004
	Tiruvannamalai	0.698	0.932	0.958	0.856	223	0.006
	Cuddalore	0.687	0.925	0.955	0.849	235	0.007
	Thiruvarur	0.616	0.949	1.000	0.841	246	0.014
	Dindigul	0.666	0.907	0.960	0.837	250	0.008
	Viluppuram	0.644	0.922	0.958	0.832	257	0.009
	Nagapattinam	0.581	0.901	0.985	0.807	295	0.015
	Ariyalur	0.539	0.937	0.958	0.792	322	0.019
	Pudukkottai	0.560	0.884	0.953	0.784	336	0.015
Telangana							
	Hyderabad	0.981	0.904	0.975	0.953	34	0.000

(continued)

Table 6.7 (continued)

State/Union Territory/District		*SL*	*ED*	*HE*	HDS	Rank in HDS	Inequality in progress
	Medchal-Malkajgiri	0.963	0.909	0.981	0.951	36	0.000
	Nalgonda	0.816	0.963	0.991	0.921	92	0.002
	Jangoan	0.812	0.976	0.977	0.919	96	0.003
	Warangal Urban	0.862	0.917	0.973	0.916	102	0.001
	Yadadri Bhuvanagiri	0.862	0.938	0.948	0.915	105	0.001
	Ranga Reddy	0.913	0.891	0.942	0.915	106	0.000
	Rajanna Sircilla	0.839	0.953	0.955	0.915	107	0.001
	Karimnagar	0.859	0.891	0.992	0.912	115	0.001
	Suryapet	0.790	0.969	0.955	0.902	140	0.003
	Peddapalli	0.831	0.875	0.989	0.897	149	0.002
	Siddipet	0.775	0.929	0.974	0.890	159	0.003
	Khammam	0.815	0.853	0.989	0.883	168	0.002
	Wanaparthy	0.767	0.951	0.936	0.881	172	0.003
	Mancherial	0.717	0.923	1.000	0.874	189	0.006
	Jagitial	0.803	0.859	0.961	0.872	193	0.002
	Nagarkurnool	0.724	0.900	0.971	0.860	211	0.005
	Mahabubnagar	0.719	0.887	0.982	0.857	218	0.005
	Nizamabad	0.765	0.825	0.991	0.856	222	0.004
	Warangal Rural	0.629	0.933	0.992	0.839	247	0.012
	Mahabubabad	0.634	0.902	0.992	0.832	256	0.011
	Sangareddy	0.620	0.938	0.966	0.829	264	0.012
	Bhadradri Kothagudem	0.732	0.808	0.939	0.823	275	0.003
	Nirmal	0.648	0.849	0.976	0.816	284	0.009
	Jayashankar Bhupalapally	0.610	0.895	0.948	0.807	298	0.011
	Jogulamba Gadwal	0.701	0.745	0.978	0.801	310	0.007
	Vikarabad	0.609	0.857	0.935	0.791	324	0.010
	Kamareddy	0.576	0.864	0.971	0.790	325	0.014
	Medak	0.512	0.925	0.954	0.776	358	0.022
	Adilabad	0.579	0.662	0.971	0.724	432	0.014
	Komaram Bheem Asifabad	0.444	0.702	0.966	0.679	503	0.025
Tripura							
	West Tripura	0.547	0.886	0.970	0.784	337	0.017
	North Tripura	0.375	0.833	0.932	0.677	508	0.036

(continued)

Table 6.7 (continued)

State/Union Territory/District		*SL*	*ED*	*HE*	HDS	Rank in HDS	Inequality in progress
	Sepahijala	0.328	0.879	0.964	0.673	514	0.051
	Gomati	0.340	0.872	0.933	0.670	518	0.045
	Dhalai	0.217	0.905	0.979	0.615	581	0.085
	Khowai	0.239	0.857	0.945	0.610	589	0.071
	South Tripura	0.225	0.860	0.953	0.603	598	0.077
	Unakoti	0.219	0.778	0.958	0.578	630	0.073
Uttar Pradesh							
	Ghaziabad	0.968	0.844	0.970	0.926	85	0.001
	Gautam Buddha Nagar	0.929	0.850	0.947	0.908	129	0.001
	Baghpat	0.824	0.814	0.951	0.861	209	0.002
	Meerut	0.886	0.739	0.943	0.853	228	0.003
	Hapur	0.856	0.777	0.930	0.852	229	0.002
	Lucknow	0.748	0.794	0.962	0.831	259	0.004
	Bulandshahr	0.765	0.786	0.939	0.827	269	0.003
	Muzaffarnagar	0.822	0.695	0.981	0.826	270	0.006
	Kanpur Nagar	0.714	0.827	0.946	0.825	271	0.004
	Agra	0.812	0.718	0.939	0.819	279	0.004
	Bijnor	0.732	0.768	0.950	0.813	290	0.004
	Varanasi	0.679	0.791	0.989	0.812	291	0.007
	Etawah	0.641	0.844	0.941	0.801	308	0.008
	Saharanpur	0.783	0.677	0.938	0.794	320	0.005
	Deoria	0.570	0.864	0.985	0.791	323	0.015
	Jhansi	0.574	0.872	0.966	0.790	326	0.014
	Mathura	0.729	0.709	0.943	0.789	329	0.005
	Moradabad	0.709	0.728	0.933	0.785	334	0.005
	Jyotiba Phule Nagar	0.709	0.706	0.943	0.780	345	0.006
	Aligarh	0.690	0.735	0.924	0.778	351	0.005
	Jaunpur	0.502	0.904	0.976	0.771	365	0.023
	Firozabad	0.634	0.768	0.912	0.765	380	0.006
	Shamli	0.779	0.594	0.944	0.762	385	0.010
	Gorakhpur	0.550	0.813	0.942	0.755	393	0.014
	Mahamaya Nagar	0.567	0.778	0.923	0.745	404	0.011
	Sant Ravidas Nagar (Bhadohi)	0.488	0.871	0.927	0.741	408	0.021
	Mau	0.516	0.784	0.970	0.739	413	0.018

(continued)

Table 6.7 (continued)

State/Union Territory/District		*SL*	*ED*	*HE*	HDS	Rank in HDS	Inequality in progress
	Ballia	0.471	0.842	0.965	0.735	417	0.024
	Chandauli	0.468	0.852	0.944	0.731	420	0.024
	Mirzapur	0.466	0.851	0.947	0.730	422	0.024
	Faizabad	0.482	0.806	0.967	0.729	423	0.022
	Azamgarh	0.448	0.875	0.933	0.725	430	0.027
	Bareilly	0.679	0.565	0.961	0.722	436	0.014
	Mainpuri	0.487	0.839	0.889	0.720	438	0.018
	Rampur	0.683	0.562	0.935	0.715	446	0.012
	Jalaun	0.466	0.785	0.956	0.712	448	0.023
	Farrukhabad	0.524	0.726	0.927	0.712	449	0.014
	Ghazipur	0.403	0.867	0.972	0.711	450	0.036
	Pratapgarh	0.430	0.834	0.956	0.711	451	0.029
	Mahrajganj	0.472	0.763	0.951	0.708	456	0.021
	Allahabad	0.468	0.771	0.935	0.704	465	0.021
	Basti	0.457	0.754	0.949	0.697	474	0.023
	Kushinagar	0.435	0.746	0.965	0.689	482	0.027
	Hamirpur	0.393	0.827	0.946	0.688	484	0.034
	Mahoba	0.386	0.822	0.952	0.685	492	0.035
	Etah	0.439	0.747	0.942	0.685	493	0.024
	Auraiya	0.410	0.796	0.931	0.683	495	0.029
	Sambhal	0.540	0.598	0.931	0.674	510	0.015
	Sultanpur	0.362	0.817	0.950	0.671	517	0.039
	Ambedkar Nagar	0.347	0.827	0.963	0.668	520	0.044
	Shahjahanpur	0.481	0.669	0.900	0.667	522	0.016
	Kanpur Dehat	0.379	0.761	0.948	0.662	529	0.034
	Sant Kabir Nagar	0.369	0.738	0.965	0.654	538	0.037
	Gonda	0.397	0.683	0.951	0.647	544	0.030
	Rae Bareli	0.347	0.758	0.954	0.646	545	0.040
	Kannauj	0.382	0.726	0.923	0.646	546	0.031
	Pilibhit	0.472	0.596	0.930	0.646	547	0.020
	Kanshiram Nagar	0.463	0.597	0.921	0.640	556	0.020
	Lalitpur	0.323	0.790	0.929	0.636	561	0.044
	Unnao	0.382	0.659	0.945	0.631	565	0.031
	Amethi	0.353	0.713	0.885	0.618	580	0.032
	Siddharthnagar	0.370	0.589	0.986	0.611	585	0.037

(continued)

Table 6.7 (continued)

State/Union Territory/District		*SL*	*ED*	*HE*	HDS	Rank in HDS	Inequality in progress
	Fatehpur	0.320	0.687	0.931	0.605	594	0.041
	Kaushambi	0.305	0.677	0.933	0.594	612	0.044
	Budaun	0.401	0.499	0.947	0.584	623	0.032
	Sonbhadra	0.254	0.706	0.960	0.581	628	0.060
	Chitrakoot	0.261	0.679	0.952	0.575	634	0.056
	Bara Banki	0.333	0.557	0.930	0.569	642	0.038
	Kheri	0.288	0.643	0.904	0.567	643	0.044
	Balrampur	0.325	0.522	0.933	0.553	655	0.040
	Banda	0.234	0.669	0.932	0.551	659	0.061
	Hardoi	0.252	0.579	0.931	0.533	669	0.054
	Sitapur	0.240	0.529	0.917	0.508	683	0.055
	Shrawasti	0.195	0.449	0.955	0.461	699	0.072
	Bahraich	0.192	0.395	0.936	0.436	705	0.072
Uttarakhand							
	Dehradun	0.950	0.892	0.967	0.936	63	0.000
	Tehri Garhwal	0.698	0.986	0.987	0.882	169	0.008
	Nainital	0.793	0.865	0.979	0.876	184	0.002
	Garhwal	0.652	0.979	0.980	0.859	215	0.011
	Rudraprayag	0.654	0.964	0.969	0.852	230	0.010
	Chamoli	0.659	0.980	0.942	0.851	232	0.010
	Udham Singh Nagar	0.803	0.795	0.940	0.844	240	0.002
	Pithoragarh	0.624	0.971	0.939	0.833	253	0.012
	Uttarkashi	0.604	0.972	0.952	0.829	265	0.014
	Bageshwar	0.564	0.969	0.983	0.820	277	0.019
	Hardwar	0.812	0.656	1.000	0.814	286	0.009
	Champawat	0.562	0.950	0.956	0.806	301	0.017
	Almora	0.481	0.974	0.979	0.782	339	0.029
West Bengal							
	Kolkata	0.882	0.884	0.987	0.917	99	0.001
	North Twenty Four Parganas	0.653	0.892	0.990	0.836	251	0.009
	Darjiling	0.589	0.908	0.991	0.814	285	0.015
	Haora	0.655	0.821	0.990	0.813	287	0.009
	Paschim Barddhaman	0.636	0.809	0.945	0.789	328	0.008
	Hugli	0.536	0.880	0.976	0.779	349	0.018
	Nadia	0.386	0.911	0.977	0.717	445	0.041

(continued)

Table 6.7 (continued)

State/Union Territory/District		*SL*	*ED*	*HE*	HDS	Rank in HDS	Inequality in progress
	Jalpaiguri	0.405	0.852	0.966	0.707	460	0.034
	Purba Barddhaman	0.377	0.866	0.975	0.699	472	0.040
	South Twenty Four Parganas	0.383	0.811	0.973	0.685	491	0.037
	Maldah	0.306	0.799	0.970	0.640	555	0.052
	Murshidabad	0.287	0.826	0.955	0.633	564	0.056
	Bankura	0.238	0.903	0.965	0.625	570	0.077
	Dakshin Dinajpur	0.252	0.850	0.986	0.625	571	0.070
	Birbhum	0.287	0.770	0.974	0.622	577	0.055
	Puruliya	0.227	0.865	0.969	0.609	591	0.078
	Purba Medinipur	0.209	0.882	0.984	0.605	595	0.087
	Paschim Medinipur	0.210	0.874	0.978	0.602	601	0.085
	Uttar Dinajpur	0.239	0.783	0.961	0.593	614	0.068
	Koch Bihar	0.201	0.875	0.956	0.591	615	0.087

Source Author

References

Anand S (2018) Recasting human development measures. Working Paper 23. London School of Economics and Political Science. International Inequality Institute, London

Chaurasia AR (2013) Social class and residential disparities in human development in Madhya Pradesh. Indian J Hum Dev 7(2):275–299

Chaurasia AR (2022) Surface measure of human development: alternative to HDI. SSRN: https://ssrn.com/abstract=4228534 or https://doi.org/10.2139/ssrn.4228534

Chaurasia AR (2023) Human development in districts of India, 2019–2021. Indian J Hum Dev 17(2):219–252

Croft TN, Aileen MJ, Marshall CK Allen et al (2018) Guide to DHS statistics. ICF, Rockville, MD

Ghislandi S, Sanderson WC, Scherbov S (2019) A simple measure of human development: the human life indicator. Popul Dev Rev 45(1):219–233

Government of India (2002) National Human Development Report 2001. Planning Commission, New Delhi

Government of India (2010) Human development in India. Analysis to action. Planning Commission, New Delhi

Government of India (2011) India Human Development Report 2011: towards social inclusion. Planning Commission. Institute of Applied Manpower Research, New Delhi

Government of India (2015) Human development towards bridging inequalities. NITI Aayog, New Delhi

Government of India (2022) National Family Health Survey (NFHS-5), 2019–21. India report. Ministry of Health and Family Welfare, New Delhi

Klugman J, Rodriguez F, Choi H-J (2011) The HDI 2010: new controversies, old critiques. Human Development Research Paper, 2011/01. United Nations Development Program, New York

Kovacevic M (2010) Review of HDI critiques and potential improvements. Human Development Research Paper No. 2010/33, United Nations Development Programme, New York

Pedersen J, Liu J (2012) Child mortality estimation: appropriate time periods for child mortality estimates from full birth histories. PLoS Med 9(8):e1001289

Rutstein SO, Johnson K (2004) The DHS Wealth Index. DHS Comparative Report No. 6. ORC Macro, Calverton, MD

United Nations (1990) Human Development Report 1990. United Nations Development Program, New York

United Nations (2010) Human Development Report 2010. The real wealth of nations. Pathways to development. United Nations Development Programme, New York

United Nations (2016) Human Development Report 2016. Technical notes. United Nations, New York

United Nations (2022) Human Development Report 2021/2022. Uncertain times, uncertain lives: shaping our future in a transforming world. United Nations Development Program, New York

Chapter 7
Characteristics of Human Development

Abstract This chapter proposes a framework for characterising human development in a population in the three-dimensional space of individual capacity, individual knowledge and individual standard of living that constitutes the core of human development and to measure the inequality or the disparity in progress in the three core dimensions of human development. The framework helps in understanding the dynamics of human development and for setting priorities for investment in human development to maximise the impact of the investment on human progress. Application of the framework to characterise human development in the districts of India highlights the variation in the dynamics of human development within the country in terms of both profiles of human development and human development inequality. The variation in the dynamics of human development within the country calls for a decentralised approach for planning and programming human development activities and for prioritising investment in human development.

Keywords India · Districts · Human development · Human development characterisation

Introduction

The human development index (HDI) is the universally accepted index to measure and monitor human development. HDI reflects development in three core dimensions of human development—individual capacity, individual knowledge and individual standard of living. Macro indicators are used to measure the progress in the three dimensions of human development. These indicators have evolved over time since 1990 when the HDI was first used to rank countries in terms of human development (United Nations 1990). The HDI in the first Human Development Report prepared by the United Nations was based on an index of health to measure the progress in individual capacity, an index of education to measure the progress in individual knowledge and an index of income to measure the progress in individual standard of living. The index of health was derived from the life expectancy at birth, the index

A. R. Chaurasia, *Population and Human Development in Districts of India*,
https://doi.org/10.1007/978-981-95-8169-6_7

of education was derived from the adult literacy rate, while the index of income was derived from the gross national product per capita (United Nations 1990). Since 2010, United Nations estimates HDI based on four indicators—life expectancy at birth to construct the health index, mean years of schooling and expected years of schooling to construct the education index and gross national income per capita in terms of purchasing power parity to construct the income index (United Nations 2010). The construction of the three indexes involves setting of goal posts for the minimum and maximum possible values of the four macro indicators and normalisation so that all the three indexes range between 0 and 1. The normalised indexes are then combined into HDI through an aggregation function which has also evolved over time. Initially, the arithmetic mean was used as the aggregation function. However, since 2010, United Nations uses the geometric mean as the aggregation function to construct HDI because of the concerns raised in using the arithmetic mean as the aggregation function (Desai 1991; Klugman et al. 2011; Kovacevic 2010). There are also problems in using the geometric mean as the aggregation function in the construction of HDI (Anand 2018). Other aggregation functions have also been suggested and used (Chaurasia 2022a; Mishra and Nathan 2018). The HDI is sensitive to the aggregation function used. For the same set of health, education and income indexes, HDI is the highest when the arithmetic mean is used as the aggregation function but the lowest when their product is used to construct the HDI (Sagar and Najam 1998).

The HDI is a composite index and, therefore, its utility in characterising human development is limited. HDI is a number that corresponds to a set of three numbers—health index, education index and income index. There is always loss of some information in using a number to reflect a set of numbers or using a scalar to reflect a vector (Chaurasia 2024). There is, however, no way of capturing the entire wealth of information embodied in a set of numbers (a vector) through one number (a scalar) (Anand and Sen 1994). This means that HDI hides more than what it reveals. HDI tells little about the three indexes on which it is based. It is possible that more than one set of the three indexes has the same HDI. Since human development is a multidimensional construct, it is obvious that HDI cannot characterise human development in its true perspective. The characterisation of human development is, however, important for understanding the dynamics of human development and for prioritising investment in human development. A population which has a high health index, but a low education index will require increased investment in the knowledge dimension of human development whereas a population which has a high education index, but a low health index will require increased investment in the individual capacity dimension of human development to maximise the impact of investment on human development. It is, therefore, imperative for human development planning and programming to go beyond HDI for characterising or profiling human development to better understand the dynamics of human development. However, measuring and monitoring of human development has largely been limited to constructing and calculating HDI and ranking populations based on HDI. Measuring and monitoring human development and ranking populations in terms of HDI is, however, imperfect as the increase in HDI does not necessarily mean increase in all the three dimensions of human development. HDI may increases even if one of the three indexes that constitute HDI increases but the other two indexes remain unchanged. The HDI may also increase even if there is a decrease in one of the three indexes but a compensating increase in the other indexes.

In view of the limitations of HDI in characterising human development, this chapter proposes an analytical framework for characterising human development in a population on the three-dimensional space of individual capacity, individual knowledge, and individual standard of living. The framework classifies a population in terms of mutually exclusive yet exhaustive human development profiles and in terms of the inequality or the disparity in progress in the three dimensions. The application of the analytical framework to districts of India reveals that the districts of the country can be grouped into eight mutually exclusive and exhaustive categories depending upon the progress in the three core dimensions of human development and each category has its own unique human development profile. This characterisation of human development may contribute to more effective planning and programming for human development by prioritising investment in human development in the cuntry.

Human development matters for India as India ranked 134 out of 193 countries of the world in 2022 in terms of HDI prepared by the United Nations (2023). An accelerated progress in human development remains a major development challenge in India. There is also evidence to suggest that progress in human development varies widely within the country and it is argued that reduction in the disparity or inequality in human development within the country may contribute significantly to accelerating progress in human development in the country. It is in this context that an understanding of the dynamics of human development in the districts of the country is important as a district is the lowest tier of the development administration system in the country. Characterising human development at the district level is also the first but the most important step towards prioritising investment in human development to maximise its impact.

This chapter presents a framework for characterising human development that goes beyond any composite index based on the progress in the three core dimensions of human development. It has four sections including this introduction. The framework for characterising human development is presented in the next section. The third section characterises human development in districts using the data from National Family Health Survey (NFHS), 2019–2021 (Government of India 2022). This characterisation allows anticipation of the variation in constellations and trajectories of human development across districts. The last section summarises the key findings of the analysis and their policy relevance.

Characterising Human Development

In its core formulation, human development is represented on the three-dimensional space of individual capacity, individual knowledge and individual standard of living or a set of three elements that reflect, respectively, expanding individual capacity, enhancing individual knowledge and improving individual standard of living. The characterisation of human development on this three-dimensional space is challenging as the progress in the three dimensions of human development may be uneven within a population and across populations. The progress, say, in individual capacity dimension may be fairly advanced but the progress in other dimension, say, individual standard of living dimension may be lagging while the progress in the individual dimension may be average. Similarly, while ranking populations, rank of a population in one dimension of human development may be different from the rank in other dimensions of human development. This implies that characterising human development on the three-dimensional space is fuzzy. Different dimensions of human development may lead to different characterisation and ranking of the same population in terms of human development. The commonly followed approach to address the problem of fuzziness is to construct a composite measure, like HDI, that aggregates the progress in the three dimensions of human development in a single index through the use of an aggregation function. However, any composite measure of human development, by its very construction, tells little about how progress in different dimensions of human development is different and how progress in different dimensions contributes to the composite index. A composite index like HDI is like a black box and, therefore, contributes little to characterising human development.

One approach to addressing the fuzziness of the set that characterises human development is to establish some relationship among the three indexes that reflect the progress in the three dimensions that constitute human development. One such relationship that may serve the basis for characterising human development is the ordering of the progress in the three dimensions of human development. If the progress in the individual capacity dimension of human development is measured through an index H ($0 \leq H \leq 1$); progress in the individual knowledge dimension is measured through and index E ($0 \leq E \leq 1$); and the progress in the standard of living dimension is measured through an index I ($0 \leq I \leq 1$), then the ordering of H, E and I can be grouped into the following 12 mutually exclusive yet exhaustive human development categories or profiles (Table 7.1).

Table 7.1 Possible categories of ordering of indexes H, E and I

Profile 1:	$H = E > I$	Profile 5:	$H > I > E$	Profile 9:	$I = H > E$
Profile 2:	$I > H = E$	Profile 6:	$E > H = I$	Profile 10:	$I > H > E$
Profile 3:	$H > E = I$	Profile 7:	$E > H > I$	Profile 11:	$I > E > H$
Profile 4:	$H > E > I$	Profile 8:	$E > I > H$	Profile 12:	$H = E = I$

Source Author

It is obvious that, based on the values of H, E and I, human development in any population can be classified into one and only one of the 12 possible categories or profiles of human development. This classification addresses the problem of fuzziness inherent in the conceptualisation of human development. It is also obvious from the above classification that two populations having the same composite index of human development such as HDI may have different human development profiles. An advantage of profiling human development by ordering the progress in different dimensions of human development is that it also captures the unevenness of human development. The human development in any population may be termed as balanced only when the progress in the three dimensions of human development is the same or when H = E = I irrespective of the level of human development. Any composite index of the progress in the three dimensions of human development does not reflect how the progress in the three dimensions is ordered. The ordering of the progress in different dimensions of human development is a simple approach to understanding the dynamics of human development in the population. This ordering has imperatives for setting up priorities for investment in human development.

The profiling of human development according to Table 7.1, however, is silent about the degree of unevenness or the inequality in progress in the three dimensions of human development. The degree of unevenness in human development can be measured in terms of the inequality in progress in the three dimensions of human development and the larger the inequality in progress the larger the unevenness in human development. It is also obvious that populations having the same human development profile may differ in terms of the unevenness in human development. Reducing and ultimately eliminating the unevenness in human development is also an important component of the human development agenda as the inequality in progress in the three dimensions of human development provides the basis for prioritising investment in human development.

The unevenness in human development or the inequality in progress across different dimensions of human development can be quantified in different ways. We measure the unevenness in human development in district *j* in terms of the unevenness index U_j which is defined as

$$U_j = 1 - \frac{GM_j}{AM_j}$$

Here GM_j is the geometric mean and AM_j is the simple arithmetic mean of the three indexes H, E and I in district *j*. It may be noticed that GM_j is always less than or equal to AM_j. $GM_j = AM_j$ only when H = E = I and, in this case the unevenness index $U_j = 0$. On the other hand, the higher the unevenness in human development in a district the larger the difference between GM_j and AM_j and the higher the unevenness index U_j. The unevenness index of human development provides yet another characterisation of human development in a population.

The characterisation of human development in terms of human development profile and unevenness in human development or the inequality in progress in the three dimensions of human development facilitates better understanding of the dynamics of human development which is not possible through calculating any composite index of human development, including the HDI used by the United Nations. Different populations having the same composite index of human development may have entirely different characterisation of human development in terms of human development profile and in terms of unevenness of human development. Characterising human development in terms of human development profile and in terms of unevenness of human development is also relevant from the perspective of accelerating the progress in human development as it facilitates in prioritising investment in human development by focussing on those dimensions of human development in which the progress is lagging or by focusing on those interventions and activities which are directed towards reducing the unevenness in human development. The conventional approach of classifying human development in a population as very low, low, medium, high and very high based on some composite index of human development or ranking populations according to some composite index of human development contributes little to prioritising investment in human development and for planning and programming for accelerating the progress in human development. At the best, the classification and ranking of populations in terms of any composite index of human development necessitates the need of exploring further the relative contribution of the progress in different dimensions of human development to the composite index of human development. The human development characterisation framework outlined above provides a simple yet straightforward approach to going beyond the composite index of human development to understand the dynamics of human development in terms of the progress in different dimensions of human development and by measuring the unevenness in human development. This understanding is important to prioritising investment in human development.

Characterising Human Development in Districts

The characterisation of human development in 707 districts of India is presented in the appendix Table 7.5 to this chapter, along with the surface measure of human development (HDS) which is proposed by Chaurasia (2022a) as an alternative to HDI. The surface measure of human development (HDS) is constructed from three indicators that reflect the progress in the individual capacity, individual knowledge and individual standard of living—(1) probability of survival in the first five years of life (HE), (2) secondary school net attendance ratio (ED), defined as the proportion of secondary school age children attending a secondary school (Croft et al. 2018), and (3) proportion of individuals having wealth index equal to or more than the second quintile of the inter-household distribution of the wealth index (SL), which is a measure of the cumulative living standard of a household which is calculated using data on household ownership of selected assets; materials used for housing construction and types of water access and sanitation facilities (Rutstein and Johnson 2004). The HDS can be calculated in absolute terms as well as in relative terms (Chaurasia 2022a). Calculation of HDS in relative terms involves constructing index

H, E and I from HE, ED and SL as follows:

$$H = \frac{(HE - 0.85)}{(1.00 - 0.85)}$$

$$E = \frac{(ED - 0.35)}{(1.00 - 0.35)}$$

$$I = SL$$

The HDS is then calculated as (Chaurasia 2022):

$$HDS = \frac{\sqrt{H \times E} + \sqrt{E \times I} + \sqrt{I \times H}}{3}$$

HDS takes into account the correlation that exists among the three dimensions of human development. The rationale and construction of HDS are discussed in detail in the previous chapter and in Chaurasia (2022a). The probability of survival in the first five years of life (HE) reflects the progress in the individual capacity dimension of human development. It is also directly related to the life expectancy at birth, which is used by the United Nations in the construction of HDI. The proportion of secondary school age children attending a secondary school (ED) reflects the progress in the individual knowledge dimension of human development. Finally, the proportion of individuals having wealth index equal to or more than the second quintiles of the inter-household distribution of the wealth index (SL) reflects the progress in the individual standard of living dimension of human development. The three indicators are proportion indicators which have advantage over the average indicators used by the United Nations for the construction of HDI as discussed in the previous chapter and in Chaurasia (2023).

Figure 7.1 depicts inter-district variation in the human development profile of 707 districts of the country as they existed at the time of NFHS, 2019–2021. Table 7.2 summarises the distribution of districts by human development profile and states and Union Territories. Based on the progress in the three dimensions of human development, 707 districts of the country can be grouped into 8 human development profiles. There is only one district—district Mahe in Puducherry—where H = E = I. In addition, there are four districts—Upper Siang and Lower Dibang Valley in Arunachal Pradesh, Kozhikode in Kerala and Lakshadweep in Lakshadweep—where E = H > I. In another 43 districts of the country, I > H > E whereas in 54 districts, E > I > H. Table 7.2 also suggests there are 63 districts in which I > E > H; 73 districts in which H > I > E; 214 districts in which E > H > I; and 255 districts in which H > E > I. The distribution of districts by different human development profiles shows that that in 473 or around two-third districts of the country, progress in the dimension of individual standard of living is comparatively the slowest among the three dimensions of human development. On the other hand, progress in the individual capacity dimension of human development is comparatively the slowest in 117 or in about 16 per cent districts whereas progress in the individual knowledge dimension of human development is also comparatively the slowest in 116 or around 16 per cent districts. It is also clear from Table 7.2 that the progress in the individual standard of living dimension of human development is comparatively the best in only 106 districts of the country. The progress is comparatively the best in the individual knowledge dimensions in 272 districts whereas the progress is comparatively the best in the individual capacity dimension in 328 districts of the country. The profiling of human development suggests that the major challenge to advancing human development in most of the districts of the country is the improvement in the individual standard of living, although there are districts, where expanding the individual capacity and enhancing the individual knowledge remains major challenge to human progress.

Table 7.2 also gives the distribution of districts in different states/Union Territories of the country by human development profile. In all districts of Assam, Jharkhand, Manipur, Meghalaya and Odisha, the progress in the individual standard of living dimension of human development is relatively the slowest. In Arunachal Pradesh,

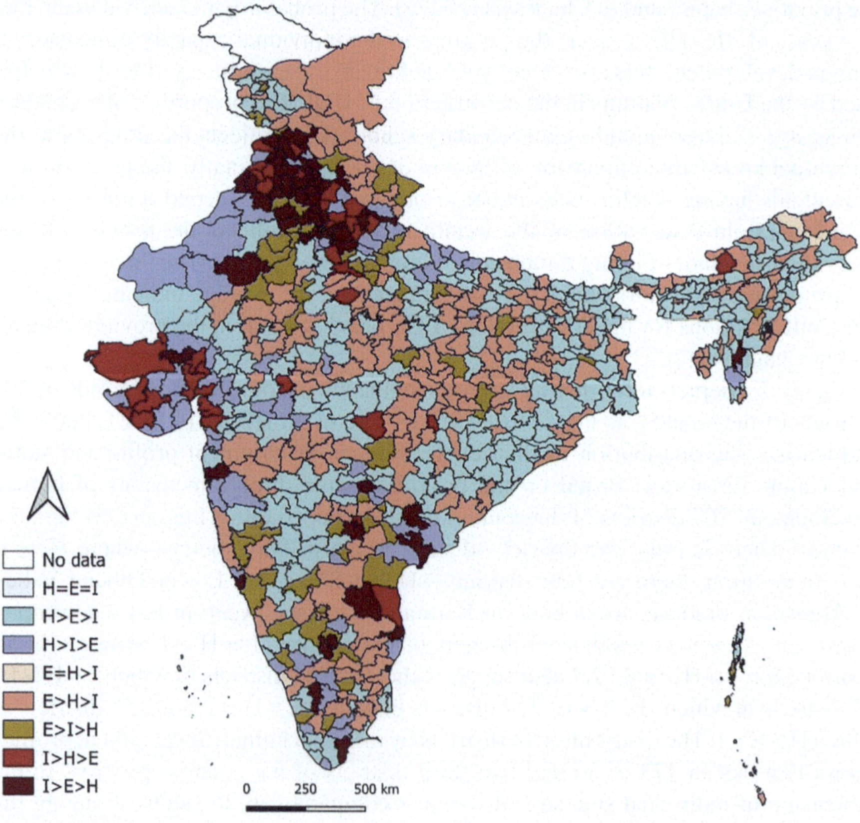

Fig. 7.1 The distribution of human development profiles across 707 districts of India, 2019–2021. *Source* Author

Table 7.2 Distribution of 707 districts by human development profile in India, 2019–2021

State/Union Territory	Human development profile								Total
	H = E = I	H > E > I	H > I > E	E = H > I	E > H > I	E > I > H	I > H > E	I > E > H	
Andaman & Nicobar Islands	0	1	0	0	1	1	0	0	3
Andhra Pradesh	0	1	2	0	6	1	1	2	13
Arunachal Pradesh	0	11	0	2	6	0	1	0	20
Assam	0	20	0	0	13	0	0	0	33
Bihar	0	19	0	0	18	1	0	0	38
Chandigarh	0	0	0	0	0	0	1	0	1
Chhattisgarh	0	10	0	0	14	2	1	0	27
Dadra & Nagar Haveli and Daman & Diu	0	0	0	0	0	0	5	6	11
Goa	0	1	1	0	0	0	1	0	3
Gujarat	0	0	1	0	0	0	0	1	2
Haryana	0	6	15	0	0	0	11	1	33
Himachal Pradesh	0	0	2	0	0	5	0	15	22
Jammu & Kashmir	0	1	1	0	5	1	1	3	12
Jharkhand	0	12	0	0	5	3	0	0	20
Karnataka	0	11	0	0	13	0	0	0	24
Kerala	0	13	2	0	6	7	0	2	30
Ladakh	0	7	2	1	3	1	0	0	14
Lakshadweep	0	0	0	0	2	0	0	0	2
Madhya Pradesh	0	0	0	1	0	0	0	0	1
Maharashtra	0	23	5	0	20	1	2	0	51

(continued)

Table 7.2 (continued)

State/Union Territory	Human development profile								Total
	H = E = I	H > E > I	H > I > E	E = H > I	E > H > I	E > I > H	I > H > E	I > E > H	
Manipur	0	13	2	0	14	3	1	3	36
Meghalaya	0	3	0	0	6	0	0	0	9
Mizoram	0	9	0	0	2	0	0	0	11
Nagaland	0	4	2	0	1	0	0	1	8
National Capital Territory of Delhi	0	7	0	0	3	1	0	0	11
Odisha	0	20	0	0	10	0	0	0	30
Puducherry	1	0	2	0	0	0	1	0	4
Punjab	0	0	2	0	0	1	6	13	22
Rajasthan	0	11	5	0	10	6	0	1	33
Sikkim	0	1	1	0	2	0	0	0	4
Tamil Nadu	0	5	7	0	14	3	0	3	32
Telangana	0	7	6	0	7	6	1	4	31
Tripura	0	2	0	0	6	0	0	0	8
Uttar Pradesh	0	23	12	0	16	8	10	6	75
Uttarakhand	0	0	2	0	7	2	0	2	13
West Bengal	0	14	1	0	4	1	0	0	20
India	1	255	73	4	214	54	43	63	707

Source Author

Bihar, Chhattisgarh, Jammu & Kashmir, Kerala, Nagaland and West Bengal also, the progress in the individual standard of living dimension of human development is relatively the slowest in all but a few districts. The progress in the individual knowledge dimension is found to be relatively the slowest in 26 of the 33 districts of Gujarat whereas the progress in the individual capacity dimension is found to be relatively the slowest in 20 of the 22 districts in Haryana, in 15 of 22 districts in Himachal Pradesh and in 14 of 22 districts of Punjab. It is obvious that the dynamics of human development is different in different districts. The diversity in the characterisation of human development across the districts has implications for prioritising investment in human development.

The unevenness in human development or the inequality in the progress in the three dimensions of human development, measured in terms of the unevenness index, U, also varies widely across districts as may be seen from Fig. 7.2 and Table 7.3. There is only one district—district Mahe in Puducherry—in which, there is no unevenness in human development whereas in 198 (28 per cent) districts, the unevenness in human development is very low ($0 < U < 0.005$). In these districts, the progress in the three dimensions of human development is nearly the same. In another 77 (11 per cent) districts, the unevenness in human development may be termed as low as $0.005 \leq U < 0.010$. However, in 178 (25 per cent) districts, the unevenness in human development is very high ($U \geq 0.050$). In these districts, progress in one of the three dimensions of human development is very advanced but, at the same time, progress lags significantly in one or both of the remaining dimensions of human development so that the progress in human development in these districts is significantly compromised by the unevenness in the progress in different dimensions of human development. Similarly, in 142 (20 per cent) districts of the country, the unevenness in human development is high ($0.020 \leq U < 0.050$). In other words in 320 or more than 45 per cent districts of the country, there is very marked imbalance in progress in different dimensions of human development. The unevenness in human development is the highest in district Bahraich of Uttar Pradesh. In this district, $H = 0.575$, $I = 0.192$ but $E = 0.069$ so that the unevenness index $U = 0.294$.

Figure 7.2 also shows that unevenness in human development appears to be relatively high in districts in the eastern parts of the country including all north-eastern states, except Mizoram, and in Gujarat. There are only 9 districts in Gujarat where unevenness in human development is either low or very low. The unevenness in human development is high or very high in all districts of Meghalaya and Tripura; in 7 of the 9 districts of Manipur; in 32 of the 33 districts of Assam, in 9 of the 11 districts of Nagaland; in 16 of the 18 districts of Arunachal Pradesh. However, there is no district in Mizoram where the unevenness in human development is either high or very high. Similarly, the unevenness in human development is found to be either high or very high in all districts of Jharkhand, in 37 of the 38 districts of Bihar; in 28 of 30 districts in Odisha; in 25 of the 27 districts of Chhattisgarh; and in 28 of the 33 districts in Gujarat (Table 7.3). There are only eight states/Union Territories—Chandigarh, Goa, Kerala, Lakshadweep, Mizoram, National Capital Territory of Delhi, Puducherry and Sikkim—in which there is no district where the unevenness of human development is found to be either high or very high. On the other

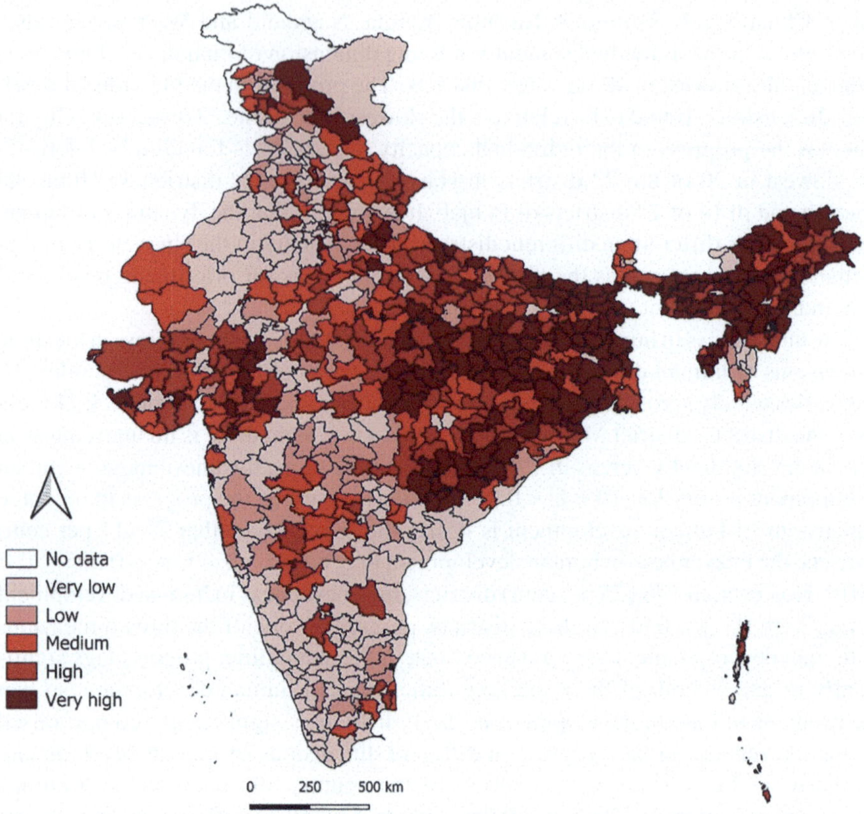

Fig. 7.2 Unevenness of human development in districts of India, 2019–2021. *Source* Author

hand, the unevenness in human development is either low or very low in 11 of the 13 districts in Andhra Pradesh, in 9 of the 11 districts in the National Capital Territory of Delhi, in 18 of the 22 districts in Haryana, in 8 of the 12 districts in Himachal Pradesh, in 18 of the 22 districts in Punjab, in 15 of the 20 districts in Jammu and Kashmir and in 27 of the 32 districts in Tamil Nadu. The variation in the unevenness in human development across districts may be attributed to district-specific factors that differentially affect the progress in the three dimensions of human development.

Table 7.4 classifies 707 districts according to both human development profile and unevenness in human development. There are 473 districts in which progress is relatively the slowest in the individual standard of living dimension of human development. Out of these 473 districts, the unevenness in human development is either high or very high in 273 or close to 60 per cent districts. Alternatively, out of 320 districts, in which the unevenness in human development or the inequality in progress in different dimensions of human development is either high or very high, the progress in the individual standard of living dimension is relatively the slowest in 273 or more than 85 per cent districts. On the other hand, there are 116 districts

Table 7.3 Distribution of 707 districts across states and Union Territories by index of unevenness in human development (U).

	No unevenness	Very low	Low	Medium	High	Very high	Total
Andaman & Nicobar Islands	0	1	0	0	2	0	3
Andhra Pradesh	0	10	1	1	1	0	13
Arunachal Pradesh	0	2	0	2	5	11	20
Assam	0	1	0	0	5	27	33
Bihar	0	1	0	4	10	23	38
Chandigarh	0	1	0	0	0	0	1
Chhattisgarh	0	2	0	6	8	11	27
Dadra & Nagar Haveli and Daman & Diu	0	1	1	0	1	0	3
Goa	0	2	0	0	0	0	2
Gujarat	0	4	1	7	16	5	33
Haryana	0	16	2	1	2	1	22
Himachal Pradesh	0	7	1	3	0	1	12
Jammu & Kashmir	0	8	3	4	4	1	20
Jharkhand	0	0	0	4	2	18	24
Karnataka	0	17	3	7	3	0	30
Kerala	0	14	0	0	0	0	14
Ladakh	0	1	0	0	0	1	2
Lakshadweep	0	1	0	0	0	0	1
Madhya Pradesh	0	9	8	8	13	13	51
Maharashtra	0	14	12	7	2	1	36
Manipur	0	0	1	1	4	3	9
Meghalaya	0	0	0	0	2	9	11
Mizoram	0	4	3	1	0	0	8
Nagaland	0	1	0	1	2	7	11
National Capital Territory of Delhi	0	7	2	2	0	0	11
Odisha	0	0	2	3	11	14	30
Puducherry	1	2	0	1	0	0	4
Punjab	0	12	6	2	2	0	22
Rajasthan	0	14	4	9	3	3	33
Sikkim	0	1	1	2	0	0	4
Tamil Nadu	0	21	6	3	2	0	32
Telangana	0	13	9	6	3	0	31
Tripura	0	0	0	1	1	6	8

(continued)

Table 7.3 (continued)

	No unevenness	Very low	Low	Medium	High	Very high	Total
Uttar Pradesh	0	7	10	17	30	11	75
Uttarakhand	0	2	1	4	6	0	13
West Bengal	0	2	0	4	2	12	20
India	1	198	77	111	142	178	707

Remarks:
No unevenness if $U = 0$
Very low unevenness if $0 < U < 0.005$
Low unevenness if $0.005 \le U < 0.010$
Medium unevenness if $0.010 \le U < 0.020$
High unevenness if $0.020 \le U < 0.050$
Very high unevenness if $U \ge 0.050$
Source Author

where the progress is relatively the slowest in the individual knowledge dimension of human development and the unevenness in human development is either high or very high in only 29 of these districts. Similarly, there are 117 districts in which the progress is relatively the slowest in the individual capacity dimension of human development and the unevenness in human development is either high or very high in only 18 of these 117 districts. It is obvious that districts where the unevenness in human development is either high or very high are mostly districts in which the progress in the individual standard of living dimension of human development is relatively the slowest. It may also be seen from Table 7.4 that there are 106 districts in which the progress is relatively the fastest in the individual standard of living dimension and in 76 or almost 72 per cent of these districts, the unevenness of human development is either low or very low. Table 7.4 suggests that the unevenness in human development in a district is primarily determined by the progress in the individual standard of living dimension of human development. It is also clear that, despite the above average progress in individual capacity and individual knowledge dimensions of human development, the progress in the individual standard of living dimension of human development lags in many districts of the country. It appears that progress in individual capacity and individual knowledge dimensions of human development does not lead to progress in individual standard of living dimension of human development. Investment in improving the standard of living of the people appears to be important to hasten the progress in human development. The pathway to accelerating human progress is through improving the standard of living of the people.

Table 7.4 Classification of 707 districts by human development profile and unevenness in human development in the district

	Unevenness in human development						Total
Human development profile	No unevenness	Very low	Low	Average	High	Very high	
H = E = I	1	0	0	0	0	0	1
H > E > I	0	36	21	39	60	99	255
H > I > E	0	36	7	6	15	9	73
E = H > I	0	2	0	0	1	1	4
E > H > I	0	36	24	42	49	63	214
E > I > H	0	30	7	6	6	5	54
I > H > E	0	22	7	9	5	0	43
I > E > H	0	36	11	9	6	1	63
Total	1	198	77	111	142	178	707

Source Author

Discussions and Conclusions

Since its introduction in 1990, the HDI and its different variants have been the mainstay of measuring and monitoring the progress in human development across countries and within countries, although the conceptual and operational framework of HDI has been subject to frequent criticism. This chapter emphasises that HDI or its different variants, by its very construction, contributes little to planning and programming for human development and for prioritising investment in human development. The construction and estimation of HDI or its different variants has, therefore, largely remained an academic exercise and for ranking populations—the higher the HDI the more advanced the human development and vice versa. United Nations is ranking countries based on HDI since 1990 as an integral component of its annual Human Development Report. A high rank in HDI rank is a matter of aggrandisement for the country whereas a low rank is a matter of development concern for the country. The HDI, however, informs little about how to accelerate the progress in human development and to prioritise investment in human development to maximise the impact of investment on advancing human development. This limitation calls for going beyond HDI and its many variants to understand the dynamics of human development. There has been little effort in this direction. There is a need to characterise human development in a manner that facilitates evidence-based planning and programming for advancing human progress.

The human development characterisation framework presented in this chapter is an attempt to characterise human development in terms of human development profile, defined in terms of the ordering the progress in the three core dimensions of human development and measuring the unevenness of human development or the inequality in progress in the three dimensions of human development. The characterisation framework may serve the basis for the understanding the dynamics of human development in a systematic and articulate manner. It helps in monitoring the progress in human development in terms of both change in the human development profile and the change the unevenness of human development or the inequality in progress in different dimensions of human development. Characterising human development following the characterisation framework presented here can be a useful support to prioritising investment in human development as priorities for investment are different in different profiles of human development. By providing an understanding of the dynamics of human development, the characterisation framework facilitates evidence-based planning and programming for human development activities to address the diversity in human development that is so pervasive in India.

Application of the characterisation framework to characterise human development in districts of India highlights the diversity in the dynamics of human development within the country. The 707 districts of the country can be categorised into 8 categories having different human development profiles and, in more than two-third districts of the country, the dynamics of human development is characterised by relatively the slowest progress in the individual standard of living dimension of human development. Most of these districts are also those districts in which the unevenness of human development or the inequality in progress in different dimensions of human development is either high or very high. In these districts, investment in human development needs to be directed towards improving the individual standard of living to accelerate the progress in human development. At the same time, there are districts where investment in human development needs to be directed towards expanding individual capacity and increasing individual knowledge. The diversity in the dynamics of human development across the districts of the country also emphasises the need of a decentralised district-based approach to foster human development. This may be a challenging proposition as development planning and programming in India has always been highly centralised and top-down in nature, despite the recognition of the highly pervasive diversity in social and economic development within the country since independence.

Human development movement in India has never been strong despite repeated commitment at the highest level of the development policy. The country could produce only two official human development reports, one in 2001 and the other in 2011 (Government of India 2002, 2011) but no institutional mechanism could be evolved within the development administration system of the country to measure and monitor the progress in human development even at national or sub-national levels if not at the district level to imbibe sensitiveness to human development issues and concerns in the development planning and programming process. Recently, the Government of India has issued a working paper that provides estimates of HDI for the country and for its selected states for the period 2017–2018 using a set of indicators that are different from the indicator set used by the United Nations in the calculation of HDI (Government of India, n.d.). The Government of India had also launched two projects, one in 2010 and the other in 2015 to shift the focus of social and economic development planning and programming in the country towards human development

(Government of India 2010, 2015). However, both these time bound projects failed to enthuse human development orientation in the development planning and programming within the country. A noble initiative that was taken under these initiatives was to prepare state and district human development reports. However, no institutional system could be evolved either at the state or at the district level and the activity was largely confined to project period only. Another limitation has been the lack of a unified framework for measuring and monitoring human development. The index of human development calculated to monitor the progress in human development was invariably based on the available indicators of individual capacity, individual knowledge and individual standard of living and, therefore, lacked comparability over time and place.

The present chapter suggests that the main pathway to accelerate the progress in human development in most of the districts of the country is to accelerate the progress in the individual standard of living dimension of human development. There are 217 districts in the country where the progress in this dimension of human development is relatively the slowest among the three dimensions of human development. The very slow progress in this dimension of human development is responsible for either very high or high unevenness in human development and slow human progress. These districts may be regarded as hotspot districts of India in terms of human progress. Investment in improving the standard of living of the people in these districts may contribute to reducing the unevenness in human progress in these districts which, in turn, may contribute to accelerating the pace of human progress in the country.

Appendix

See Table 7.5.

Table 7.5 Characterisation of human development in 707 districts of India as they existed at the time of National Family Health Survey, 2019–2021

State	District	Index of progress in			Human development		HDS
		Individual capacity H	Individual knowledge E	Individual standard of living I	Profile	Unevenness	
Andaman & Nicobar Islands							
	Nicobar	0.723	1.000	0.523	E > H > I	0.026	0.730
	North & Middle Andaman	0.956	0.860	0.526	H > E > I	0.023	0.763
	South Andaman	0.861	0.927	0.919	E > I > H	0.000	0.902
Andhra Pradesh							
	Srikakulam	0.863	0.923	0.676	E > H > I	0.006	0.815
	Vizianagaram	0.734	0.910	0.583	E > H > I	0.012	0.733
	Visakhapatnam	0.742	0.800	0.735	E > H > I	0.001	0.759
	East Godavari	0.849	0.810	0.762	H > E > I	0.001	0.807
	West Godavari	0.875	0.801	0.867	H > I > E	0.001	0.847
	Krishna	0.722	0.731	0.755	I > E > H	0.000	0.736
	Guntur	0.841	0.719	0.811	H > I > E	0.002	0.789
	Prakasam	0.687	0.799	0.788	E > I > H	0.002	0.757
	Sri Potti Sriramulu Nellore	0.668	0.650	0.785	I > H > E	0.003	0.699
	Y.S.R	0.522	0.802	0.864	I > E > H	0.017	0.717
	Kurnool	0.695	0.699	0.649	E > H > I	0.000	0.681
	Anantapur	0.792	0.812	0.711	E > H > I	0.001	0.771
	Chittoor	0.802	0.841	0.768	E > H > I	0.001	0.803

(continued)

Table 7.5 (continued)

State	District	Index of progress in			Human development		HDS
		Individual capacity H	Individual knowledge E	Individual standard of living I	Profile	Unevenness	
Arunachal Pradesh							
	Tawang	0.992	1.000	0.824	E > H > I	0.003	0.936
	West Kameng	0.818	0.780	0.850	I > H > E	0.000	0.815
	East Kameng	0.710	0.780	0.358	E > H > I	0.039	0.592
	Papum Pare	0.990	0.829	0.686	H > E > I	0.008	0.828
	Upper Subansiri	0.968	0.692	0.192	H > E > I	0.128	0.538
	Upper Siang	1.000	1.000	0.367	E = H > I	0.066	0.737
	Changlang	1.000	0.890	0.350	H > E > I	0.065	0.698
	Lower Subansiri	0.918	0.780	0.534	H > E > I	0.018	0.731
	Dibang Valley	0.881	1.000	0.412	E > H > I	0.048	0.728
	Lower Dibang Valley	1.000	1.000	0.550	E = H > I	0.026	0.828
	Anjaw	0.880	1.000	0.179	E > H > I	0.146	0.586
	East Siang	1.000	0.808	0.584	H > E > I	0.018	0.783
	Kra Daadi	0.829	1.000	0.443	E > H > I	0.040	0.727
	Kurung Kumey	0.943	0.744	0.239	H > E > I	0.100	0.578
	Lohit	1.000	0.692	0.553	H > E > I	0.023	0.731
	Longding	1.000	0.615	0.162	H > E > I	0.154	0.501
	Namsai	0.902	0.615	0.233	H > E > I	0.096	0.527
	Siang	0.914	1.000	0.279	E > H > I	0.093	0.663

(continued)

Table 7.5 (continued)

State	District	Index of progress in			Human development		HDS
		Individual capacity H	Individual knowledge E	Individual standard of living I	Profile	Unevenness	
	Tirap	0.761	0.692	0.394	H > E > I	0.028	0.599
	West Siang	0.844	0.808	0.540	H > E > I	0.014	0.720
Assam							
	Kokrajhar	0.740	0.794	0.308	E > H > I	0.056	0.580
	Goalpara	0.796	0.692	0.251	H > E > I	0.077	0.535
	Barpeta	0.815	0.753	0.252	H > E > I	0.081	0.557
	Morigaon	0.653	0.668	0.214	E > H > I	0.080	0.471
	Lakhimpur	0.840	0.753	0.283	H > E > I	0.070	0.582
	Dhemaji	0.841	0.865	0.181	E > H > I	0.131	0.547
	Tinsukia	0.583	0.663	0.354	E > H > I	0.025	0.520
	Dibrugarh	0.846	0.595	0.413	H > E > I	0.031	0.599
	Golaghat	0.805	0.669	0.261	H > E > I	0.072	0.537
	Dima Hasao	0.937	0.732	0.392	H > E > I	0.044	0.657
	Cachar	0.833	0.622	0.259	H > E > I	0.075	0.529
	Karimganj	0.691	0.601	0.216	H > E > I	0.077	0.464
	Hailakandi	0.318	0.644	0.203	E > H > I	0.083	0.356
	Bongaigaon	0.839	0.793	0.349	H > E > I	0.049	0.628
	Chirang	0.754	0.772	0.230	E > H > I	0.088	0.534
	Kamrup	0.881	0.739	0.411	H > E > I	0.035	0.653

(continued)

Table 7.5 (continued)

State	District	Index of progress in			Human development		HDS
		Individual capacity H	Individual knowledge E	Individual standard of living I	Profile	Unevenness	
	Kamrup Metropolitan	0.814	0.821	0.776	E > H > I	0.000	0.804
	Nalbari	0.915	0.765	0.329	H > E > I	0.061	0.629
	Baksa	0.810	0.753	0.190	H > E > I	0.115	0.517
	Darrang	0.825	0.673	0.232	H > E > I	0.088	0.526
	Udalguri	0.661	0.698	0.218	E > H > I	0.081	0.483
	Biswanath	0.525	0.704	0.251	E > H > I	0.060	0.464
	Charaideo	0.673	0.552	0.246	H > E > I	0.058	0.462
	Dhubri	0.750	0.696	0.214	H > E > I	0.091	0.503
	Hojai	0.807	0.772	0.361	H > E > I	0.043	0.619
	Jorhat	0.621	0.796	0.466	E > H > I	0.017	0.617
	Karbi Anglong	0.790	0.796	0.294	E > H > I	0.065	0.586
	Majuli	0.722	0.838	0.183	E > H > I	0.121	0.511
	Nagaon	0.574	0.593	0.243	E > H > I	0.052	0.445
	Sivasagar	0.861	0.692	0.463	H > E > I	0.023	0.657
	Sonitpur	0.725	0.604	0.320	H > E > I	0.040	0.528
	South Salmara Mancachar	0.873	0.615	0.113	H > E > I	0.182	0.437
	West Karbi Anglong	0.794	0.792	0.121	H > E > I	0.172	0.471
Bihar							

(continued)

Table 7.5 (continued)

State	District	Index of progress in			Human development		HDS
		Individual capacity H	Individual knowledge E	Individual standard of living I	Profile	Unevenness	
	Pashchim Champaran	0.374	0.473	0.220	E > H > I	0.035	0.343
	Purba Champaran	0.606	0.548	0.277	H > E > I	0.039	0.459
	Sheohar	0.570	0.627	0.207	E > H > I	0.073	0.434
	Sitamarhi	0.452	0.541	0.212	E > H > I	0.052	0.381
	Madhubani	0.702	0.600	0.220	H > E > I	0.077	0.468
	Supaul	0.712	0.445	0.081	H > E > I	0.198	0.331
	Araria	0.627	0.445	0.135	H > E > I	0.118	0.355
	Kishanganj	0.636	0.344	0.172	H > E > I	0.096	0.347
	Purnia	0.481	0.403	0.179	H > E > I	0.057	0.334
	Katihar	0.695	0.479	0.155	H > E > I	0.114	0.393
	Madhepura	0.650	0.450	0.101	H > E > I	0.159	0.337
	Saharsa	0.505	0.497	0.153	H > E > I	0.087	0.351
	Darbhanga	0.658	0.560	0.272	H > E > I	0.047	0.474
	Muzaffarpur	0.740	0.770	0.315	E > H > I	0.052	0.577
	Gopalganj	0.728	0.663	0.367	H > E > I	0.030	0.568
	Siwan	0.597	0.772	0.442	E > H > I	0.019	0.592
	Saran	0.603	0.692	0.358	E > H > I	0.027	0.536
	Vaishali	0.522	0.709	0.329	E > H > I	0.035	0.502
	Samastipur	0.611	0.643	0.174	E > H > I	0.099	0.429

(continued)

Table 7.5 (continued)

State	District	Index of progress in			Human development		HDS
		Individual capacity H	Individual knowledge E	Individual standard of living I	Profile	Unevenness	
	Begusarai	0.669	0.688	0.304	E > H > I	0.045	0.529
	Khagaria	0.666	0.578	0.247	H > E > I	0.058	0.468
	Bhagalpur	0.576	0.730	0.462	E > H > I	0.013	0.581
	Banka	0.731	0.672	0.239	H > E > I	0.074	0.507
	Munger	0.784	0.712	0.516	H > E > I	0.011	0.663
	Lakhisarai	0.750	0.666	0.377	H > E > I	0.030	0.580
	Sheikhpura	0.511	0.736	0.392	E > H > I	0.025	0.533
	Nalanda	0.669	0.608	0.446	H > E > I	0.011	0.568
	Patna	0.698	0.619	0.608	H > E > I	0.001	0.641
	Bhojpur	0.755	0.756	0.456	E > H > I	0.019	0.643
	Buxar	0.601	0.727	0.466	E > H > I	0.012	0.591
	Kaimur (Bhabua)	0.490	0.714	0.300	E > H > I	0.044	0.479
	Rohtas	0.371	0.823	0.518	E > I > H	0.040	0.548
	Aurangabad	0.686	0.756	0.390	E > H > I	0.028	0.594
	Gaya	0.748	0.640	0.328	H > E > I	0.041	0.549
	Nawada	0.807	0.716	0.404	H > E > I	0.030	0.623
	Jamui	0.509	0.617	0.258	E > H > I	0.045	0.440
	Jehanabad	0.681	0.819	0.358	E > H > I	0.041	0.594
	Arwal	0.497	0.777	0.301	E > H > I	0.053	0.497

(continued)

Table 7.5 (continued)

State	District	Index of progress in			Human development		HDS
		Individual capacity H	Individual knowledge E	Individual standard of living I	Profile	Unevenness	
Chandigarh							
	Chandigarh	0.886	0.805	0.967	I > H > E	0.002	0.884
Chhattisgarh							
	Koriya	0.217	0.831	0.352	E > I > H	0.113	0.414
	Jashpur	0.675	0.684	0.146	E > H > I	0.130	0.437
	Raigarh	0.390	0.775	0.425	E > I > H	0.037	0.510
	Korba	0.484	0.641	0.463	E > H > I	0.008	0.525
	Janjgir - Champa	0.799	0.861	0.460	E > H > I	0.026	0.688
	Kabeerdham	0.833	0.824	0.504	H > E > I	0.019	0.707
	Rajnandgaon	0.648	0.871	0.550	E > H > I	0.014	0.680
	Mahasamund	0.760	0.678	0.404	H > E > I	0.025	0.598
	Dhamtari	0.906	0.859	0.618	H > E > I	0.010	0.786
	Uttar Bastar Kanker	0.853	0.895	0.342	E > H > I	0.059	0.655
	Narayanpur	0.744	0.457	0.217	H > E > I	0.083	0.433
	Bijapur	0.450	0.310	0.148	H > E > I	0.068	0.282
	Balod	0.944	0.945	0.573	E > H > I	0.019	0.805
	Baloda Bazar	0.677	0.711	0.467	E > H > I	0.012	0.611
	Balrampur	0.582	0.612	0.146	E > H > I	0.114	0.396
	Bastar	0.501	0.417	0.269	H > E > I	0.024	0.386

(continued)

Table 7.5 (continued)

State	District	Index of progress in			Human development		HDS
		Individual capacity H	Individual knowledge E	Individual standard of living I	Profile	Unevenness	
	Bemetara	0.768	0.797	0.527	E > H > I	0.012	0.689
	Bilaspur	0.696	0.712	0.474	E > H > I	0.012	0.620
	Dantewada	0.683	0.571	0.252	H > E > I	0.058	0.472
	Durg	0.858	0.796	0.867	I > H > E	0.001	0.840
	Gariyaband	0.682	0.730	0.348	E > H > I	0.036	0.565
	Kodagaon	0.583	0.595	0.169	E > H > I	0.095	0.406
	Mungeli	0.624	0.738	0.370	E > H > I	0.029	0.561
	Raipur	0.854	0.758	0.757	H > E > I	0.001	0.789
	Sukma	0.524	0.423	0.092	H > E > I	0.145	0.296
	Surajpur	0.771	0.752	0.228	H > E > I	0.089	0.531
	Surguja	0.565	0.703	0.254	E > H > I	0.059	0.477
Dadra & Nagar Haveli and Daman & Diu							
	Diu	0.899	0.823	0.963	I > H > E	0.002	0.894
	Daman	0.729	0.868	0.961	I > E > H	0.005	0.849
	Dadra & Nagar Haveli	0.857	0.849	0.960	I > H > E	0.001	0.888
Goa							
	North Goa	0.792	0.735	0.928	I > H > E	0.004	0.815
	South Goa	0.797	0.841	0.990	I > E > H	0.003	0.873

(continued)

Table 7.5 (continued)

State	District	Index of progress in			Human development		HDS
		Individual capacity H	Individual knowledge E	Individual standard of living I	Profile	Unevenness	
Gujarat							
	Kachchh	0.893	0.895	0.987	I > E > H	0.001	0.924
	Banas Kantha	0.673	0.798	0.987	I > E > H	0.009	0.812
	Patan	0.651	0.896	0.989	I > E > H	0.011	0.836
	Mahesana	0.897	0.888	0.972	I > H > E	0.001	0.919
	Gandhinagar	0.934	0.898	0.981	I > H > E	0.000	0.937
	Porbandar	0.754	0.825	0.969	I > E > H	0.004	0.846
	Amreli	1.000	0.615	0.955	H > I > E	0.016	0.843
	Anand	0.717	0.692	0.867	I > H > E	0.004	0.756
	Dohad	0.669	0.530	0.520	H > E > I	0.005	0.570
	Narmada	1.000	0.946	0.966	H > I > E	0.000	0.971
	Bharuch	0.893	0.908	0.971	I > E > H	0.000	0.923
	The Dangs	0.745	0.405	0.821	I > H > E	0.032	0.636
	Navsari	0.663	0.355	0.479	H > I > E	0.024	0.487
	Valsad	0.401	0.447	0.648	I > E > H	0.016	0.491
	Surat	0.579	0.526	0.665	I > H > E	0.003	0.588
	Tapi	0.715	0.670	0.816	I > H > E	0.003	0.732
	Ahmadabad	0.880	0.567	0.896	I > H > E	0.015	0.769
	Aravali	0.868	0.410	0.781	H > I > E	0.036	0.662

(continued)

Table 7.5 (continued)

State	District	Index of progress in			Human development		HDS
		Individual capacity H	Individual knowledge E	Individual standard of living I	Profile	Unevenness	
	Bhavnagar	0.564	0.501	0.697	I > H > E	0.007	0.583
	Botad	0.823	0.376	0.184	H > E > I	0.126	0.403
	Chhota Udaipur	0.691	0.487	0.305	H > E > I	0.040	0.475
	Devbhumi Dwarka	0.649	0.497	0.711	I > H > E	0.008	0.614
	Gir Somnath	0.835	0.341	0.227	H > E > I	0.111	0.415
	Jamnagar	0.879	0.580	0.729	H > I > E	0.011	0.722
	Junagadh	0.847	0.700	0.725	H > I > E	0.003	0.755
	Kheda	0.827	0.598	0.870	I > H > E	0.010	0.758
	Mahisagar	0.728	0.444	0.404	H > E > I	0.026	0.512
	Morbi	0.743	0.590	0.934	I > H > E	0.013	0.746
	Panch Mahals	0.741	0.518	0.458	H > E > I	0.016	0.563
	Rajkot	0.865	0.471	0.717	H > I > E	0.022	0.669
	Sabar Kantha	0.867	0.378	0.810	H > I > E	0.044	0.654
	Surendranagar	0.508	0.306	0.329	H > I > E	0.020	0.373
	Vadodara	0.807	0.154	0.776	H > I > E	0.143	0.497
Haryana							
	Panchkula	0.839	0.557	0.751	H > I > E	0.011	0.708
	Ambala	0.859	0.559	0.915	I > H > E	0.016	0.765
	Yamunanagar	0.743	0.575	0.817	I > H > E	0.008	0.706

(continued)

Table 7.5 (continued)

State	District	Index of progress in			Human development		HDS
		Individual capacity H	Individual knowledge E	Individual standard of living I	Profile	Unevenness	
	Kurukshetra	0.561	0.350	0.555	H > I > E	0.016	0.480
	Kaithal	0.803	0.483	0.394	H > E > I	0.035	0.540
	Karnal	0.928	0.466	0.868	H > I > E	0.031	0.730
	Panipat	0.592	0.409	0.474	H > I > E	0.009	0.488
	Sonipat	0.864	0.734	0.934	I > H > E	0.004	0.841
	Jind	0.785	0.425	0.587	H > I > E	0.023	0.585
	Fatehabad	0.779	0.455	0.721	H > I > E	0.019	0.639
	Sirsa	0.844	0.537	0.831	H > I > E	0.015	0.726
	Hisar	0.735	0.870	0.968	I > E > H	0.005	0.854
	Rohtak	0.546	0.829	0.939	I > E > H	0.019	0.757
	Jhajjar	0.781	0.816	0.918	I > E > H	0.002	0.837
	Mahendragarh	0.719	0.793	0.914	I > E > H	0.004	0.806
	Rewari	0.693	0.900	0.914	I > E > H	0.006	0.831
	Gurgaon	0.814	0.834	0.912	I > E > H	0.001	0.853
	Mewat	0.730	0.887	0.897	I > E > H	0.003	0.835
	Faridabad	0.765	0.903	0.940	I > E > H	0.003	0.867
	Palwal	0.629	0.916	0.906	E > I > H	0.011	0.808
	Bhiwani	0.738	0.894	0.885	E > I > H	0.003	0.837
	Charkhi Dadri	0.722	0.800	0.879	I > E > H	0.002	0.799

(continued)

Table 7.5 (continued)

State	District	Index of progress in			Human development		HDS
		Individual capacity H	Individual knowledge E	Individual standard of living I	Profile	Unevenness	
Himachal Pradesh							
	Chamba	0.810	0.862	0.910	I > E > H	0.001	0.860
	Kangra	0.833	0.881	0.927	I > E > H	0.001	0.880
	Lahul & Spiti	0.943	0.955	0.954	E > I > H	0.000	0.951
	Kullu	0.764	0.954	0.851	E > I > H	0.003	0.853
	Mandi	0.796	0.902	0.926	I > E > H	0.002	0.873
	Hamirpur	0.771	0.863	0.940	I > E > H	0.002	0.856
	Una	0.606	0.271	0.561	H > I > E	0.041	0.459
	Bilaspur	0.778	0.804	0.949	I > E > H	0.003	0.841
	Solan	0.874	0.755	0.805	H > I > E	0.001	0.810
	Sirmaur	0.527	0.889	0.891	I > E > H	0.020	0.753
	Shimla	0.782	0.975	0.935	E > I > H	0.003	0.894
	Kinnaur	0.802	0.911	0.673	E > H > I	0.006	0.791
Jammu & Kashmir							
	Kupwara	0.702	0.829	0.855	I > E > H	0.003	0.793
	Badgam	0.970	1.000	0.481	E > H > I	0.037	0.787
	Punch	0.890	0.933	0.661	E > H > I	0.008	0.821
	Rajouri	0.857	0.919	0.820	E > H > I	0.001	0.865
	Kathua	0.814	0.846	0.872	I > E > H	0.000	0.844

(continued)

Table 7.5 (continued)

State	District	Index of progress in			Human development		HDS
		Individual capacity H	Individual knowledge E	Individual standard of living I	Profile	Unevenness	
	Baramula	0.868	0.819	0.900	I > H > E	0.001	0.862
	Bandipore	0.948	0.825	0.899	H > I > E	0.001	0.890
	Srinagar	0.604	0.846	0.902	I > E > H	0.011	0.775
	Ganderbal	0.867	0.860	0.824	H > E > I	0.000	0.850
	Pulwama	0.768	0.911	0.837	E > I > H	0.002	0.837
	Shupiyan	0.789	1.000	0.695	E > H > I	0.009	0.821
	Anantnag	0.743	0.868	0.581	E > H > I	0.010	0.723
	Kulgam	0.969	0.884	0.773	H > E > I	0.003	0.873
	Doda	0.795	0.846	0.742	E > H > I	0.001	0.794
	Ramban	0.837	1.000	0.388	E > H > I	0.053	0.703
	Kishtwar	0.817	0.922	0.581	E > H > I	0.013	0.763
	Udhampur	0.926	0.808	0.621	H > E > I	0.010	0.777
	Reasi	0.874	0.939	0.847	E > H > I	0.001	0.886
	Jammu	0.837	0.700	0.648	H > E > I	0.004	0.725
	Samba	0.792	0.766	0.584	H > E > I	0.007	0.709
Jharkhand							
	Garhwa	0.921	0.968	0.947	E > I > H	0.000	0.946
	Chatra	0.915	0.752	0.687	H > E > I	0.005	0.780
	Kodarma	0.827	0.923	0.869	E > I > H	0.001	0.872

(continued)

Table 7.5 (continued)

State	District	Index of progress in			Human development		HDS
		Individual capacity H	Individual knowledge E	Individual standard of living I	Profile	Unevenness	
	Giridih	1.000	0.860	0.803	H > E > I	0.003	0.885
	Deoghar	1.000	0.928	0.830	H > E > I	0.002	0.917
	Godda	0.885	0.886	0.542	E > H > I	0.018	0.757
	Sahibganj	0.870	0.841	0.514	H > E > I	0.020	0.727
	Pakur	0.956	0.792	0.370	H > E > I	0.053	0.669
	Dhanbad	0.930	0.856	0.522	H > E > I	0.022	0.753
	Bokaro	0.870	0.923	0.606	E > H > I	0.012	0.790
	Lohardaga	0.905	0.842	0.506	H > E > I	0.023	0.734
	Purbi Singhbhum	0.902	0.971	0.952	E > I > H	0.000	0.941
	Palamu	0.938	0.917	0.876	H > E > I	0.000	0.910
	Latehar	0.682	0.734	0.173	E > H > I	0.115	0.469
	Hazaribagh	0.680	0.649	0.196	H > E > I	0.091	0.462
	Ramgarh	0.754	0.864	0.399	E > H > I	0.037	0.648
	Dumka	0.708	0.702	0.331	H > E > I	0.040	0.557
	Jamtara	0.782	0.574	0.194	H > E > I	0.101	0.465
	Ranchi	0.530	0.573	0.213	E > H > I	0.061	0.412
	Khunti	0.790	0.458	0.161	H > E > I	0.127	0.410
	Gumla	0.728	0.420	0.117	H > E > I	0.157	0.356
	Simdega	0.841	0.765	0.555	H > E > I	0.011	0.712

(continued)

Table 7.5 (continued)

State	District	Index of progress in			Human development		HDS
		Individual capacity H	Individual knowledge E	Individual standard of living I	Profile	Unevenness	
	Pashchimi Singhbhum	0.716	0.805	0.541	E > H > I	0.010	0.680
	Saraikela-Kharsawan	0.724	0.789	0.234	E > H > I	0.086	0.532
Karnataka							
	Belgaum	0.679	0.801	0.551	E > H > I	0.009	0.671
	Bagalkot	0.624	0.715	0.235	E > H > I	0.072	0.487
	Bijapur	0.753	0.655	0.094	H > E > I	0.190	0.406
	Bidar	0.694	0.819	0.434	E > H > I	0.025	0.633
	Raichur	0.728	0.786	0.484	E > H > I	0.016	0.656
	Koppal	0.605	0.670	0.173	E > H > I	0.102	0.434
	Gadag	0.810	0.683	0.197	H > E > I	0.106	0.504
	Dharwad	0.718	0.832	0.546	E > H > I	0.011	0.691
	Uttara Kannada	0.740	0.576	0.112	H > E > I	0.163	0.398
	Haveri	0.556	0.726	0.108	E > H > I	0.165	0.387
	Bellary	0.760	0.712	0.119	H > E > I	0.166	0.442
	Chitradurga	0.699	0.631	0.091	H > E > I	0.186	0.386
	Davanagere	0.528	0.759	0.298	E > H > I	0.050	0.502
	Shimoga	0.831	0.857	0.670	E > H > I	0.004	0.783
	Udupi	0.817	0.779	0.533	H > E > I	0.013	0.701
	Chikmagalur	0.872	0.714	0.520	H > E > I	0.016	0.690

(continued)

Table 7.5 (continued)

State	District	Index of progress in			Human development		HDS
		Individual capacity H	Individual knowledge E	Individual standard of living I	Profile	Unevenness	
	Tumkur	0.793	0.818	0.539	E > H > I	0.012	0.708
	Bangalore	0.638	0.598	0.497	H > E > I	0.004	0.575
	Mandya	0.715	0.680	0.448	H > E > I	0.015	0.605
	Hassan	0.891	0.848	0.567	H > E > I	0.014	0.758
	Dakshina Kannada	0.730	0.888	0.762	E > I > H	0.003	0.791
	Kodagu	0.757	0.833	0.786	E > I > H	0.001	0.791
	Mysore	0.768	0.734	0.582	H > E > I	0.005	0.691
	Chamarajanagar	0.549	0.788	0.643	E > I > H	0.008	0.655
	Gulbarga	0.716	0.826	0.700	E > H > I	0.002	0.746
	Yadgir	0.845	0.786	0.743	H > E > I	0.001	0.790
	Kolar	0.857	0.880	0.833	E > H > I	0.000	0.857
	Chikkaballapura	0.900	0.942	0.837	E > H > I	0.001	0.892
	Bangalore Rural	0.797	0.908	0.826	E > I > H	0.001	0.842
	Ramanagara	0.503	0.986	0.823	E > I > H	0.027	0.750
Kerala							
	Kasaragod	0.878	0.921	0.969	I > E > H	0.001	0.922
	Kannur	1.000	0.926	0.864	H > E > I	0.001	0.929
	Wayanad	0.930	0.915	0.812	H > E > I	0.001	0.885
	Kozhikode	0.947	0.924	0.903	H > E > I	0.000	0.924

(continued)

Table 7.5 (continued)

State	District	Index of progress in			Human development		HDS
		Individual capacity H	Individual knowledge E	Individual standard of living I	Profile	Unevenness	
	Malappuram	0.829	0.887	0.868	E > I > H	0.000	0.861
	Palakkad	0.910	0.830	0.858	H > I > E	0.001	0.866
	Thrissur	0.777	0.826	0.748	E > H > I	0.001	0.783
	Ernakulam	0.854	0.658	0.554	H > E > I	0.012	0.680
	Idukki	0.719	0.592	0.458	H > E > I	0.012	0.582
	Kottayam	0.917	0.911	0.892	H > E > I	0.000	0.907
	Alappuzha	0.851	0.778	0.849	H > I > E	0.001	0.826
	Pathanamthitta	0.791	0.894	0.891	E > I > H	0.001	0.858
	Kollam	0.609	0.848	0.901	I > E > H	0.010	0.778
	Thiruvananthapuram	0.965	0.960	0.916	H > E > I	0.000	0.947
Ladakh							
	Leh (Ladakh)	1.000	0.950	0.968	H > I > E	0.000	0.972
	Kargil	0.916	0.929	0.815	E > H > I	0.001	0.886
Lakshadweep							
	Lakshadweep	1.000	1.000	0.980	E = H > I	0.000	0.993
Madhya Pradesh							
	Sheopur	0.969	1.000	0.966	E > H > I	0.000	0.978
	Morena	1.000	0.981	0.888	H > E > I	0.001	0.955
	Bhind	1.000	0.976	0.971	H > E > I	0.000	0.982

(continued)

Table 7.5 (continued)

State	District	Index of progress in			Human development		HDS
		Individual capacity H	Individual knowledge E	Individual standard of living I	Profile	Unevenness	
	Gwalior	0.927	0.985	0.982	E > I > H	0.000	0.964
	Datia	0.926	0.950	0.832	E > H > I	0.001	0.902
	Shivpuri	1.000	0.954	0.956	H > I > E	0.000	0.970
	Tikamgarh	1.000	0.980	0.968	H > E > I	0.000	0.983
	Chhatarpur	1.000	0.984	0.934	H > E > I	0.000	0.972
	Panna	1.000	0.989	0.943	H > E > I	0.000	0.977
	Sagar	1.000	0.989	0.925	H > E > I	0.000	0.971
	Damoh	1.000	1.000	0.981	E = H > I	0.000	0.993
	Satna	0.559	0.474	0.323	H > E > I	0.018	0.444
	Rewa	0.573	0.664	0.545	E > H > I	0.003	0.592
	Umaria	0.714	0.775	0.573	E > H > I	0.006	0.683
	Neemuch	0.814	0.770	0.828	I > H > E	0.000	0.804
	Mandsaur	0.612	0.695	0.516	E > H > I	0.005	0.605
	Ratlam	0.569	0.567	0.394	H > E > I	0.010	0.505
	Ujjain	0.737	0.689	0.476	H > E > I	0.013	0.626
	Dewas	0.559	0.574	0.413	E > H > I	0.008	0.511
	Dhar	0.338	0.577	0.234	E > H > I	0.051	0.363
	Indore	0.737	0.738	0.421	E > H > I	0.023	0.617

(continued)

Table 7.5 (continued)

State	District	Index of progress in			Human development		HDS
		Individual capacity H	Individual knowledge E	Individual standard of living I	Profile	Unevenness	
	Khargone (West Nimar)	0.605	0.628	0.307	E > H > I	0.035	0.495
	Barwani	0.637	0.602	0.391	H > E > I	0.016	0.534
	Rajgarh	0.519	0.498	0.246	H > E > I	0.038	0.405
	Vidisha	0.526	0.573	0.310	E > H > I	0.025	0.458
	Bhopal	0.752	0.728	0.657	H > E > I	0.001	0.711
	Sehore	0.599	0.729	0.576	E > H > I	0.004	0.632
	Raisen	0.686	0.559	0.592	H > I > E	0.003	0.611
	Betul	0.841	0.720	0.667	H > E > I	0.004	0.740
	Harda	0.682	0.695	0.575	E > H > I	0.003	0.649
	Hoshangabad	0.820	0.586	0.478	H > E > I	0.019	0.616
	Katni	0.804	0.768	0.895	I > H > E	0.002	0.821
	Jabalpur	0.724	0.551	0.647	H > I > E	0.005	0.638
	Narsimhapur	0.788	0.342	0.371	H > I > E	0.057	0.472
	Dindori	0.715	0.683	0.350	H > E > I	0.034	0.563
	Mandla	0.615	0.515	0.460	H > E > I	0.005	0.527
	Chhindwara	0.893	0.691	0.850	H > I > E	0.004	0.808
	Seoni	0.601	0.746	0.495	E > H > I	0.011	0.608
	Balaghat	0.700	0.706	0.466	E > H > I	0.013	0.616

(continued)

Table 7.5 (continued)

State	District	Index of progress in			Human development		HDS
		Individual capacity H	Individual knowledge E	Individual standard of living I	Profile	Unevenness	
	Guna	0.763	0.688	0.391	H > E > I	0.029	0.596
	Ashoknagar	0.752	0.660	0.597	H > E > I	0.003	0.667
	Shahdol	0.679	0.639	0.531	H > E > I	0.004	0.614
	Anuppur	0.309	0.584	0.348	E > I > H	0.030	0.401
	Sidhi	1.000	0.606	0.395	H > E > I	0.052	0.632
	Singrauli	0.841	0.593	0.438	H > E > I	0.026	0.607
	Jhabua	0.634	0.660	0.146	E > H > I	0.124	0.421
	Alirajpur	0.708	0.720	0.297	E > H > I	0.052	0.545
	Khandwa (East Nimar)	0.675	0.641	0.462	H > E > I	0.010	0.587
	Burhanpur	0.668	0.623	0.305	H > E > I	0.040	0.511
	Agar Malwa	0.698	0.790	0.285	E > H > I	0.062	0.554
	Shajapur	0.663	0.531	0.428	H > E > I	0.012	0.535
Maharashtra							
	Nandurbar	0.714	0.581	0.331	H > E > I	0.035	0.523
	Dhule	0.393	0.658	0.280	E > H > I	0.046	0.423
	Jalgaon	0.645	0.689	0.286	E > H > I	0.049	0.513
	Buldana	0.539	0.574	0.232	E > H > I	0.052	0.425
	Akola	0.586	0.613	0.320	E > H > I	0.029	0.492
	Washim	0.555	0.472	0.181	H > E > I	0.072	0.374

(continued)

Table 7.5 (continued)

State	District	Index of progress in			Human development		HDS
		Individual capacity H	Individual knowledge E	Individual standard of living I	Profile	Unevenness	
	Amravati	0.801	0.293	0.171	H > E > I	0.147	0.359
	Wardha	0.823	0.697	0.533	H > E > I	0.012	0.677
	Nagpur	0.641	0.518	0.612	H > I > E	0.003	0.589
	Bhandara	0.870	0.672	0.522	H > E > I	0.016	0.677
	Gondiya	0.602	0.695	0.526	E > H > I	0.005	0.605
	Gadchiroli	0.614	0.543	0.362	H > E > I	0.018	0.497
	Chandrapur	0.896	0.637	0.594	H > E > I	0.013	0.700
	Yavatmal	0.868	0.727	0.711	H > E > I	0.003	0.766
	Nanded	0.792	0.902	0.667	E > H > I	0.006	0.783
	Hingoli	0.792	0.819	0.776	E > H > I	0.000	0.795
	Parbhani	0.724	0.812	0.590	E > H > I	0.006	0.704
	Jalna	0.873	0.870	0.754	H > E > I	0.002	0.831
	Aurangabad	0.896	0.837	0.836	H > E > I	0.000	0.856
	Nashik	0.931	0.925	0.932	I > H > E	0.000	0.929
	Mumbai Suburban	0.714	0.876	0.677	E > H > I	0.005	0.752
	Mumbai	0.642	0.936	0.607	E > H > I	0.015	0.718
	Raigarh	0.760	0.812	0.507	E > H > I	0.015	0.683
	Pune	0.687	0.928	0.760	E > I > H	0.006	0.787
	Ahmadnagar	0.668	0.819	0.644	E > H > I	0.004	0.707

(continued)

Table 7.5 (continued)

State	District	Index of progress in			Human development		HDS
		Individual capacity H	Individual knowledge E	Individual standard of living I	Profile	Unevenness	
	Bid	0.760	0.719	0.597	H > E > I	0.004	0.689
	Latur	0.768	0.804	0.608	E > H > I	0.005	0.723
	Osmanabad	0.799	0.767	0.559	H > E > I	0.009	0.702
	Solapur	0.732	0.767	0.577	E > H > I	0.006	0.688
	Satara	0.845	0.850	0.740	E > H > I	0.002	0.810
	Ratnagiri	0.654	0.715	0.640	E > H > I	0.001	0.669
	Sindhudurg	0.237	0.947	0.979	I > E > H	0.113	0.640
	Kolhapur	0.887	0.932	0.968	I > E > H	0.000	0.929
	Sangli	0.867	0.767	0.814	H > I > E	0.001	0.815
	Palghar	1.000	0.769	0.882	H > I > E	0.004	0.880
	Thane	0.827	0.838	0.724	E > H > I	0.002	0.795
Manipur							
	Senapati	0.864	0.817	0.577	H > E > I	0.011	0.744
	Tamenglong	0.814	0.808	0.686	H > E > I	0.002	0.768
	Churachandpur	0.965	0.862	0.667	H > E > I	0.009	0.824
	Bishnupur	0.873	0.794	0.661	H > E > I	0.005	0.772
	Thoubal	1.000	0.820	0.755	H > E > I	0.005	0.854
	Imphal West	0.742	0.928	0.586	E > H > I	0.013	0.742
	Imphal East	0.844	0.961	0.687	E > H > I	0.007	0.825

(continued)

Table 7.5 (continued)

State	District	Index of progress in			Human development		HDS
		Individual capacity H	Individual knowledge E	Individual standard of living I	Profile	Unevenness	
	Ukhrul	0.777	0.935	0.817	E > I > H	0.002	0.841
	Chandel	0.804	0.930	0.813	E > I > H	0.002	0.848
Meghalaya							
	South Garo Hills	0.919	0.682	0.675	H > E > I	0.008	0.753
	Ribhoi	0.862	0.882	0.896	I > E > H	0.000	0.880
	East Khasi Hills	0.904	0.872	0.302	H > E > I	0.074	0.641
	East Garo Hills	0.666	0.846	0.189	E > H > I	0.115	0.502
	East Jantia Hills	0.837	0.799	0.422	H > E > I	0.032	0.665
	North Garo Hills	0.816	0.872	0.430	E > H > I	0.033	0.683
	South West Garo Hills	0.809	0.836	0.422	E > H > I	0.032	0.667
	South West Khasi Hills	0.940	0.897	0.697	H > E > I	0.006	0.840
	West Garo Hills	0.716	0.858	0.597	E > H > I	0.008	0.718
	West Jaintia Hills	0.773	0.846	0.164	E > H > I	0.138	0.512
	West Khasi Hills	0.745	0.780	0.410	E > H > I	0.028	0.627
Mizoram							
	Mamit	0.963	0.910	0.268	H > E > I	0.095	0.646
	Kolasib	0.759	0.599	0.261	H > E > I	0.064	0.505
	Aizawl	1.000	0.661	0.472	H > E > I	0.035	0.686
	Champhai	0.938	0.795	0.330	H > E > I	0.063	0.644

(continued)

Table 7.5 (continued)

State	District	Index of progress in			Human development		HDS
		Individual capacity H	Individual knowledge E	Individual standard of living I	Profile	Unevenness	
	Serchhip	0.648	0.532	0.274	H > E > I	0.044	0.463
	Lunglei	0.825	0.882	0.277	E > H > I	0.080	0.608
	Lawngtlai	0.822	0.732	0.301	H > E > I	0.061	0.581
	Saiha	0.623	0.712	0.203	E > H > I	0.088	0.467
Nagaland							
	Mon	0.951	0.908	0.480	H > E > I	0.032	0.755
	Mokokchung	0.670	0.365	0.236	H > E > I	0.067	0.395
	Zunheboto	0.529	0.465	0.112	H > E > I	0.125	0.322
	Wokha	0.863	0.780	0.655	H > E > I	0.005	0.763
	Dimapur	0.943	0.846	0.820	H > E > I	0.001	0.868
	Phek	0.800	0.840	0.935	I > E > H	0.002	0.857
	Tuensang	0.936	0.846	0.813	H > E > I	0.001	0.864
	Longleng	0.870	0.780	0.840	H > I > E	0.001	0.829
	Kiphire	1.000	0.744	0.754	H > I > E	0.007	0.826
	Kohima	0.704	0.615	0.472	H > E > I	0.010	0.591
	Peren	0.857	1.000	0.707	E > H > I	0.007	0.848
NCT of Delhi							
	Central	0.878	0.692	0.124	H > E > I	0.172	0.468
	East	0.891	0.860	0.534	H > E > I	0.018	0.748

(continued)

Table 7.5 (continued)

State	District	Index of progress in			Human development		HDS
		Individual capacity H	Individual knowledge E	Individual standard of living I	Profile	Unevenness	
	New Delhi	0.840	0.692	0.240	H > E > I	0.086	0.540
	North	0.719	1.000	0.467	E > H > I	0.034	0.704
	North East	0.908	0.907	0.794	H > E > I	0.001	0.868
	North West	0.832	0.744	0.225	H > E > I	0.096	0.543
	Shahdara	0.783	0.599	0.187	H > E > I	0.106	0.467
	South	0.689	0.615	0.130	H > E > I	0.140	0.411
	South East	0.445	0.692	0.159	E > H > I	0.111	0.384
	South West	0.894	0.897	0.655	E > H > I	0.008	0.809
	West	0.336	0.692	0.345	E > I > H	0.045	0.437
Odisha							
	Bargarh	0.814	0.730	0.370	H > E > I	0.039	0.613
	Jharsuguda	0.597	0.863	0.511	E > H > I	0.019	0.645
	Sambalpur	0.776	0.744	0.380	H > E > I	0.035	0.611
	Debagarh	0.718	0.704	0.247	H > E > I	0.072	0.516
	Sundargarh	0.503	0.661	0.411	E > H > I	0.014	0.517
	Kendujhar	0.636	0.611	0.290	H > E > I	0.041	0.491
	Mayurbhanj	0.685	0.536	0.154	H > E > I	0.114	0.406
	Baleshwar	0.809	0.817	0.362	E > H > I	0.045	0.633
	Bhadrak	0.712	0.850	0.343	E > H > I	0.049	0.604

(continued)

Table 7.5 (continued)

State	District	Index of progress in			Human development		HDS
		Individual capacity H	Individual knowledge E	Individual standard of living I	Profile	Unevenness	
	Kendrapara	0.592	0.855	0.430	E > H > I	0.029	0.607
	Jagatsinghapur	0.810	0.822	0.515	E > H > I	0.016	0.704
	Cuttack	0.886	0.671	0.583	H > E > I	0.012	0.705
	Jajapur	0.606	0.689	0.412	E > H > I	0.017	0.559
	Dhenkanal	0.766	0.704	0.383	H > E > I	0.031	0.599
	Anugul	0.897	0.534	0.418	H > E > I	0.039	0.592
	Nayagarh	0.957	0.826	0.452	H > E > I	0.035	0.719
	Khordha	0.926	0.832	0.699	H > E > I	0.005	0.815
	Puri	0.861	0.886	0.570	E > H > I	0.014	0.761
	Ganjam	0.835	0.695	0.639	H > E > I	0.005	0.719
	Gajapati	0.574	0.441	0.283	H > E > I	0.030	0.420
	Kandhamal	0.682	0.607	0.206	H > E > I	0.082	0.457
	Baudh	0.659	0.565	0.329	H > E > I	0.030	0.502
	Subarnapur	0.608	0.714	0.433	E > H > I	0.015	0.576
	Balangir	0.729	0.666	0.362	H > E > I	0.032	0.567
	Nuapada	0.680	0.660	0.220	H > E > I	0.078	0.479
	Kalahandi	0.589	0.454	0.247	H > E > I	0.044	0.411
	Rayagada	0.744	0.391	0.260	H > E > I	0.070	0.433
	Nabarangapur	0.715	0.289	0.132	H > E > I	0.157	0.319

(continued)

Table 7.5 (continued)

State	District	Index of progress in			Human development		HDS
		Individual capacity H	Individual knowledge E	Individual standard of living I	Profile	Unevenness	
	Koraput	0.766	0.438	0.220	H > E > I	0.087	0.434
	Malkangiri	0.258	0.435	0.189	E > H > I	0.045	0.281
Puducherry							
	Yanam	0.730	0.692	0.972	I > H > E	0.009	0.791
	Puducherry	1.000	0.863	0.914	H > I > E	0.001	0.925
	Mahe	1.000	1.000	1.000	H = E = I	0.000	1.000
	Karaikal	0.908	0.835	0.839	H > I > E	0.001	0.860
Punjab							
	Kapurthala	0.795	0.853	0.949	I > E > H	0.002	0.864
	Jalandhar	0.785	0.863	0.985	I > E > H	0.003	0.875
	Hoshiarpur	0.929	0.956	0.964	I > E > H	0.000	0.950
	Shahid Bhagat Singh Nagar	0.977	0.922	0.965	H > I > E	0.000	0.954
	Fatehgarh Sahib	0.903	0.838	0.970	I > H > E	0.001	0.903
	Ludhiana	0.732	0.816	0.955	I > E > H	0.005	0.831
	Moga	0.772	0.758	0.929	I > H > E	0.003	0.817
	Muktsar	0.569	0.753	0.894	I > E > H	0.013	0.730
	Faridkot	0.703	0.666	0.901	I > H > E	0.007	0.751
	Bathinda	0.762	0.672	0.907	I > H > E	0.006	0.776

(continued)

Table 7.5 (continued)

State	District	Index of progress in			Human development		HDS
		Individual capacity H	Individual knowledge E	Individual standard of living I	Profile	Unevenness	
	Mansa	0.734	0.817	0.878	I > E > H	0.002	0.808
	Patiala	0.781	0.814	0.963	I > E > H	0.003	0.850
	Amritsar	0.835	0.795	0.956	I > H > E	0.002	0.860
	Tarn Taran	0.913	0.723	0.873	H > I > E	0.004	0.833
	Rupnagar	0.832	0.870	0.957	I > E > H	0.001	0.885
	Sahibzada Ajit Singh Nagar	0.582	0.914	0.967	I > E > H	0.018	0.806
	Sangrur	0.465	0.869	0.960	I > E > H	0.034	0.739
	Barnala	0.667	0.868	0.947	I > E > H	0.008	0.821
	Fazilka	0.672	0.706	0.820	I > E > H	0.003	0.731
	Firozpur	0.735	0.700	0.901	I > H > E	0.005	0.775
	Gurdaspur	0.729	0.907	0.937	I > E > H	0.004	0.854
	Pathankot	0.778	0.973	0.962	E > I > H	0.004	0.901
Rajasthan							
	Ganganagar	0.810	0.767	0.778	H > I > E	0.000	0.785
	Hanumangarh	0.835	0.782	0.786	H > I > E	0.000	0.801
	Bikaner	0.845	0.682	0.754	H > I > E	0.003	0.758
	Churu	0.808	0.807	0.700	H > E > I	0.002	0.770
	Jhunjhunun	0.777	0.854	0.835	E > I > H	0.001	0.821

(continued)

Table 7.5 (continued)

State	District	Index of progress in			Human development		HDS
		Individual capacity H	Individual knowledge E	Individual standard of living I	Profile	Unevenness	
	Alwar	0.734	0.697	0.675	H > E > I	0.000	0.702
	Bharatpur	0.833	0.657	0.568	H > E > I	0.009	0.680
	Dhaulpur	0.650	0.744	0.479	E > H > I	0.012	0.616
	Karauli	0.818	0.751	0.437	H > E > I	0.026	0.651
	Sawai Madhopur	0.521	0.732	0.525	E > I > H	0.010	0.587
	Dausa	0.785	0.869	0.567	E > H > I	0.012	0.732
	Jaipur	0.773	0.885	0.836	E > I > H	0.001	0.830
	Sikar	0.762	0.816	0.804	E > I > H	0.000	0.794
	Nagaur	0.725	0.772	0.780	I > E > H	0.000	0.759
	Jodhpur	0.825	0.681	0.743	H > I > E	0.002	0.748
	Jaisalmer	0.815	0.562	0.637	H > I > E	0.009	0.665
	Barmer	0.944	0.690	0.588	H > E > I	0.015	0.730
	Jalor	0.848	0.701	0.652	H > E > I	0.005	0.730
	Sirohi	0.731	0.630	0.560	H > E > I	0.004	0.637
	Pali	0.841	0.847	0.839	E > H > I	0.000	0.842
	Ajmer	0.591	0.824	0.821	E > I > H	0.009	0.739
	Tonk	0.685	0.770	0.532	E > H > I	0.009	0.657
	Bundi	0.632	0.775	0.513	E > H > I	0.011	0.633
	Bhilwara	0.736	0.788	0.605	E > H > I	0.005	0.706

(continued)

Table 7.5 (continued)

State	District	Index of progress in			Human development		HDS
		Individual capacity H	Individual knowledge E	Individual standard of living I	Profile	Unevenness	
	Rajsamand	0.794	0.787	0.652	H > E > I	0.003	0.742
	Dungarpur	0.920	0.824	0.361	H > E > I	0.054	0.664
	Banswara	0.682	0.652	0.282	H > E > I	0.051	0.511
	Chittaurgarh	0.592	0.696	0.592	E > H > I	0.002	0.625
	Kota	0.708	0.849	0.835	E > I > H	0.002	0.795
	Baran	0.724	0.748	0.544	E > H > I	0.007	0.667
	Jhalawar	0.683	0.710	0.441	E > H > I	0.016	0.601
	Udaipur	0.868	0.858	0.507	H > E > I	0.021	0.729
	Pratapgarh	0.690	0.708	0.286	E > H > I	0.054	0.531
Sikkim							
	North District	0.697	1.000	0.667	E > H > I	0.013	0.778
	West District	0.899	1.000	0.635	E > H > I	0.013	0.834
	South District	1.000	0.872	0.739	H > E > I	0.006	0.865
	East District	0.946	0.846	0.869	H > I > E	0.001	0.886
Tamil Nadu							
	Thiruvallur	0.875	0.894	0.908	I > E > H	0.000	0.892
	Chennai	1.000	0.959	0.976	H > I > E	0.000	0.978
	Kancheepuram	0.923	0.829	0.862	H > I > E	0.001	0.871
	Vellore	0.868	0.963	0.854	E > H > I	0.001	0.894

(continued)

Table 7.5 (continued)

State	District	Index of progress in			Human development		HDS
		Individual capacity H	Individual knowledge E	Individual standard of living I	Profile	Unevenness	
	Tiruvannamalai	0.720	0.895	0.698	E > H > I	0.005	0.767
	Viluppuram	0.721	0.879	0.644	E > H > I	0.006	0.743
	Salem	0.802	0.914	0.778	E > H > I	0.002	0.830
	Namakkal	0.806	0.905	0.846	E > I > H	0.001	0.852
	Erode	0.963	0.757	0.852	H > I > E	0.004	0.854
	The Nilgiris	0.903	0.960	0.849	E > H > I	0.001	0.903
	Dindigul	0.736	0.856	0.666	E > H > I	0.004	0.750
	Karur	0.851	0.899	0.713	E > H > I	0.004	0.818
	Tiruchirappalli	0.941	0.914	0.739	H > E > I	0.004	0.861
	Perambalur	0.855	0.906	0.734	E > H > I	0.003	0.829
	Ariyalur	0.722	0.902	0.539	E > H > I	0.016	0.710
	Cuddalore	0.699	0.885	0.687	E > H > I	0.005	0.753
	Nagapattinam	0.901	0.848	0.581	H > E > I	0.013	0.767
	Thiruvarur	1.000	0.922	0.616	H > E > I	0.015	0.833
	Thanjavur	0.809	0.917	0.741	E > H > I	0.003	0.820
	Pudukkottai	0.685	0.822	0.560	E > H > I	0.009	0.683
	Sivaganga	0.770	0.882	0.756	E > H > I	0.002	0.801
	Madurai	1.000	0.810	0.876	H > I > E	0.003	0.893
	Theni	0.734	0.908	0.853	E > I > H	0.003	0.829

(continued)

Table 7.5 (continued)

State	District	Index of progress in			Human development		HDS
		Individual capacity H	Individual knowledge E	Individual standard of living I	Profile	Unevenness	
	Virudhunagar	1.000	0.723	0.763	H > I > E	0.008	0.822
	Ramanathapuram	0.941	0.913	0.751	H > E > I	0.004	0.865
	Thoothukkudi	0.820	0.857	0.871	I > E > H	0.000	0.849
	Tirunelveli	0.951	0.876	0.831	H > E > I	0.001	0.885
	Kanniyakumari	1.000	0.928	0.938	H > I > E	0.000	0.955
	Dharmapuri	0.812	0.946	0.759	E > H > I	0.003	0.836
	Krishnagiri	0.786	0.838	0.808	E > I > H	0.000	0.810
	Coimbatore	0.712	0.887	0.896	I > E > H	0.004	0.828
	Tiruppur	0.958	0.873	0.880	H > I > E	0.001	0.903
Telangana							
	Adilabad	0.806	0.480	0.579	H > I > E	0.018	0.611
	Bhadradri Kothagudem	0.591	0.705	0.732	I > E > H	0.003	0.674
	Hyderabad	0.833	0.852	0.981	I > E > H	0.002	0.887
	Jagitial	0.738	0.783	0.803	I > E > H	0.000	0.774
	Jangoan	0.846	0.963	0.812	E > H > I	0.002	0.872
	Jayashankar Bhupalapally	0.654	0.838	0.610	E > H > I	0.007	0.696
	Jogulamba Gadwal	0.850	0.607	0.701	H > I > E	0.007	0.714

(continued)

Table 7.5 (continued)

State	District	Index of progress in			Human development		HDS
		Individual capacity H	Individual knowledge E	Individual standard of living I	Profile	Unevenness	
	Kamareddy	0.808	0.791	0.576	H > E > I	0.008	0.719
	Karimnagar	0.943	0.832	0.859	H > I > E	0.001	0.877
	Khammam	0.927	0.774	0.815	H > I > E	0.002	0.837
	Komaram Bheem Asifabad	0.772	0.542	0.444	H > E > I	0.020	0.574
	Mahabubabad	0.948	0.849	0.634	H > E > I	0.010	0.802
	Mahabubnagar	0.878	0.825	0.719	H > E > I	0.003	0.805
	Mancherial	1.000	0.882	0.717	H > E > I	0.007	0.860
	Medak	0.695	0.885	0.512	E > H > I	0.018	0.684
	Medchal-Malkajgiri	0.873	0.860	0.963	I > H > E	0.001	0.898
	Nagarkurnool	0.806	0.846	0.724	E > H > I	0.002	0.791
	Nalgonda	0.939	0.943	0.816	E > H > I	0.002	0.898
	Nirmal	0.843	0.768	0.648	H > E > I	0.004	0.750
	Nizamabad	0.941	0.730	0.765	H > I > E	0.005	0.808
	Peddapalli	0.930	0.808	0.831	H > I > E	0.001	0.855
	Rajanna Sircilla	0.701	0.928	0.839	E > I > H	0.005	0.819
	Ranga Reddy	0.611	0.832	0.913	I > E > H	0.010	0.777
	Sangareddy	0.772	0.904	0.620	E > H > I	0.009	0.759
	Siddipet	0.829	0.891	0.775	E > H > I	0.001	0.831

(continued)

Table 7.5 (continued)

State	District	Index of progress in			Human development		HDS
		Individual capacity H	Individual knowledge E	Individual standard of living I	Profile	Unevenness	
	Suryapet	0.703	0.953	0.790	E > I > H	0.006	0.810
	Vikarabad	0.568	0.780	0.609	E > I > H	0.007	0.648
	Wanaparthy	0.573	0.924	0.767	E > I > H	0.014	0.744
	Warangal Rural	0.946	0.897	0.629	H > E > I	0.011	0.815
	Warangal Urban	0.820	0.872	0.862	E > I > H	0.000	0.851
	Yadadri Bhuvanagiri	0.654	0.904	0.862	E > I > H	0.007	0.801
Tripura							
	Dhalai	0.862	0.853	0.217	H > E > I	0.110	0.574
	Gomati	0.551	0.803	0.340	E > H > I	0.043	0.540
	Khowai	0.632	0.780	0.239	E > H > I	0.078	0.508
	North Tripura	0.546	0.744	0.375	E > H > I	0.028	0.539
	Sepahijala	0.762	0.814	0.328	E > H > I	0.053	0.601
	South Tripura	0.686	0.785	0.225	E > H > I	0.088	0.516
	Unakoti	0.718	0.658	0.219	H > E > I	0.082	0.488
	West Tripura	0.798	0.824	0.547	E > H > I	0.012	0.714
Uttar Pradesh							
	Saharanpur	0.583	0.504	0.783	I > H > E	0.013	0.615
	Bijnor	0.664	0.644	0.732	I > H > E	0.001	0.679
	Rampur	0.566	0.327	0.683	I > H > E	0.033	0.508

(continued)

Table 7.5 (continued)

State	District	Index of progress in			Human development		HDS
		Individual capacity H	Individual knowledge E	Individual standard of living I	Profile	Unevenness	
	Jyotiba Phule Nagar	0.617	0.548	0.709	I > H > E	0.004	0.622
	Meerut	0.622	0.598	0.886	I > H > E	0.012	0.693
	Baghpat	0.672	0.713	0.824	I > E > H	0.003	0.734
	Gautam Buddha Nagar	0.646	0.769	0.929	I > E > H	0.008	0.775
	Bulandshahr	0.595	0.671	0.765	I > E > H	0.004	0.674
	Aligarh	0.492	0.593	0.690	I > E > H	0.007	0.587
	Mahamaya Nagar	0.484	0.658	0.567	E > I > H	0.006	0.566
	Mathura	0.618	0.552	0.729	I > H > E	0.005	0.630
	Agra	0.590	0.567	0.812	I > H > E	0.010	0.650
	Firozabad	0.410	0.644	0.634	E > I > H	0.015	0.554
	Mainpuri	0.257	0.753	0.487	E > I > H	0.065	0.467
	Bareilly	0.743	0.331	0.679	H > I > E	0.041	0.560
	Pilibhit	0.536	0.379	0.472	H > I > E	0.007	0.459
	Shahjahanpur	0.336	0.490	0.481	E > I > H	0.011	0.431
	Kheri	0.363	0.450	0.288	E > H > I	0.012	0.362
	Sitapur	0.446	0.276	0.240	H > E > I	0.028	0.312
	Hardoi	0.539	0.352	0.252	H > E > I	0.036	0.367
	Unnao	0.633	0.475	0.382	H > E > I	0.016	0.489
	Lucknow	0.749	0.684	0.748	H > I > E	0.001	0.727

(continued)

Table 7.5 (continued)

State	District	Index of progress in			Human development		HDS
		Individual capacity H	Individual knowledge E	Individual standard of living I	Profile	Unevenness	
	Farrukhabad	0.515	0.579	0.524	E > I > H	0.001	0.539
	Kannauj	0.485	0.578	0.382	E > H > I	0.011	0.477
	Etawah	0.607	0.760	0.641	E > I > H	0.004	0.667
	Auraiya	0.540	0.686	0.410	E > H > I	0.016	0.536
	Kanpur Dehat	0.656	0.633	0.379	H > E > I	0.021	0.544
	Kanpur Nagar	0.643	0.734	0.714	E > I > H	0.001	0.696
	Jalaun	0.704	0.669	0.466	H > E > I	0.012	0.606
	Jhansi	0.773	0.803	0.574	E > H > I	0.008	0.711
	Lalitpur	0.527	0.676	0.323	E > H > I	0.032	0.492
	Hamirpur	0.638	0.734	0.393	E > H > I	0.024	0.574
	Mahoba	0.677	0.727	0.386	E > H > I	0.026	0.581
	Banda	0.543	0.491	0.234	H > E > I	0.044	0.404
	Chitrakoot	0.681	0.507	0.261	H > E > I	0.052	0.458
	Fatehpur	0.540	0.519	0.320	H > E > I	0.019	0.451
	Pratapgarh	0.704	0.745	0.430	E > H > I	0.021	0.613
	Kaushambi	0.550	0.502	0.305	H > E > I	0.023	0.442
	Allahabad	0.567	0.648	0.468	E > H > I	0.007	0.557
	Bara Banki	0.532	0.318	0.333	H > I > E	0.022	0.386
	Faizabad	0.777	0.701	0.482	H > E > I	0.015	0.644

(continued)

Table 7.5 (continued)

State	District	Index of progress in			Human development		HDS
		Individual capacity H	Individual knowledge E	Individual standard of living I	Profile	Unevenness	
	Ambedkar Nagar	0.751	0.733	0.347	H > E > I	0.041	0.586
	Bahraich	0.575	0.069	0.192	H > I > E	0.227	0.215
	Shrawasti	0.697	0.153	0.195	H > I > E	0.170	0.289
	Balrampur	0.550	0.265	0.325	H > I > E	0.037	0.366
	Gonda	0.671	0.512	0.397	H > E > I	0.017	0.518
	Siddharthnagar	0.906	0.368	0.370	H > I > E	0.072	0.509
	Basti	0.660	0.622	0.457	H > E > I	0.009	0.574
	Sant Kabir Nagar	0.768	0.597	0.369	H > E > I	0.032	0.560
	Mahrajganj	0.676	0.636	0.472	H > E > I	0.009	0.590
	Gorakhpur	0.614	0.712	0.550	E > H > I	0.004	0.623
	Kushinagar	0.768	0.609	0.435	H > E > I	0.020	0.592
	Deoria	0.899	0.791	0.570	H > E > I	0.013	0.743
	Azamgarh	0.555	0.807	0.448	E > H > I	0.023	0.590
	Mau	0.799	0.668	0.516	H > E > I	0.012	0.653
	Ballia	0.765	0.758	0.471	H > E > I	0.017	0.653
	Jaunpur	0.843	0.853	0.502	E > H > I	0.021	0.718
	Ghazipur	0.812	0.795	0.403	H > E > I	0.034	0.647
	Chandauli	0.628	0.772	0.468	E > H > I	0.015	0.613
	Varanasi	0.928	0.678	0.679	H > I > E	0.009	0.755

(continued)

Table 7.5 (continued)

State	District	Index of progress in			Human development		HDS
		Individual capacity H	Individual knowledge E	Individual standard of living I	Profile	Unevenness	
	Sant Ravidas Nagar	0.512	0.801	0.488	E > H > I	0.020	0.589
	Mirzapur	0.644	0.770	0.466	E > H > I	0.015	0.617
	Sonbhadra	0.737	0.548	0.254	H > E > I	0.063	0.481
	Etah	0.612	0.610	0.439	H > E > I	0.009	0.549
	Kanshiram Nagar	0.472	0.380	0.463	H > I > E	0.003	0.437
	Amethi	0.236	0.558	0.353	E > I > H	0.045	0.365
	Budaun	0.646	0.229	0.401	H > I > E	0.062	0.399
	Ghaziabad	0.800	0.760	0.968	I > H > E	0.004	0.839
	Hapur	0.531	0.657	0.856	I > E > H	0.014	0.671
	Moradabad	0.552	0.581	0.709	I > E > H	0.004	0.611
	Muzaffarnagar	0.872	0.530	0.822	H > I > E	0.017	0.729
	Rae Bareli	0.696	0.628	0.347	H > E > I	0.031	0.540
	Sambhal	0.539	0.381	0.540	I > H > E	0.009	0.482
	Shamli	0.628	0.375	0.779	I > H > E	0.032	0.575
	Sultanpur	0.664	0.718	0.362	E > H > I	0.031	0.564
Uttarakhand							
	Uttarkashi	0.682	0.957	0.604	E > H > I	0.015	0.737
	Chamoli	0.616	0.969	0.659	E > I > H	0.016	0.736
	Rudraprayag	0.792	0.945	0.654	E > H > I	0.008	0.791

(continued)

Table 7.5 (continued)

State	District	Index of progress in			Human development		HDS
		Individual capacity H	Individual knowledge E	Individual standard of living I	Profile	Unevenness	
	Tehri Garhwal	0.916	0.978	0.698	E > H > I	0.008	0.857
	Dehradun	0.781	0.833	0.950	I > E > H	0.003	0.853
	Garhwal	0.864	0.968	0.652	E > H > I	0.010	0.820
	Pithoragarh	0.591	0.955	0.624	E > I > H	0.018	0.710
	Bageshwar	0.885	0.952	0.564	E > H > I	0.018	0.786
	Almora	0.857	0.960	0.481	E > H > I	0.030	0.743
	Champawat	0.705	0.923	0.562	E > H > I	0.015	0.719
	Nainital	0.860	0.792	0.793	H > I > E	0.001	0.814
	Udham Singh Nagar	0.600	0.685	0.803	I > E > H	0.005	0.692
	Hardwar	1.000	0.471	0.812	H > I > E	0.034	0.735
West Bengal							
	Darjiling	0.938	0.858	0.589	H > E > I	0.014	0.784
	Jalpaiguri	0.773	0.772	0.405	H > E > I	0.030	0.630
	Koch Bihar	0.710	0.807	0.201	E > H > I	0.105	0.513
	Uttar Dinajpur	0.737	0.666	0.239	H > E > I	0.075	0.506
	Dakshin Dinajpur	0.904	0.768	0.252	H > E > I	0.090	0.584
	Maldah	0.799	0.691	0.306	H > E > I	0.055	0.566
	Murshidabad	0.702	0.732	0.287	E > H > I	0.056	0.541
	Birbhum	0.828	0.646	0.287	H > E > I	0.063	0.550

(continued)

Table 7.5 (continued)

State	District	Index of progress in			Human development		HDS
		Individual capacity H	Individual knowledge E	Individual standard of living I	Profile	Unevenness	
	Nadia	0.845	0.863	0.386	E > H > I	0.044	0.667
	North Twenty Four Parganas	0.931	0.834	0.653	H > E > I	0.008	0.800
	Hugli	0.842	0.816	0.536	H > E > I	0.015	0.720
	Bankura	0.767	0.851	0.238	E > H > I	0.092	0.562
	Puruliya	0.792	0.792	0.227	H > E > I	0.094	0.547
	Haora	0.934	0.724	0.655	H > E > I	0.009	0.765
	Kolkata	0.916	0.822	0.882	H > I > E	0.001	0.873
	South Twenty Four Parganas	0.820	0.709	0.383	H > E > I	0.036	0.614
	Paschim Medinipur	0.851	0.806	0.210	H > E > I	0.110	0.554
	Purba Medinipur	0.894	0.818	0.209	H > E > I	0.115	0.567
	Paschim Barddhaman	0.632	0.707	0.636	E > I > H	0.001	0.658
	Purba Barddhaman	0.830	0.794	0.377	H > E > I	0.041	0.639

Source Author's calculations based on data from National Family Health Survey, 2019–2021

References

Anand S (2018) Recasting human development measures. Working Paper No. 23. The London School of Economics and Political Sciences. International Inequalities Institute

Anand S, Sen AK (1994) Human Development Index: methodology and measurement. Human Development Report Office Occasional Paper 1. United Nations Development Programme, New York

Chaurasia AR (2022a) Surface measure of human development: alternative to HDI. https://ssrn.com/abstract=4228534 or https://doi.org/10.2139/ssrn.4228534 [Accessed on 25 April 2025].

Chaurasia AR (2022b) An index to measure progress towards universal social protection with application of India. Indian J Hum Dev 16:34–54

Chaurasia AR (2023) Human development in districts of India, 2019–2021. Indian J Hum Dev 17:219–252

Chaurasia AR (2024) Fertility transition in India, 1985–2020: selection of aggregation function matters. https://ssrn.com/abstract=4901450

Croft TN, Marshall AMJ, Allen CCK, Arnold F, Assaf S, Balian S (2018) Guide to DHS statistics. ICF

Desai MJ (1991) Human development concepts and measures. Eur Econ Rev 35(2–3):350–357

Government of India (2002) National Human Development Report 2001. Planning Commission, New Delhi

Government of India (2010) Human development in India. Analysis to action. Planning Commission, New Delhi

Government of India (2011) India Human Development Report 2011: towards social inclusion. Institute of Applied Manpower Research, New Delhi

Government of India (2015) Human development towards bridging inequalities. NITI Aayog, New Delhi

Government of India (2022) National Family Health Survey (NFHS-5) India report. Ministry of Health, and Family Welfare, New Delhi

Government of India (n.d.) Gendering human development. A working paper for computing HDI, GDI and GII for states of India. Ministry of Statistics and Programme Implementation. National Statistical Office, Social Statistics Division, New Delhi

Klugman J, Rodriguez F, Choi H-J (2011) The HDI 2010: new controversies, old critiques. Human Development Research Paper, 2011/01. United Nations Development Program, New York

Kovacevic M (2010) Review of HDI critiques and potential improvements. Human Development Research Paper No. 2010/33. United Nations Development Programme, New York

Mishra S, Nathan HSK (2018) A MANUSH or HUMANS characterisation of human development index. J Hum Dev Capab 19(3):398–415

Rutstein SO, Johnson K (2004) The DHS Wealth Index. DHS Comparative Report No. 6. ORC Macro

Sagar AD, Najam A (1998) The human development index: a critical review. Ecol Econ 25(3):249–264

United Nations (1990) Human Development Report 1990. United Nations Development Program, New York

United Nations. (2010) *Human Development Report 2010. The real wealth of nations. Pathways to development*. United Nations Development Programme, New York

United Nations (2023) Human Development Report 2023/2024. United Nations Development Programme, New York

Chapter 8
Human Development and Fertility

Abstract This chapter analyses the association between human development and fertility based on the variation in the characteristics of human development and the level of fertility across districts. The analysis reveals that, although there is a statistically significant negative relationship between inter-district variation in total fertility rate and inter-district variation in the surface measure of human development, yet the relationship is not strong. Inter-district variation in the surface measure of human development explains only a part of inter-district variation in total fertility rate. The analysis suggests that there are factors other than individual capacity, individual knowledge and individual standard of living which have more dominating effect on inter-district variation in the level of fertility than inter-district variation in the three core dimensions of human development. Reduction in inter-district variation in human development, however, contributes to reduction in inter-district variation in fertility.

Keywords India · Districts · Fertility · Human development

Introduction

The evidence presented in this monograph has highlighted persistent and pervasive variation in both demography and human development across the districts of India. In nearly half of the districts of the country, fertility now appears to have decreased below the replacement level, but it remains above the replacement level in around half of the districts and exceptionally high in some of the districts. Similarly, the progress in human development varies widely across districts not only in terms of the surface measure of human development but also in terms of the characteristics of human development—human development profile and unevenness of human development or the inequality in progress in the three dimensions of human development. These variations are expected because of the mesmerising social, cultural and economic diversity of the country India. India is quite often christened as the country

A. R. Chaurasia, *Population and Human Development in Districts of India*,
https://doi.org/10.1007/978-981-95-8169-6_8

of countries from the perspective of both demography and development. This diversity also emphasises the need of a decentralised approach to development planning and programming to meet the development and welfare needs of the people of the country. The inter-district diversity in both demography and human development, elaborated in this monograph, also calls for examining the linkages between demographic dynamics and the state of human development at the district level, the lowest tier of the public administration system. Such analyses may provide the evidence for integrating demographic factors in planning and programming for human development at the district level. Such integration is necessary to ensure that progress in human development leads to demographic transition and, in turn, demographic transition contributes to accelerating the progress in human development. Given the vastness and the diversity of the country India, it is obvious that there can be no universal approach to both demographic transition and progress in human development that can be applied uniformly to all districts of the country. The challenges of addressing demographic issues and advancing human development are different in different districts of the country. The need of the time is to identify uniformities in the diversity in both demography and human development within the country.

Attempts to explore how human development contributes to fertility transition have a long history (Luci and Thévenon 2010). The demographic transition theory asserts that progress in the three core dimensions of human development—individual capacity, individual knowledge and individual standard of living—are related to demographic transition, especially transition in fertility (Bryant 2007; Notestein 1945; Davis 1945). This relationship suggests that fertility and human development are inversely related—the higher the human development the lower the fertility and vice versa. However, the relationship between fertility and human development does not appear to be linear. At low and medium levels, human development appears to promote fertility transition but at advanced level of human development, fertility and human development relationship is observed to be weak or even reversed in many studies (Wilson and Airey 1999; Myrskylä et al. 2009; Harttgen and Vollmer 2014; Fox et al. 2019), although the association remains negative (Myrskylä et al. 2009; Furuoka 2009; Myrskyläet al. 2011; Esping-Andersen and Billari 2015; Goldscheider et al. 2015). In India, both the level of human development and the level of fertility vary widely across districts. There are districts in which fertility has decreased to very low levels. At the same time, there are districts in which fertility still remains more than 4 births per woman of reproductive age. Similarly, there are districts where human development is at a very advanced stage, whereas there are districts where human development remains unacceptably low. Inter-district variation in the level of fertility and inter-district variation in human development within India provides an opportunity to examine fertility and human development relationship in the Indian context. This chapter attempts to examine this relationship based on the district level data available through the latest round, 2019–2021, of the National Family Health Survey. Both bivariate and multivariate tools have been used in the analyses. The National Family Health Survey, 2019–2021 provides data for 707 districts. The number of districts in the country has, however, increased substantially after 2019–2021 round of the National Family Health Survey.

Fertility and Human Development

The bivariate relationship between fertility and human development across the districts of the country is depicted in Fig. 8.1 in terms of the scatter plot of inter-district variation in the total fertility rate (TFR) and inter-district variation in the surface measure of human development (HDS). The figure suggests that inter-district variation in human development and inter-district variation in fertility are inversely related—the higher the level of human development in a district the lower the fertility in that district and vice versa. This linear relationship, however, is not strong as there are districts where human development is well advanced, but fertility remains high. Similarly, there are districts where human development is either low or very low, but, at the same time, fertility is also either low or very low. More specifically, there are 144 districts where human development is well advanced (HDS $\geq$ 0.900), but TFR is below the replacement level (TFR = 2.1) in only 117 of these districts. Similarly, there are 102 districts in which human development is either low or very low (HDS < 0.600) but, in 12 of these districts, TFR is below the replacement level. Figure 8.1 suggests that human development contributes to fertility transition but there are factors other than human development which influence fertility transition at the district level.

The relatively weak relationship between the level of human development and fertility transition across the districts is confirmed by the ordinary least square regression of TFR on HDS (Table 8.1). The regression coefficient of HDS is statistically significant and negative which confirms that advancement in human development contributes to the transition in fertility. However, inter-district variation in HDS explains less than 35 per cent of the inter-district variation in TFR which confirms

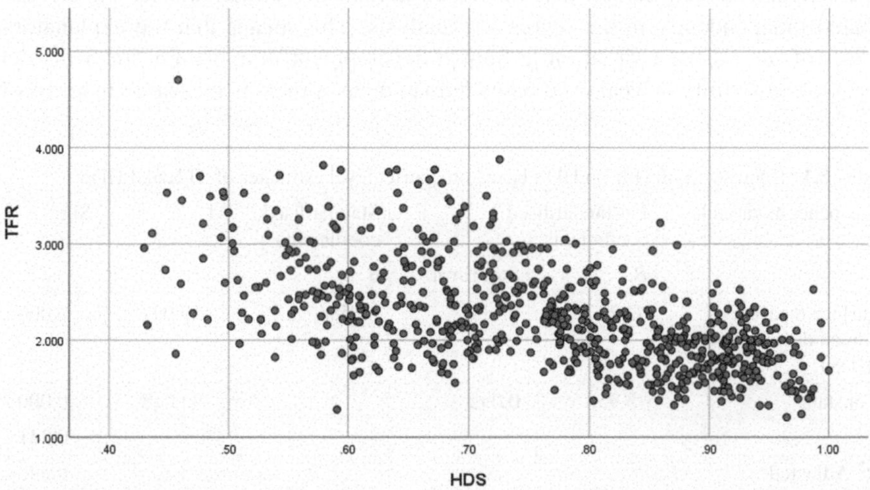

Fig. 8.1 Association between the level of fertility (TFR) and human development (HDS) across 707 districts of India. *Source* Author

that fertility in a district is influenced by factors other than the three core dimensions of human development—individual capacity, individual knowledge and individual standard of living. TFR is not the lowest in that district in which HDS is the highest. Similarly, TFR is not the highest in that district in which HDS is the lowest. Inter-district variation in TFR and HDS confirms that fertility is negatively associated with human development, but human development is not the only determinant of fertility.

It may be noted that HDS is the composite measure that encapsulates the probability of survival in the first five years of life (HE) reflecting individual capacity, secondary school net attendance ratio (ED) reflecting individual knowledge and proportion of households having the wealth index equal to or more than the second quintile of the inter-household distribution of wealth index (SL) reflecting individual standard of living. Regression of TFR on HDS does not reveal the relative influence of the three dimensions of human development on TFR. Table 8.2 presents results of the regression of TFR on HE, ED and SL. Inter-district variation in all the three dimensions of human development is statistically significantly negatively associated with the inter-district variation in TFR. The table also suggests that, among the three dimensions of human development, the influence of inter-district variation in the individual knowledge dimension of human development on inter-district variation in fertility is relatively the largest while the influence of inter-district variation in the individual capacity dimension of human development is relatively the smallest. The relationship, however, remains not strong as inter-district variation in HE, ED and SL explains only about 40 per cent of the inter-district variation in TFR which is, however, higher than the variation explained by HDS.

A comparison of Tables 8.1 and 8.2 also suggests that the inter-district variation in TFR explained by the inter-district variation in (HDS) based on the bivariate regression of TFR on HDS is smaller than the inter-district variation in TFR explained by the inter-district variation when the three dimensions of human development are treated independently in the regression analysis. This means that the explanatory power of inter-district variation in human development in explaining inter-district variation in fertility is weakened when human development is measured in terms of

Table 8.1 Regression of TFR on HDS based on district level estimates of TFR and HDS

Independent variable	Unstandardised coefficients		Standardised coefficients	t	Sig.
	B	Std. Error	*β*		
Surface measure of human development (HDS)	–2.267	0.119	–0.584	–19.097	0.000
Constant	3.956	0.092		43.018	0.000
R^2					0.341
R^2 Adjusted					0.340
N					707

Source Author

Table 8.2 Results of the regression of TFR on HE, ED and SL based on district level data

Independent variables	Unstandardised coefficients		Standardised coefficients	t	Sig.
	B	Std. Error	*β*		
Probability of survival in the first five years of life (HE)	–4.240	0.792	–0.178	–5.353	0.000
Secondary school net attendance ratio (ED)	–1.575	0.174	–0.319	–9.054	0.000
Proportion of households having at least second quintile of wealth index distribution (SL)	–0.613	0.071	–0.296	–8.693	0.000
Constant	7.970	0.718		11.095	0.000
R^2					0.404
R^2 Adjusted					0.401
N					707

Source Author

the composite index of human development compared to when different dimensions of human development are treated independently in exploring fertility and human development relationship. This loss of explanatory power is expected as there is always some loss of information when a set of indicators (vector) is represented by a single indicator (scalar) through some aggregation function that combines the indicators in the set into a single indicator. Table 8.2 suggests that it is more appropriate to analyse the relationship between fertility and human development by treating different dimensions of human development independently rather than by combining them into a single composite index.

Classification of Districts

The foregoing discussions show that fertility in a district is influenced by the progress in the three dimensions of human development—individual capacity, individual knowledge and individual standard of living—and the influence of the progress in the three dimensions is not the same. On average, the influence of the individual knowledge dimension of human development on fertility is relatively the largest whereas the influence of the individual capacity dimension is relatively the smallest. This means that the defining characteristics of human development in a district may also influence the level of fertility in the district. As discussed in the previous chapter, human development may be characterised in terms of human development profile or the relative ordering of the progress in the three dimensions of human development and the unevenness of human development or the inequality in progress in the three dimensions of human development in the same district. It has been shown in the previous chapter that human development in the districts of the country can be characterised into 8 human development profiles depending upon whether the progress is relatively the most advanced in the individual capacity dimension or the individual knowledge dimension or the individual standard of living dimension of human development. On the other hand, the unevenness of human development has also been found to vary widely across the districts. Any analysis of the relationship between inter-district variation in fertility and human development, therefore, should also consider inter-district variation in these characteristics of human development.

The classification modelling approach (Han et al. 2012; Tan et al. 2014) has been used to classify districts into mutually exclusive yet exhaustive groups or clusters of districts to analyse how fertility varies in different groups or clusters of districts having distinct human development characteristics defined in terms of HE, ED and SL, in terms of human development profile and in terms of inequality in progress in human development. There are different methods of classification available for the purpose. These include support vector machine (Urso et al. 2019), decision or classification tree (Liu et al. 2016), logistic regression (Leon 1998), nearest neighbour (Parihari et al. 2023), imbalanced classification (Sun et al. 2009), random forest (Ho 1998), multiclass classification (Venkatesan 2016), neural network (Muratbek and Bektemisova 2024) and others. The decision or classification tree method is a powerful statistical method that is popularly used for the purpose of classification. It is a non-parametric, recursive partitioning method which does not require any assumption about the data distribution and is robust to outliers. It is easy to interpret and understand and can be used even when the underlying data are heavily skewed.

Different statistical techniques have been proposed to construct a decision or classification tree. The popular ones are classification and regression trees (CART) (Brieman et al. 1984), Chi-squared automatic interaction detection (CHAID) (Kass 1980) and quick, unbiased, efficient, statistical tree (QUEST) (Loh and Shih 1997). A comparison of these techniques is given elsewhere (Song and Lu 2015). The classification and regression trees (CART) technique is used here to classify districts into mutually exclusive and exhaustive groups or clusters in such a way that within-group homogeneity in the level of fertility is the maximum. A cluster or a group in which the level of fertility is the same for all districts in the cluster or the group is termed as 'pure'. If a cluster is not pure then the impurity in the cluster can be measured through the Gini index. The dependent variable in CART can be both categorical and continuous. When the dependent variable is a categorical one, CART produces the classification tree with the proportionate distribution of data in different categories in each identified cluster. When the dependent variable is a continuous one, CART produces the regression tree with estimates of arithmetic mean and standard deviation of the dependent variable in each identified cluster. In the present case, the dependent variable is the level of fertility in the district, measured in terms of TFR, which is a continuous variable. The CART, therefore, has generated the regression tree with estimates of mean fertility and associated standard deviation in each identified cluster.

The independent or explanatory variables used in the classification modelling exercise are related to the defining characteristics of human development in each district. Five indicators have been used to characterise human development in the district: (1) probability of survival in the first five years of life (HE) which reflects the progress in the individual capacity dimension of human development; (2) secondary school net attendance ratio (ED) which reflects the progress in the individual knowledge dimension of human development; (3) proportion of households having wealth index at least second quintile of the inter-household distribution of wealth index (SL) which reflects the progress in the individual standard of living dimension of human development; (4) human development profile (HDP) which captures the relative ordering of progress in the three dimensions of human development in the district; and (5) unevenness in human development (HDU) or the inequality in progress in the three dimensions of human development. The indicators HDP and HDU are estimated from the normalised values H, E and I of the indicators HE, ED and SL respectively as discussed in the previous chapter. Out of the five independent variables, four are continuous variables while the fifth is a categorical variable. The advantage of CART is that it can handle both continuous and categorical variable simultaneously.

Results of the classification exercise are presented in Fig. 8.2 and Table 8.3. The mean TFR across 707 districts is estimated to be 2.227 ± 0.526 births per woman of childbearing age. The classification exercise suggests that 707 districts of the country can be divided into 17 mutually exclusive and exhaustive groups (Nodes) and both defining characteristics of human development and average fertility in each group are different. Table 8.3 gives the characteristics of human development in each of the 17 groups (Nodes) in terms of probability of survival in the first five years of life (HE), secondary school net attendance ratio (ED), proportion of households having wealth index at least second quintile of inter-household distribution of wealth index (SL), human development profile (HDP) and unevenness of human development (HDU) or the inequality in progress in different dimensions of human development. Among 17 groups, mean TFR is the highest (3.113 ± 0.617) in Node 17 comprising of 31 districts. The probability of survival in the first five years of life (HE) in districts of this group is less than or equal to 0.951, the secondary school net attendance ratio (ED) is less than or equal to 0.731 and the unevenness in human development is large (HDU > 0.041) which means that there is substantial difference in progress in the three dimensions of human development in the districts of this group. On the other hand, average TFR is found to be the lowest (1.714 ± 0.250) in Node 14 comprising of 80 districts. In these districts, HE>0.980, ED > 0.842 and SL > 0.687. The human development profile (HDP) and the unevenness in human development (HDU) in these districts do not matter in deciding the level of fertility. The average TFR is also found to be very low (1.717 ± 0.258) in Node 32 comprising of 25 districts. The probability of survival in the first five years of life (HE) in the districts of this group ranges between 0.969 and 0.980, the secondary net attendance rate (ED) is at least 0.842, and proportion of households having wealth index equal to or higher than the second quintile of the inter-household distribution of the wealth index (SL) is equal to or more than 0.687. In all districts of this group, the progress in the individual opportunity dimension of human development is better than the progress in either the individual capacity dimension or the individual standard of living dimension of human development. The unevenness in human development does not influence fertility in districts of this group.

The classification exercise also highlights the importance of unevenness in human development or the inequality in progress in different dimensions of human development (HDU) and human development profile (HDP) in determining the level of fertility in different groups of districts (Nodes). For example, there are 114 districts in which SL $\leq$ 0.687 while ED > 0.871 (Node 12). In 81of these districts, the index of unevenness in human development, HDU $\leq$ 0.046 (Node 23) whereas in 33 districts, HDU > 0.046 (Node 24) and the average TFR of districts of Node 23 is markedly lower than the average TFR of districts of Node 24. Similarly, there are 47 districts in which HE $\leq$ 0.982, ED $\leq$ 0.842 and SL > 0.581 (Node 21). Out of these 47 districts, human development profile (HDP) is defined by either H > E > I or H > I > E in 31 districts (Node 27) whereas HDP is defined by either E > H > I or I > E > H or I > E > H in 16 districts (Node 28). The average TFR of districts of Node 28 is below the replacement level whereas average TFR of districts of Node 27 is well above the replacement level. There are 124 districts in which HE $\leq$ 0.980, ED > 0.842 and SL > 0.687 (Node 13). In 59 of these districts, the human development profile is defined by either E > H > I or E > I > H (Node 25) whereas in 65 districts, human development profile is defined by either H > E > I or H > I > E or I > E > H or I > H > E (Node 26) and there is marked difference in the average TFR between the two groups of districts. In districts where the progress in the individual opportunity dimension of human development is the most advanced relative to the progress in the individual capacity and individual standard of living dimensions of human development, fertility, on average, is lower than the average fertility of those districts where the progress in either the health dimension of the standard of living dimension of human development is relatively the most advanced. The importance of the individual opportunity dimension of human development in deciding the level of TFR is very much obvious from the classification exercise.

In each group of districts (Nodes) identified through the classification exercise, districts vary in terms of the level of fertility. The within-group variation in TFR is the largest in Node 17 but the smallest in Node 28 as may be seen from the value of the standard deviation in Table 8.3. The within-group (Node) distribution of districts is, however, different in different groups. In 14 out of 17 Nodes, the within-group (Node) distribution of TFR is negatively skewed Meaning that the mean TFR is less than the median TFR in these groups (Nodes). There are, however, three groups (Nodes) in which the distribution of TFR is positively skewed or the mean TFR is less than the median TFR. A negative skewness indicates that there are some outlier districts in the group in which TFR is exceptionally high. On the other hand, a positive skewness means that in some districts of the group (Node), the TFR is exceptionally low as may be seen from Fig. 8.4. The number of outlier districts in different groups (Nodes), districts having exceptionally high or exceptionally low values of TFR is, however, small. There are only 8 groups (Nodes) in which, there are outlier districts

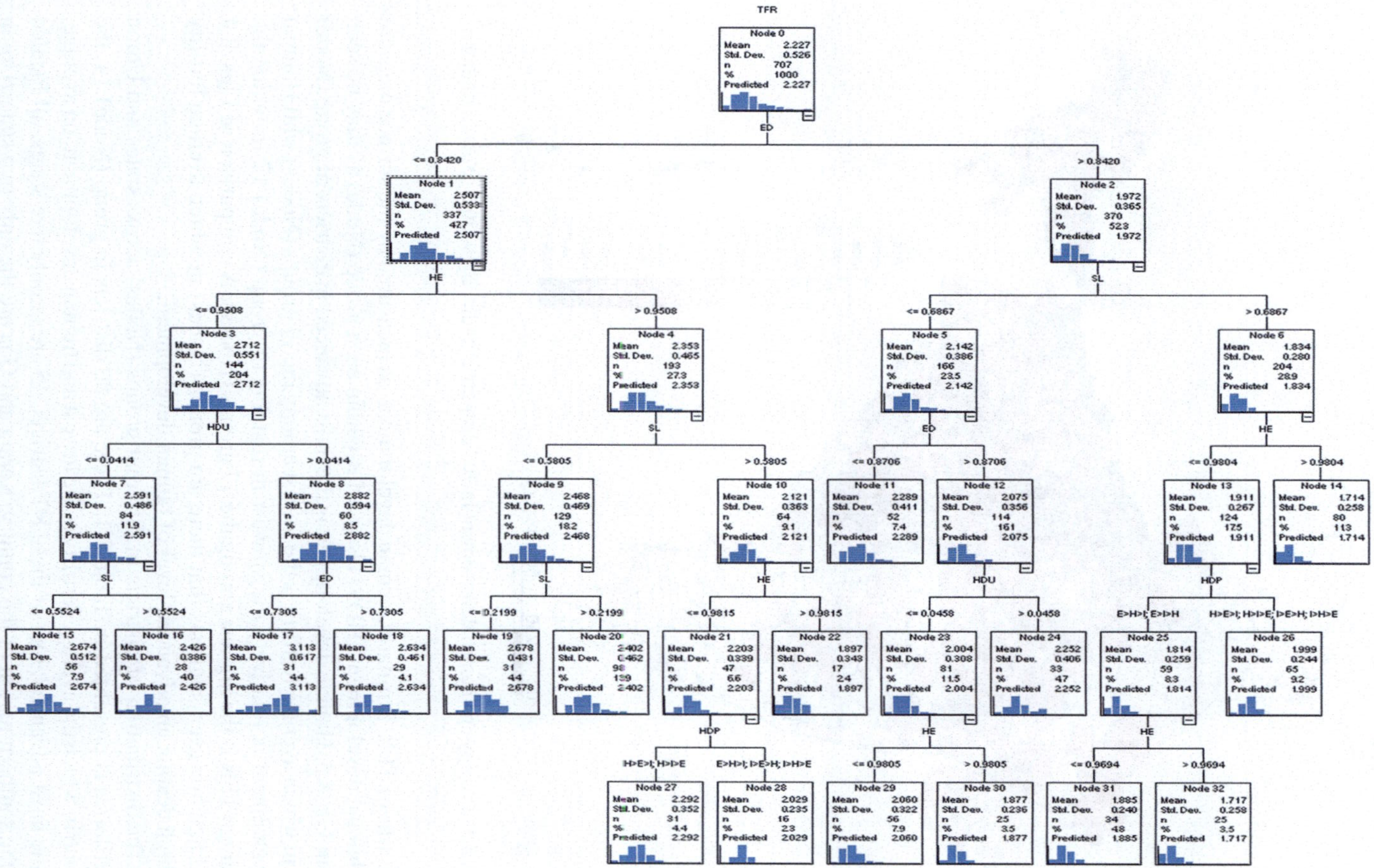

Fig. 8.2 The classification tree showing the segmentation of districts in terms of fertility and characteristics of human development. *Source* Author

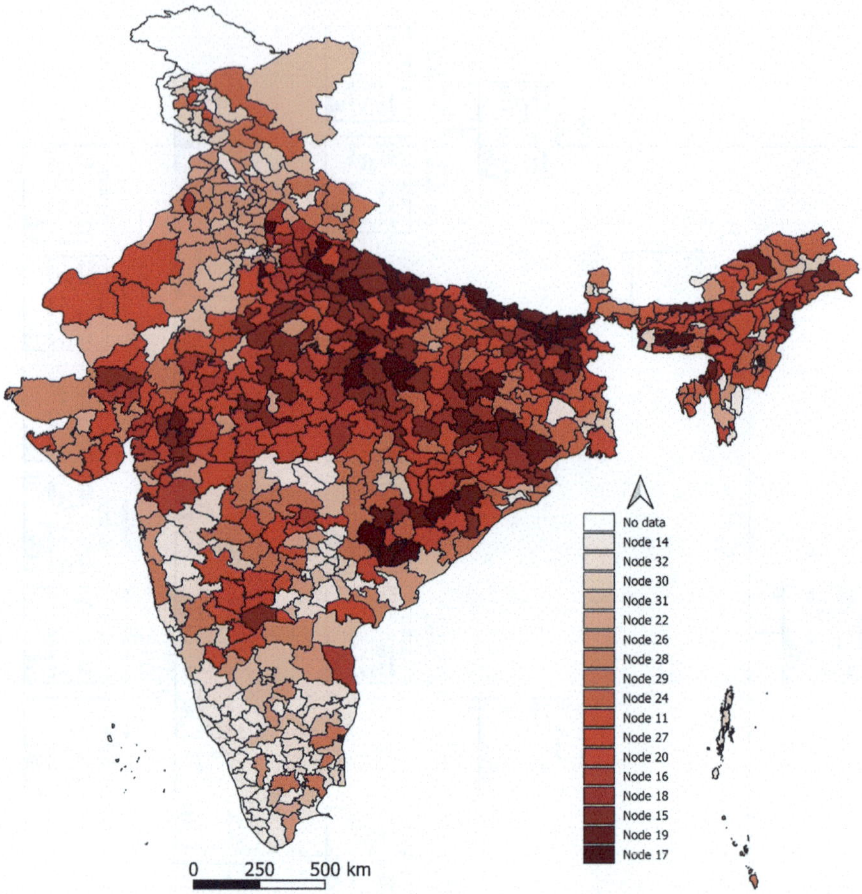

Fig. 8.3 Fertility and human development in 707 districts of India, 2019–2021. *Source* Author

in terms of TFR. Similarly, human development, as measured by the surface measure of human development (HDS) also varies within each group (Node) as may be seen from Fig. 8.5. Human development is the most advanced in districts of group (Node) 14 in which average fertility is the lowest among the 17 groups (Nodes) while human development is relatively the slowest in districts of group (Node) 17 in which the average TFR is the highest among the 17 groups (Nodes). A comparison of Figs. 8.4 and 8.5, however, suggests that there are groups (Nodes) in which average fertility is high despite an advanced level of human development.

The distribution of districts in different groups (Nodes) across states and Union Territories is presented in Table 8.4. There are 80 districts in group (Node) 14, the group in which the average fertility is the lowest and human development, on average, is the most advanced. Out of these 80 districts, 45 are in only five states, all located in the southern region of the country. Kerala and Goa are the only two states/Union

Table 8.3 Average fertility (TFR) and characteristics of human development in mutually exclusive and exhaustive clusters of districts identified through classification modelling exercise

Node	Predictor variables					Total fertility rate (TFR)				Surface measure of human development (HDS)		Number of districts
	HE	ED	SL	HDP	HDU	Mean	SD	Median	IQR	Mean	SD	
17	≤0.951	≤0.731			>0.041	3.113	0.617	3.167	0.804	0.385	0.073	31
19	≤0.951	≤0.842	≤0.220			2.678	0.431	2.734	0.717	0.432	0.060	31
15	>0.951	≤0.842	≤0.552		≤0.041	2.674	0.512	2.727	0.649	0.512	0.071	56
18	≤0.951	0.731–0.842			>0.041	2.634	0.461	2.554	0.729	0.458	0.048	29
16	>0.951	≤0.842	>0.552		≤0.041	2.426	0.386	2.482	0.395	0.632	0.063	28
20	>0.951	≤0.842	0.220–0.581			2.402	0.462	2.352	0.528	0.577	0.064	98
27	>0.982	>0.842	>0.581	H > E > E H > I > E		2.292	0.352	2.260	0.476	0.691	0.068	31
11		0.842–0.871	>0.687			2.289	0.411	2.286	0.577	0.633	0.075	52
24		>0.871	≤0.687		>0.046	2.252	0.406	2.198	0.453	0.617	0.084	33
29	≤0.981	>0.842	>0.687		≤0.046	2.060	0.322	2.043	0.465	0.721	0.046	56
28	0.982	≤0.842	>0.581	E > H > I I > E > H I > H > E		2.029	0.235	1.992	0.210	0.750	0.049	16
26	≤0.980	>0.842	>0.687	H > E > I H > I > E I > E > H I > H > E		1.999	0.244	1.982	0.327	0.822	0.045	65
22	>0.982	≤0.842	>0.581			1.897	0.343	1.863	0.638	0.770	0.046	17

(continued)

Table 8.3 (continued)

Node	Predictor variables					Total fertility rate (TFR)				Surface measure of human development (HDS)		Number of districts
	HE	ED	SL	HDP	HDU	Mean	SD	Median	IQR	Mean	SD	
31	≤0.969	>0.842	>0.687	E > H > I E > I > H		1.885	0.240	1.849	0.339	0.803	0.040	34
30	>0.981	>0.842	>0.687		≤0.046	1.877	0.236	1.868	0.365	0.787	0.032	25
32	0.969–0.980	>0.842	>0.687	E > H > I E > I > H		1.717	0.258	1.712	0.379	0.840	0.029	25
14	>0.980	>0.842	>0.687			1.714	0.258	1.660	0.333	0.907	0.047	80

Source Author

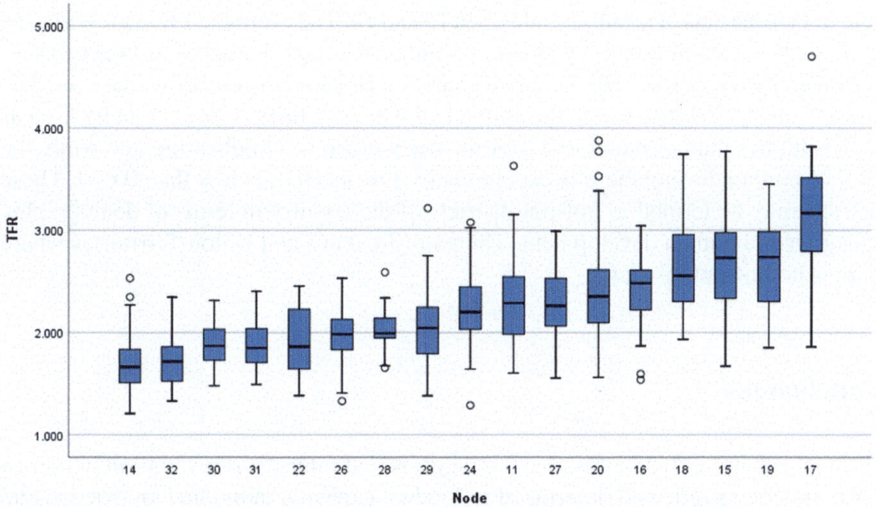

Fig. 8.4 Variation in TFR within different groups (Nodes) identified through the classification exercise. *Source* Author

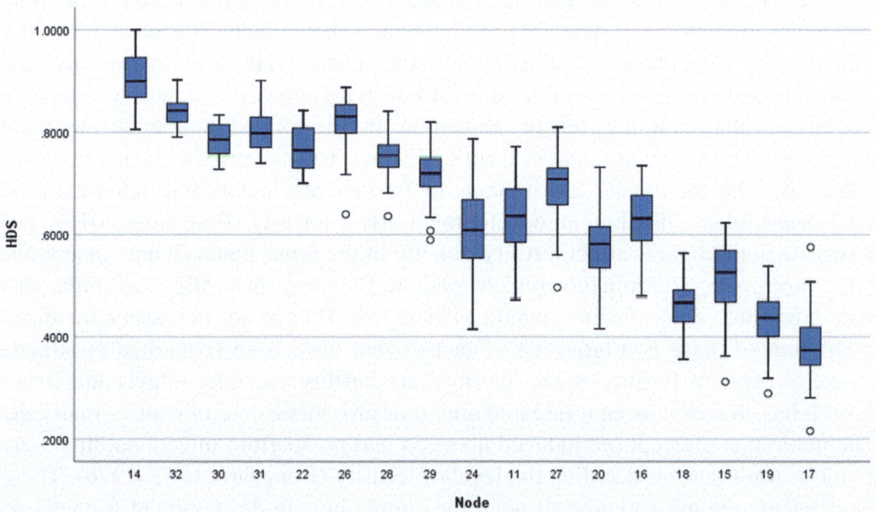

Fig. 8.5 Variation in HDS within groups (Nodes) identified through classification exercise. *Source* Author

Territories, all districts of which belong to this group (Node). On the other hand, there are 31 districts in group (Node) 17 in which fertility is, on average, the highest among all groups (Nodes) while the level of human development is relatively the lowest on average. Out of these 31 districts, 22 districts are in only three states—Bihar, Uttar Pradesh and Madhya Pradesh. Fertility and human development scenario is also quite poor in 116 districts of groups (Nodes) 19, 15 and 18. The average TFR in these groups is more than 2.5 children per women of childbearing age while the surface measure of human development (HDS) is less than 0.550. More specifically, there are 124 districts out of 707 districts in the country in which fertility is exceptionally high as TFR is higher than or equal to 2.5 births per women of childbearing age while the level of human development is exceptionally low as HDS is less than 0.600. These districts may be termed as hotspot districts of the country in terms of demographic transition and human development. There are 14 states and Union Territories where there is no hotspot district.

Conclusions

The analysis presented in this chapter suggests that inter-district variation in human development, as reflected in terms of individual capacity, individual knowledge and individual standard of living, explains inter-district variation in fertility, as measured in terms of TFR, only partly. The progress in human development in a district influences the level of fertility in the district but the influence does not appear to be strong. There are district level factors other than individual capacity, individual knowledge and individual standard of living which also influence the level of fertility in the district. The enhancement in the individual capacity, increase in the individual knowledge and improvement in the standard of living, however, creates a conducive environment that facilitates fertility transition. In this context, human development is, at best, the distal determinant of fertility distinct from the proximate determinants of fertility. The proximate determinants of fertility are factors through which the distal determinants like human development affect fertility (Bongaarts 2015). The proximate determinants affect fertility directly in the sense that a change in any one of the proximate determinants always leads to a change in fertility assuming that other proximate determinants remain unchanged. This is not necessary for distal determinants. There is a large list of factors that have been termed as proximate or determinants of fertility or the intermediate fertility variables (Davis and Bruce 1956) It has, however, been found that only four proximate determinants—marriage/cohabitation, contraception, induced abortion and postpartum infecundability—are the most important for deciding the level of fertility (Bongaarts 1972, 1978). These determinants are not captured through the simple human development framework, although progress in human development influences these proximate determinants also. The present analysis suggests that because of the variation in these proximate determinants, fertility in a district may be at deviance from that determined by the level of human development in the district. It may, however, be concluded that a reduction in the inter-district variation in human development may contribute to the reduction in the inter-district variation in fertility.

Table 8.4 Distribution of districts in different groups (Nodes) across states and Union Territories

State/Union Territory	Group (Node)																	Total
	14	32	30	31	22	26	28	29	24	11	27	20	16	18	15	19	17	
Andaman & Nicobar Islands	0	1	1	0	0	0	0	1	0	0	0	0	0	0	0	0	0	3
Andhra Pradesh	1	1	0	3	0	2	2	2	0	0	1	0	1	0	0	0	0	13
Arunachal Pradesh	1	0	3	0	0	1	0	1	6	2	0	4	0	0	0	2	0	20
Assam	0	1	0	0	0	0	0	0	2	8	0	12	0	5	1	4	0	33
Bihar	0	0	0	0	0	0	0	0	2	3	1	8	0	6	5	3	10	38
Chandigarh	1	0	0	0	0	0	0	0	0	0	0	0	0	0	0	0	0	1
Chhattisgarh	0	0	2	0	0	2	0	3	2	2	0	6	0	3	3	2	2	27
Delhi	4	0	0	0	0	6	1	0	0	0	0	0	0	0	0	0	0	11
Dadra & Nagar Haveli and Daman & Diu	0	0	0	0	1	0	1	0	0	0	0	0	0	0	1	0	0	3
Goa	2	0	0	0	0	0	0	0	0	0	0	0	0	0	0	0	0	2
Gujarat	0	0	0	0	3	0	7	0	0	0	9	5	5	0	3	1	0	33
Haryana	1	0	0	4	0	15	0	0	0	0	1	0	0	0	0	0	1	22
Himachal Pradesh	1	1	1	2	0	5	0	1	1	0	0	0	0	0	0	0	0	12
Jammu & Kashmir	7	1	5	0	1	0	0	3	0	2	1	0	0	0	0	0	0	20
Jharkhand	0	0	0	0	0	0	0	4	1	4	0	1	0	3	0	10	1	24
Karnataka	7	3	1	6	0	3	0	2	0	2	1	4	0	0	1	0	0	30
Kerala	14	0	0	0	0	0	0	0	0	0	0	0	0	0	0	0	0	14
Ladakh	0	0	0	1	0	0	0	0	1	0	0	0	0	0	0	0	0	2
Lakshadweep	1	0	0	0	0	0	0	0	0	0	0	0	0	0	0	0	0	1
Madhya Pradesh	0	0	0	0	1	2	0	0	0	2	5	16	2	5	13	1	4	51
Maharashtra	6	4	1	3	2	3	0	10	0	3	2	0	1	0	1	0	0	36
Manipur	1	0	0	0	0	0	0	3	3	2	0	0	0	0	0	0	0	9
Meghalaya	0	0	1	0	0	0	0	0	2	1	0	3	0	1	0	0	3	11
Mizoram	3	1	0	0	1	1	0	0	0	1	0	1	0	0	0	0	0	8
Nagaland	1	0	2	0	0	0	0	0	1	0	0	2	0	2	0	3	0	11
Odisha	1	0	0	0	1	0	0	4	3	0	1	9	0	1	5	3	2	30
Puducherry	3	0	0	0	0	0	1	0	0	0	0	0	0	0	0	0	0	4
Punjab	3	0	0	1	1	12	4	0	0	0	0	0	1	0	0	0	0	22

(continued)

Table 8.4 (continued)

State/Union Territory	Group (Node)																	Total
	14	32	30	31	22	26	28	29	24	11	27	20	16	18	15	19	17	
Rajasthan	0	1	0	5	1	4	0	2	1	4	5	7	1	0	2	0	0	33
Sikkim	2	0	1	0	0	0	0	1	0	0	0	0	0	0	0	0	0	4
Tamil Nadu	11	7	2	5	1	2	0	4	0	0	0	0	0	0	0	0	0	32
Telangana	7	4	2	4	1	3	0	3	0	3	1	2	1	0	0	0	0	31
Tripura	0	0	0	0	0	0	0	1	3	2	0	0	0	0	1	1	0	8
Uttar Pradesh	0	0	0	0	1	2	0	3	0	7	3	13	14	3	20	1	8	75
Uttarakhand	1	0	1	0	1	2	0	7	0	0	0	0	1	0	0	0	0	13
West Bengal	1	0	2	0	1	0	0	1	5	4	0	5	1	0	0	0	0	20
India	80	25	25	34	17	65	16	56	33	52	31	98	28	29	56	31	31	707

Source Author

Appendix

See Table 8.5.

Table 8.5 Total fertility rate (TFR), index of individual capacity (HE), index of individual opportunity (ED), index of individual standard of living (SL), human development profile (HDP), human development inequality (HDE) and surface measure of human development (HDS) in districts of India

State/ Union Territory	District	TFR	HE	ED	SL	HDS	HDP	HDU	Node
Andaman & Nicobar Islands	Nicobar	1.377	0.959	1.000	0.523	0.804	E > H > I	0.034	29
	North & Middle Andaman	1.784	0.993	0.909	0.526	0.788	H > E > I	0.031	30
	South Andaman	1.327	0.979	0.952	0.919	0.950	E > I > H	0.001	32
Andhra Pradesh	Srikakulam	1.565	0.979	0.950	0.676	0.860	E > H > I	0.009	29
	Vizianagaram	1.852	0.960	0.941	0.583	0.813	E > H > I	0.016	29
	Visakhapatnam	1.540	0.961	0.870	0.735	0.852	E > H > I	0.001	31
	East Godavari	1.898	0.977	0.876	0.762	0.869	H > E > I	0.001	26
	West Godavari	1.793	0.981	0.871	0.867	0.905	H > I > E	0.001	14

(continued)

Table 8.5 (continued)

State/ Union Territory	District	TFR	HE	ED	SL	HDS	HDP	HDU	Node
	Krishna	1.650	0.958	0.825	0.755	0.843	I > E > H	0.000	28
	Guntur	1.652	0.976	0.817	0.811	0.866	H > I > E	0.002	27
	Prakasam	1.704	0.953	0.869	0.788	0.868	E > I > H	0.002	31
	Sri Potti Sriramulu Nellore	1.534	0.950	0.772	0.785	0.833	I > H > E	0.004	16
	Y.S.R.	1.985	0.928	0.872	0.864	0.888	I > E > H	0.023	26
	Kurnool	1.936	0.954	0.804	0.649	0.795	E > H > I	0.001	28
	Anantapur	1.774	0.969	0.878	0.711	0.847	E > H > I	0.002	31
	Chittoor	1.631	0.970	0.897	0.768	0.875	E > H > I	0.001	32
Arunachal Pradesh	Tawang	2.347	0.999	1.000	0.824	0.938	E > H > I	0.004	14
	West Kameng	1.703	0.973	0.857	0.850	0.892	I > H > E	0.001	26
	East Kameng	2.527	0.956	0.857	0.358	0.681	E > H > I	0.053	11
	Papum Pare	2.283	0.998	0.889	0.686	0.850	H > E > I	0.011	30
	Upper Subansiri	2.734	0.995	0.800	0.192	0.574	H > E > I	0.182	19
	Upper Siang	2.235	1.000	1.000	0.367	0.737	E = H > I	0.092	24
	Changlang	2.070	1.000	0.929	0.350	0.709	H > E > I	0.092	24
	Lower Subansiri	2.067	0.988	0.857	0.534	0.774	H > E > I	0.024	11
	Dibang Valley	2.013	0.982	1.000	0.412	0.756	E > H > I	0.067	24
	Lower Dibang Valley	2.114	1.000	1.000	0.550	0.828	E – H > I	0.036	30
	Anjaw	1.820	0.982	1.000	0.179	0.612	E > H > I	0.213	24
	East Siang	1.807	1.000	0.875	0.584	0.805	H > E > I	0.024	30
	Kra Daadi	2.294	0.974	1.000	0.443	0.770	E > H > I	0.055	24
	Kurung Kumey	2.347	0.991	0.833	0.239	0.614	H > E > I	0.141	20
	Longding	2.312	1.000	0.750	0.162	0.539	H > E > I	0.217	19
	Lohit	2.213	1.000	0.800	0.553	0.768	H > E > I	0.030	20
	Namsai	2.810	0.985	0.750	0.233	0.586	H > E > I	0.133	20
	Siang	2.506	0.987	1.000	0.279	0.682	E > H > I	0.132	24
	Tirap	2.477	0.964	0.800	0.394	0.685	H > E > I	0.038	20
	West Siang	2.150	0.977	0.875	0.540	0.779	H > E > I	0.019	29
Assam	Kokrajhar	1.953	0.961	0.866	0.308	0.658	E > H > I	0.079	11
	Goalpara	2.199	0.969	0.800	0.251	0.607	H > E > I	0.108	20
	Barpeta	2.475	0.972	0.840	0.252	0.620	H > E > I	0.115	20
	Morigaon	2.431	0.948	0.784	0.214	0.574	E > H > I	0.114	18
	Lakhimpur	1.889	0.976	0.840	0.283	0.640	H > E > I	0.099	20
	Dhemaji	2.372	0.976	0.913	0.181	0.591	E > H > I	0.191	24
	Tinsukia	1.715	0.937	0.781	0.354	0.653	E > H > I	0.034	15

(continued)

Table 8.5 (continued)

State/ Union Territory	District	TFR	HE	ED	SL	HDS	HDP	HDU	Node
	Dibrugarh	2.095	0.977	0.736	0.413	0.678	H > E > I	0.042	20
	Golaghat	2.193	0.971	0.785	0.261	0.609	H > E > I	0.101	20
	Dima Hasao	2.142	0.991	0.826	0.392	0.699	H > E > I	0.061	20
	Cachar	2.420	0.975	0.754	0.259	0.601	H > E > I	0.104	20
	Karimganj	2.474	0.954	0.741	0.216	0.565	H > E > I	0.109	19
	Hailakandi	2.428	0.898	0.768	0.203	0.551	E > H > I	0.108	18
	Bongaigaon	2.060	0.976	0.865	0.349	0.684	H > E > I	0.069	11
	Chirang	2.033	0.963	0.852	0.230	0.607	E > H > I	0.126	11
	Kamrup	1.968	0.982	0.831	0.411	0.707	H > E > I	0.048	20
	Kamrup Metropolitan	1.411	0.972	0.884	0.776	0.874	E > H > I	0.000	32
	Nalbari	1.773	0.987	0.847	0.329	0.671	H > E > I	0.084	11
	Baksa	1.850	0.971	0.840	0.190	0.578	H > E > I	0.165	19
	Darrang	2.457	0.974	0.788	0.232	0.593	H > E > I	0.124	20
	Udalguri	2.159	0.949	0.804	0.218	0.583	E > H > I	0.115	18
	Biswanath	2.201	0.929	0.808	0.251	0.600	E > H > I	0.082	18
	Charaideo	1.954	0.951	0.709	0.246	0.574	H > E > I	0.081	20
	Dhubri	2.763	0.962	0.802	0.214	0.582	H > E > I	0.129	19
	West Karbi Anglong	2.217	0.969	0.865	0.121	0.527	H > E > I	0.255	11
	Hojai	2.379	0.971	0.852	0.361	0.685	H > E > I	0.060	11
	Jorhat	1.882	0.943	0.867	0.466	0.735	E > H > I	0.023	11
	Karbi Anglong	2.370	0.968	0.867	0.294	0.652	E > H > I	0.091	11
	Majuli	2.445	0.958	0.895	0.183	0.583	E > H > I	0.174	24
	Nagaon	2.486	0.936	0.735	0.243	0.576	E > H > I	0.073	18
	Sivasagar	1.901	0.979	0.800	0.463	0.723	H > E > I	0.031	20
	Sonitpur	2.474	0.959	0.743	0.320	0.629	H > E > I	0.055	20
	South Salmara Mancachar	3.067	0.981	0.750	0.113	0.494	H > E > I	0.264	19
Bihar	Pashchim Champaran	3.256	0.906	0.658	0.220	0.533	E > H > I	0.048	17
	Purba Champaran	3.171	0.941	0.706	0.277	0.590	H > E > I	0.053	17
	Sheohar	3.253	0.936	0.757	0.207	0.559	E > H > I	0.103	18
	Sitamarhi	3.715	0.918	0.702	0.212	0.543	E > H > I	0.072	17
	Madhubani	3.363	0.955	0.740	0.220	0.567	H > E > I	0.108	19
	Supaul	2.959	0.957	0.639	0.081	0.429	H > E > I	0.286	19
	Araria	3.704	0.944	0.639	0.135	0.476	H > E > I	0.166	17
	Kishanganj	3.493	0.945	0.573	0.172	0.485	H > E > I	0.127	17
	Purnia	3.491	0.922	0.612	0.179	0.496	H > E > I	0.080	17

(continued)

Table 8.5 (continued)

State/ Union Territory	District	TFR	HE	ED	SL	HDS	HDP	HDU	Node
	Katihar	3.324	0.954	0.661	0.155	0.500	H > E > I	0.159	19
	Madhepura	3.532	0.947	0.643	0.101	0.448	H > E > I	0.228	17
	Saharsa	3.539	0.926	0.673	0.153	0.495	H > E > I	0.124	17
	Darbhanga	3.154	0.949	0.714	0.272	0.591	H > E > I	0.065	17
	Muzaffarpur	3.155	0.961	0.851	0.315	0.658	E > H > I	0.073	11
	Gopalganj	3.025	0.959	0.781	0.367	0.665	H > E > I	0.042	20
	Siwan	2.711	0.940	0.852	0.442	0.718	E > H > I	0.025	11
	Saran	2.923	0.940	0.800	0.358	0.661	E > H > I	0.037	15
	Vaishali	3.370	0.928	0.811	0.329	0.646	E > H > I	0.047	18
	Samastipur	3.345	0.942	0.768	0.174	0.540	E > H > I	0.141	18
	Begusarai	3.745	0.950	0.797	0.304	0.634	E > H > I	0.062	18
	Khagaria	3.817	0.950	0.726	0.247	0.579	H > E > I	0.082	17
	Bhagalpur	3.198	0.936	0.825	0.462	0.718	E > H > I	0.017	15
	Banka	3.282	0.960	0.787	0.239	0.594	H > E > I	0.105	20
	Munger	2.804	0.968	0.813	0.516	0.747	H > E > I	0.015	20
	Lakhisarai	3.771	0.962	0.783	0.377	0.671	H > E > I	0.041	20
	Sheikhpura	3.470	0.927	0.829	0.392	0.683	E > H > I	0.033	15
	Nalanda	3.254	0.950	0.745	0.446	0.690	H > E > I	0.014	15
	Patna	2.483	0.955	0.753	0.608	0.762	H > E > I	0.002	27
	Bhojpur	2.919	0.963	0.842	0.456	0.728	E > H > I	0.026	20
	Buxar	3.315	0.940	0.823	0.466	0.720	E > H > I	0.016	15
	Kaimur (Bhabua)	2.701	0.923	0.814	0.300	0.629	E > H > I	0.059	18
	Rohtas	2.956	0.906	0.885	0.518	0.753	E > I > H	0.052	24
	Aurangabad	3.318	0.953	0.842	0.390	0.693	E > H > I	0.039	20
	Gaya	3.529	0.962	0.766	0.328	0.641	H > E > I	0.057	20
	Nawada	3.386	0.971	0.815	0.404	0.697	H > E > I	0.042	20
	Jamui	3.202	0.926	0.751	0.258	0.588	E > H > I	0.062	18
	Jehanabad	3.024	0.952	0.882	0.358	0.688	E > H > I	0.056	24
	Arwal	2.913	0.925	0.855	0.301	0.641	E > H > I	0.070	11
Chandigarh	Chandigarh	1.675	0.983	0.873	0.967	0.940	I > H > E	0.003	14
Chhattisgarh	Koriya	2.086	0.883	0.890	0.352	0.668	E > I > H	0.145	24
	Jashpur	2.070	0.951	0.795	0.146	0.528	E > H > I	0.188	19
	Raigarh	1.893	0.908	0.854	0.425	0.701	E > I > H	0.048	11
	Korba	2.180	0.923	0.767	0.463	0.697	E > H > I	0.011	15
	Janjgir—Champa	1.872	0.970	0.909	0.460	0.751	E > H > I	0.036	29
	Kabeerdham	2.185	0.975	0.886	0.504	0.766	H > E > I	0.026	29
	Rajnandgaon	1.800	0.947	0.916	0.550	0.788	E > H > I	0.018	29

(continued)

Table 8.5 (continued)

State/ Union Territory	District	TFR	HE	ED	SL	HDS	HDP	HDU	Node
	Mahasamund	1.753	0.964	0.791	0.404	0.687	H > E > I	0.035	20
	Dhamtari	1.753	0.986	0.908	0.618	0.825	H > E > I	0.014	30
	Narayanpur	2.148	0.962	0.647	0.217	0.540	H > E > I	0.113	19
	Bijapur	1.860	0.917	0.552	0.148	0.456	H > E > I	0.093	17
	Balod	1.560	0.992	0.964	0.573	0.825	E > H > I	0.026	30
	Baloda Bazar	2.250	0.952	0.812	0.467	0.721	E > H > I	0.017	20
	Balrampur	2.466	0.937	0.748	0.146	0.513	E > H > I	0.164	18
	Bastar	2.316	0.925	0.621	0.269	0.555	H > E > I	0.032	15
	Uttar Bastar Kanker	1.868	0.978	0.932	0.342	0.699	E > H > I	0.083	24
	Bemetara	2.351	0.965	0.868	0.527	0.768	E > H > I	0.017	11
	Bilaspur	1.975	0.954	0.813	0.474	0.725	E > H > I	0.016	20
	Dantewada	1.985	0.952	0.721	0.252	0.581	H > E > I	0.081	20
	Durg	1.810	0.979	0.868	0.867	0.903	I > H > E	0.001	26
	Gariyaband	1.855	0.952	0.824	0.348	0.666	E > H > I	0.050	20
	Kodagaon	2.219	0.937	0.737	0.169	0.527	E > H > I	0.135	18
	Mungeli	2.831	0.944	0.830	0.370	0.677	E > H > I	0.040	15
	Raipur	2.036	0.978	0.843	0.757	0.856	H > E > I	0.002	26
	Sukma	2.163	0.929	0.625	0.092	0.432	H > E > I	0.211	17
	Surajpur	2.221	0.966	0.839	0.228	0.602	H > E > I	0.127	20
	Surguja	2.258	0.935	0.807	0.254	0.603	E > H > I	0.082	18
Delhi	Central	1.584	0.985	0.885	0.963	0.944	I > H > E	0.002	14
	East	1.831	0.959	0.914	0.961	0.945	I > E > H	0.006	26
	New Delhi	2.291	0.979	0.902	0.960	0.947	I > H > E	0.002	26
	North	2.587	0.969	0.827	0.928	0.907	I > H > E	0.005	28
	North East	2.108	0.970	0.897	0.990	0.951	I > E > H	0.004	26
	North West	2.002	0.984	0.931	0.987	0.967	I > E > H	0.001	14
	Shahdara	1.982	0.951	0.869	0.987	0.935	I > E > H	0.012	26
	South	2.127	0.948	0.932	0.989	0.956	I > E > H	0.015	26
	South East	1.815	0.985	0.927	0.972	0.961	I > H > E	0.001	14
	South West	1.559	0.990	0.934	0.981	0.968	I > H > E	0.001	14
	West	2.101	0.963	0.886	0.969	0.939	I > E > H	0.005	26
Dadra & Nagar Haveli and Daman & Diu	Diu	1.991	1.000	0.750	0.955	0.897	H > I > E	0.022	22
	Daman	2.036	0.958	0.800	0.867	0.873	I > H > E	0.005	28
	Dadra & Nagar Haveli	1.935	0.950	0.694	0.520	0.705	H > E > I	0.007	15
Goa	North Goa	1.239	1.000	0.965	0.966	0.977	H > I > E	0.000	14
	South Goa	1.206	0.984	0.940	0.971	0.965	I > E > H	0.001	14
	Kachchh	2.336	0.962	0.613	0.821	0.789	I > H > E	0.044	28

(continued)

Table 8.5 (continued)

State/ Union Territory	District	TFR	HE	ED	SL	HDS	HDP	HDU	Node
	Banas Kantha	2.756	0.949	0.581	0.479	0.648	H > I > E	0.032	15
	Patan	2.369	0.910	0.641	0.648	0.725	I > E > H	0.022	16
	Mahesana	2.739	0.937	0.692	0.665	0.758	I > H > E	0.005	16
	Gandhinagar	2.255	0.957	0.785	0.816	0.850	I > H > E	0.003	28
	Porbandar	1.926	0.982	0.719	0.896	0.860	I > H > E	0.021	22
	Amreli	2.146	0.980	0.616	0.781	0.782	H > I > E	0.049	27
	Anand	2.229	0.935	0.676	0.697	0.763	I > H > E	0.009	16
	Dohad	2.931	0.973	0.594	0.184	0.505	H > E > I	0.166	19
	Narmada	2.225	0.954	0.667	0.305	0.596	H > E > I	0.053	20
	Bharuch	2.201	0.947	0.673	0.711	0.770	I > H > E	0.011	16
	The Dangs	2.555	0.975	0.571	0.227	0.526	H > E > I	0.142	20
	Navsari	1.679	0.982	0.727	0.729	0.806	H > I > E	0.014	22
	Valsad	1.550	0.977	0.805	0.725	0.831	H > I > E	0.003	27
	Surat	2.039	0.974	0.739	0.870	0.857	I > H > E	0.013	28
	Tapi	1.820	0.959	0.639	0.404	0.638	H > E > I	0.034	20
	Ahmadabad	1.983	0.961	0.733	0.934	0.872	I > H > E	0.017	28
	Aravali	2.613	0.961	0.687	0.458	0.679	H > E > I	0.021	20
	Bhavnagar	2.062	0.980	0.656	0.717	0.775	H > I > E	0.030	27
	Botad	2.549	0.980	0.596	0.810	0.783	H > I > E	0.062	27
	Chhota Udaipur	2.355	0.926	0.549	0.329	0.563	H > I > E	0.026	15
	Devbhumi Dwarka	2.538	0.971	0.450	0.776	0.707	H > I > E	0.208	27
	Gir Somnath	2.013	0.976	0.712	0.751	0.807	H > I > E	0.014	27
	Jamnagar	1.806	0.979	0.713	0.915	0.863	I > H > E	0.022	28
	Junagadh	2.002	0.961	0.724	0.817	0.830	I > H > E	0.011	28
	Kheda	2.195	0.934	0.577	0.555	0.673	H > I > E	0.023	16
	Mahisagar	2.383	0.970	0.664	0.394	0.644	H > E > I	0.045	20
	Morbi	1.811	0.989	0.653	0.868	0.828	H > I > E	0.043	22
	Panch Mahals	2.165	0.939	0.616	0.474	0.656	H > I > E	0.012	15
	Rajkot	2.161	0.980	0.827	0.934	0.912	I > H > E	0.005	28
	Sabar Kantha	2.170	0.968	0.626	0.587	0.713	H > I > E	0.030	27
	Surendranagar	2.363	0.967	0.646	0.721	0.769	H > I > E	0.026	27
	Vadodara	1.877	0.977	0.699	0.831	0.830	H > I > E	0.021	27
	Panchkula	2.190	0.960	0.915	0.968	0.948	I > E > H	0.006	26
	Ambala	1.755	0.932	0.889	0.939	0.920	I > E > H	0.025	26
	Yamunanagar	1.930	0.967	0.880	0.918	0.921	I > E > H	0.002	26
	Kurukshetra	1.958	0.958	0.865	0.914	0.912	I > E > H	0.005	26
	Kaithal	2.289	0.954	0.935	0.914	0.934	I > E > H	0.008	26

(continued)

Table 8.5 (continued)

State/ Union Territory	District	TFR	HE	ED	SL	HDS	HDP	HDU	Node
	Karnal	2.391	0.972	0.892	0.912	0.925	I > E > H	0.001	26
	Panipat	2.189	0.959	0.926	0.897	0.928	I > E > H	0.004	26
	Sonipat	1.953	0.965	0.937	0.940	0.947	I > E > H	0.004	26
	Jind	2.132	0.944	0.946	0.906	0.932	E > I > H	0.015	31
	Fatehabad	1.815	0.961	0.931	0.885	0.925	E > I > H	0.004	31
	Sirsa	2.182	0.958	0.870	0.879	0.902	I > E > H	0.003	26
	Hisar	2.028	0.971	0.910	0.910	0.930	I > E > H	0.001	26
	Rohtak	1.980	0.975	0.922	0.927	0.941	I > E > H	0.001	26
	Jhajjar	1.933	0.991	0.971	0.954	0.972	E > I > H	0.000	14
	Mahendragarh	2.355	0.965	0.970	0.851	0.927	E > I > H	0.004	31
	Rewari	2.030	0.969	0.936	0.926	0.944	I > E > H	0.002	26
	Gurgaon	2.007	0.966	0.911	0.940	0.939	I > E > H	0.003	26
	Mewat	3.621	0.941	0.526	0.561	0.658	H > I > E	0.058	17
	Faridabad	2.077	0.967	0.873	0.949	0.929	I > E > H	0.004	26
	Palwal	2.989	0.981	0.841	0.805	0.873	H > I > E	0.002	27
	Bhiwani	2.328	0.929	0.928	0.891	0.916	I > E > H	0.028	26
	Charkhi Dadri	2.245	0.967	0.984	0.935	0.962	E > I > H	0.004	31
Gujarat	Chamba	2.166	0.970	0.942	0.673	0.854	E > H > I	0.008	29
	Kangra	1.751	0.955	0.889	0.855	0.899	I > E > H	0.004	26
	Lahul & Spiti	2.111	0.996	1.000	0.481	0.794	E > H > I	0.051	24
	Kullu	1.890	0.984	0.957	0.661	0.857	E > H > I	0.011	30
	Mandi	1.972	0.979	0.947	0.820	0.913	E > H > I	0.001	32
	Hamirpur	2.007	0.972	0.900	0.872	0.914	I > E > H	0.000	26
	Una	2.099	0.980	0.882	0.900	0.920	I > H > E	0.001	26
	Bilaspur	1.998	0.992	0.886	0.899	0.925	H > I > E	0.002	14
	Solan	1.936	0.941	0.900	0.902	0.914	I > E > H	0.015	26
	Sirmaur	2.490	0.980	0.909	0.824	0.903	H > E > I	0.000	26
Haryana	Shimla	1.777	0.965	0.942	0.837	0.913	E > I > H	0.002	31
	Kinnaur	2.055	0.968	1.000	0.695	0.879	E > H > I	0.011	31
Jammu & Kashmir	Anantnag	2.270	1.000	0.953	0.830	0.926	H > E > I	0.003	14
	Bandipore	1.938	0.969	0.848	0.584	0.787	H > E > I	0.009	11
	Punch	2.137	0.972	0.949	0.581	0.818	E > H > I	0.018	29
	Rajouri	2.196	0.989	0.875	0.621	0.817	H > E > I	0.013	30
	Kishtwar	1.673	0.990	0.906	0.522	0.785	H > E > I	0.030	30
	Doda	1.952	0.981	0.897	0.514	0.775	H > E > I	0.027	30
	Baramula	1.882	0.976	0.805	0.648	0.801	H > E > I	0.006	27
	Srinagar	1.479	0.988	0.979	0.947	0.972	E > I > H	0.000	14
	Jammu	2.102	0.985	0.981	0.952	0.973	E > I > H	0.000	14

(continued)

Table 8.5 (continued)

State/ Union Territory	District	TFR	HE	ED	SL	HDS	HDP	HDU	Node
Himachal Pradesh	Pulwama	1.752	0.974	0.950	0.869	0.930	E > I > H	0.001	32
	Shupiyan	2.083	1.000	0.909	0.803	0.901	H > E > I	0.004	14
	Badgam	1.841	0.995	0.925	0.773	0.894	H > E > I	0.004	14
	Kupwara	1.964	0.961	0.914	0.581	0.805	E > H > I	0.013	29
	Ganderbal	2.250	0.987	0.839	0.687	0.831	H > E > I	0.007	22
	Ramban	2.272	0.993	0.865	0.370	0.700	H > E > I	0.073	11
	Kulgam	1.868	0.983	0.926	0.542	0.798	E > H > I	0.025	30
	Udhampur	2.228	0.980	0.950	0.606	0.832	E > H > I	0.016	29
	Reasi	2.174	0.986	0.897	0.506	0.773	H > E > I	0.031	30
	Kathua	1.494	0.981	0.961	0.847	0.928	E > H > I	0.001	14
	Samba	1.894	0.991	0.946	0.876	0.937	H > E > I	0.000	14
Jharkhand	Garhwa	2.921	0.952	0.827	0.173	0.557	E > H > I	0.165	19
	Chatra	2.806	0.952	0.772	0.196	0.560	H > E > I	0.130	19
	Kodarma	2.645	0.963	0.912	0.399	0.720	E > H > I	0.051	24
	Giridih	2.612	0.956	0.806	0.331	0.653	H > E > I	0.056	20
	Deoghar	2.885	0.967	0.723	0.194	0.548	H > E > I	0.142	19
	Godda	2.726	0.930	0.723	0.213	0.552	E > H > I	0.085	17
	Sahibganj	3.169	0.969	0.648	0.161	0.504	H > E > I	0.174	19
	Pakur	2.590	0.959	0.623	0.117	0.460	H > E > I	0.218	19
	Dhanbad	2.300	0.976	0.847	0.555	0.777	H > E > I	0.015	11
	Bokaro	2.296	0.957	0.873	0.541	0.774	E > H > I	0.013	29
	Lohardaga	2.326	0.959	0.863	0.234	0.611	E > H > I	0.122	11
	Purbi Singhbhum	1.990	0.952	0.871	0.551	0.776	E > H > I	0.012	29
	Palamu	2.679	0.944	0.814	0.235	0.595	E > H > I	0.101	18
	Latehar	2.851	0.963	0.776	0.094	0.479	H > E > I	0.282	19
	Hazaribagh	2.563	0.954	0.883	0.434	0.727	E > H > I	0.034	29
	Ramgarh	2.959	0.959	0.861	0.484	0.745	E > H > I	0.021	11
	Dumka	2.668	0.941	0.785	0.173	0.544	E > H > I	0.145	18
	Jamtara	2.511	0.972	0.794	0.197	0.571	H > E > I	0.152	19
	Ranchi	2.027	0.958	0.891	0.546	0.782	E > H > I	0.015	29
	Khunti	2.217	0.961	0.725	0.112	0.483	H > E > I	0.238	19
	Gumla	2.599	0.933	0.822	0.108	0.497	E > H > I	0.240	18
	Simdega	2.288	0.964	0.813	0.119	0.511	H > E > I	0.245	19
	Pashchimi Singhbhum	2.284	0.955	0.760	0.091	0.470	H > E > I	0.277	19
	Saraikela-Kharsawan	1.965	0.929	0.844	0.298	0.638	E > H > I	0.068	11
Karnataka	Belgaum	2.193	0.975	0.907	0.670	0.843	E > H > I	0.006	29

(continued)

Table 8.5 (continued)

State/ Union Territory	District	TFR	HE	ED	SL	HDS	HDP	HDU	Node
	Bagalkot	2.197	0.972	0.857	0.533	0.770	H > E > I	0.017	11
	Bijapur	2.461	0.981	0.814	0.520	0.753	H > E > I	0.022	20
	Bidar	2.164	0.969	0.882	0.539	0.779	E > H > I	0.017	29
	Raichur	2.348	0.946	0.739	0.497	0.709	H > E > I	0.005	15
	Koppal	1.906	0.957	0.792	0.448	0.707	H > E > I	0.021	20
	Gadag	2.032	0.984	0.901	0.567	0.801	H > E > I	0.019	30
	Dharwad	1.489	0.960	0.927	0.762	0.880	E > I > H	0.004	31
	Uttara Kannada	1.608	0.963	0.892	0.786	0.878	E > I > H	0.001	31
	Haveri	1.790	0.965	0.827	0.582	0.779	H > E > I	0.007	27
	Bellary	1.813	0.932	0.862	0.643	0.805	E > I > H	0.011	11
	Chitradurga	1.693	0.957	0.887	0.700	0.843	E > H > I	0.003	31
	Davanagere	1.703	0.977	0.861	0.743	0.856	H > E > I	0.001	26
	Shimoga	1.476	0.979	0.922	0.833	0.910	E > H > I	0.000	32
	Udupi	1.503	0.985	0.963	0.837	0.926	E > H > I	0.001	14
	Chikmagalur	1.605	0.969	0.940	0.826	0.910	E > I > H	0.002	32
	Tumkur	1.729	0.926	0.991	0.823	0.911	E > I > H	0.037	31
	Bangalore	1.475	0.982	0.949	0.969	0.966	I > E > H	0.001	14
	Mandya	1.618	1.000	0.952	0.864	0.937	H > E > I	0.002	14
	Hassan	1.529	0.990	0.945	0.812	0.913	H > E > I	0.002	14
	Dakshina Kannada	1.768	0.992	0.950	0.903	0.948	H > E > I	0.000	14
	Kodagu	1.653	0.974	0.927	0.868	0.922	E > I > H	0.000	32
	Mysore	1.815	0.987	0.889	0.858	0.910	H > I > E	0.001	14
	Chamarajanagar	1.788	0.966	0.887	0.748	0.864	E > H > I	0.001	31
	Gulbarga	2.109	0.978	0.777	0.554	0.755	H > E > I	0.016	20
	Yadgir	2.196	0.958	0.735	0.458	0.694	H > E > I	0.017	20
	Kolar	1.802	0.988	0.942	0.892	0.940	H > E > I	0.000	14
	Chikkaballapura	1.810	0.978	0.856	0.849	0.893	H > I > E	0.001	26
	Bangalore Rural	1.703	0.969	0.931	0.891	0.930	E > I > H	0.002	31
	Ramanagara	1.641	0.941	0.901	0.901	0.914	I > E > H	0.014	26
Kerala	Kasaragod	2.006	0.995	0.974	0.916	0.961	H > E > I	0.000	14
	Kannur	1.733	1.000	0.967	0.968	0.978	H > I > E	0.000	14
	Wayanad	1.995	0.987	0.954	0.815	0.916	E > H > I	0.002	14
	Kozhikode	1.823	1.000	1.000	0.980	0.994	E = H > I	0.000	14
	Malappuram	2.531	0.995	1.000	0.966	0.987	E > H > I	0.000	14
	Palakkad	1.812	1.000	0.988	0.888	0.958	H > E > I	0.001	14
	Thrissur	1.410	1.000	0.984	0.971	0.985	H > E > I	0.000	14
	Ernakulam	1.498	0.989	0.990	0.982	0.987	E > I > H	0.000	14

(continued)

Table 8.5 (continued)

State/ Union Territory	District	TFR	HE	ED	SL	HDS	HDP	HDU	Node
	Idukki	1.640	0.989	0.967	0.832	0.927	E > H > I	0.002	14
	Kottayam	1.583	1.000	0.970	0.956	0.975	H > I > E	0.000	14
	Alappuzha	1.456	1.000	0.987	0.968	0.985	H > E > I	0.000	14
	Pathanamthitta	1.395	1.000	0.989	0.934	0.974	H > E > I	0.000	14
	Kollam	1.499	1.000	0.993	0.943	0.978	H > E > I	0.000	14
	Thiruvananthapuram	1.469	1.000	0.993	0.925	0.972	H > E > I	0.001	14
Ladakh	Leh (Ladakh)	1.934	0.969	0.900	0.742	0.866	E > H > I	0.001	31
	Kargil	2.098	0.975	1.000	0.388	0.742	E > H > I	0.073	24
Lakshadweep	Lakshadweep	2.182	1.000	1.000	0.981	0.994	E = H > I	0.000	14
Madhya Pradesh	Sheopur	2.502	0.934	0.658	0.323	0.598	H > E > I	0.025	15
	Morena	2.979	0.936	0.782	0.545	0.741	E > H > I	0.004	15
	Bhind	2.741	0.957	0.854	0.573	0.781	E > H > I	0.008	11
	Gwalior	2.402	0.972	0.851	0.828	0.882	I > H > E	0.000	26
	Datia	2.648	0.942	0.802	0.516	0.737	E > H > I	0.007	15
	Shivpuri	2.769	0.935	0.718	0.394	0.653	H > E > I	0.014	15
	Tikamgarh	2.538	0.960	0.798	0.476	0.722	H > E > I	0.018	20
	Chhatarpur	2.344	0.934	0.723	0.413	0.663	E > H > I	0.011	15
	Panna	2.469	0.901	0.725	0.234	0.560	E > H > I	0.067	17
	Sagar	2.925	0.961	0.830	0.421	0.707	E > H > I	0.032	20
	Damoh	2.766	0.941	0.758	0.307	0.621	E > H > I	0.048	18
	Satna	2.716	0.946	0.741	0.391	0.661	H > E > I	0.022	15
	Rewa	3.073	0.928	0.674	0.246	0.559	H > E > I	0.052	17
	Umaria	2.450	0.929	0.723	0.310	0.610	E > H > I	0.034	15
	Neemuch	2.250	0.963	0.823	0.657	0.807	H > E > I	0.002	27
	Mandsaur	2.273	0.940	0.824	0.576	0.768	E > H > I	0.005	16
	Ratlam	2.388	0.953	0.714	0.592	0.742	H > I > E	0.004	27
	Ujjain	2.404	0.976	0.818	0.667	0.813	H > E > I	0.005	27
	Dewas	2.310	0.952	0.802	0.575	0.764	E > H > I	0.004	20
	Dhar	2.031	0.973	0.731	0.478	0.706	H > E > I	0.025	20
	Indore	2.022	0.971	0.849	0.895	0.904	I > H > E	0.002	26
	Khargone (West Nimar)	2.199	0.959	0.708	0.647	0.763	H > I > E	0.006	27
	Barwani	2.471	0.968	0.572	0.371	0.601	H > I > E	0.072	20
	Rajgarh	2.359	0.957	0.794	0.350	0.659	H > E > I	0.048	20
	Vidisha	2.802	0.942	0.684	0.460	0.674	H > E > I	0.007	15
	Bhopal	1.791	0.984	0.799	0.850	0.875	H > I > E	0.006	22
	Sehore	2.913	0.940	0.835	0.495	0.737	E > H > I	0.014	15

(continued)

Table 8.5 (continued)

State/Union Territory	District	TFR	HE	ED	SL	HDS	HDP	HDU	Node
	Jhabua	3.529	0.933	0.657	0.181	0.513	H > E > I	0.101	17
	Raisen	2.796	0.955	0.809	0.466	0.720	E > H > I	0.018	20
	Betul	1.952	0.965	0.797	0.391	0.683	H > E > I	0.039	20
	Harda	2.799	0.963	0.779	0.597	0.769	H > E > I	0.004	27
	Hoshangabad	2.680	0.952	0.765	0.531	0.734	H > E > I	0.005	20
	Katni	2.321	0.896	0.730	0.348	0.624	E > I > H	0.039	15
	Jabalpur	1.735	1.000	0.744	0.395	0.678	H > E > I	0.069	20
	Narsimhapur	2.163	0.976	0.735	0.438	0.690	H > E > I	0.035	20
	Dindori	2.554	0.945	0.779	0.146	0.523	E > H > I	0.179	18
	Mandla	2.105	0.956	0.818	0.297	0.637	E > H > I	0.073	20
	Chhindwara	1.813	0.951	0.766	0.462	0.704	H > E > I	0.013	20
	Seoni	2.036	0.950	0.755	0.305	0.622	H > E > I	0.055	18
	Balaghat	1.725	0.955	0.863	0.285	0.642	E > H > I	0.087	11
	Guna	2.797	0.950	0.695	0.428	0.665	H > E > I	0.016	15
	Ashoknagar	2.848	0.957	0.728	0.331	0.629	H > E > I	0.048	20
	Shahdol	1.932	0.909	0.778	0.280	0.604	E > H > I	0.060	18
	Anuppur	2.303	0.947	0.798	0.286	0.622	E > H > I	0.069	18
	Sidhi	2.859	0.931	0.723	0.232	0.565	E > H > I	0.073	17
	Singrauli	3.340	0.938	0.748	0.320	0.625	E > H > I	0.040	15
	Alirajpur	3.210	0.970	0.540	0.171	0.479	H > E > I	0.188	19
	Khandwa (East Nimar)	2.583	0.973	0.803	0.533	0.753	H > E > I	0.016	20
	Burhanpur	2.265	0.946	0.687	0.612	0.739	H > I > E	0.004	16
	Agar Malwa	2.558	0.981	0.787	0.522	0.745	H > E > I	0.021	20
	Shajapur	2.493	0.940	0.802	0.526	0.740	E > H > I	0.006	15
Maharashtra	Nandurbar	2.176	0.942	0.703	0.362	0.634	H > E > I	0.024	15
	Dhule	2.406	0.984	0.764	0.594	0.769	H > E > I	0.017	22
	Jalgaon	2.065	0.980	0.823	0.711	0.833	H > E > I	0.004	27
	Buldana	1.881	0.969	0.937	0.667	0.849	E > H > I	0.008	29
	Akola	2.026	0.969	0.883	0.776	0.873	E > H > I	0.000	31
	Washim	2.333	0.959	0.878	0.590	0.796	E > H > I	0.009	29
	Amravati	1.703	0.981	0.915	0.754	0.879	H > E > I	0.002	14
	Wardha	1.622	0.984	0.894	0.836	0.903	H > E > I	0.001	14
	Nagpur	1.636	0.990	0.951	0.932	0.957	I > H > E	0.000	14
	Bhandara	1.785	0.957	0.919	0.677	0.844	E > H > I	0.006	29
	Gondiya	1.786	0.946	0.958	0.607	0.824	E > H > I	0.019	29
	Gadchiroli	1.672	0.964	0.878	0.507	0.762	E > H > I	0.020	29
	Chandrapur	1.841	0.953	0.953	0.760	0.885	E > I > H	0.008	31
	Yavatmal	1.751	0.950	0.882	0.644	0.817	E > H > I	0.006	29
	Nanded	2.232	0.964	0.817	0.597	0.782	H > E > I	0.005	27

(continued)

Table 8.5 (continued)

State/ Union Territory	District	TFR	HE	ED	SL	HDS	HDP	HDU	Node
	Hingoli	2.123	0.965	0.873	0.608	0.804	E > H > I	0.007	29
	Parbhani	2.262	0.970	0.848	0.559	0.777	H > E > I	0.012	11
	Jalna	2.342	0.960	0.849	0.577	0.782	E > H > I	0.008	11
	Aurangabad	2.095	0.977	0.902	0.740	0.869	E > H > I	0.002	32
	Nashik	2.603	0.948	0.815	0.640	0.793	E > H > I	0.001	16
	Mumbai Suburban	1.326	0.886	0.966	0.979	0.943	I > E > H	0.163	26
	Mumbai	1.476	0.983	0.956	0.968	0.969	I > E > H	0.001	14
	Raigarh	2.202	0.980	0.848	0.814	0.879	H > I > E	0.001	26
	Pune	1.646	1.000	0.850	0.882	0.909	H > I > E	0.006	14
	Ahmadnagar	1.862	0.974	0.895	0.724	0.859	E > H > I	0.002	32
	Bid	2.190	0.980	0.881	0.577	0.798	H > E > I	0.015	29
	Latur	2.512	0.972	0.875	0.686	0.838	H > E > I	0.003	29
	Osmanabad	2.314	0.995	0.910	0.667	0.849	H > E > I	0.012	30
	Solapur	2.744	0.981	0.866	0.661	0.828	H > E > I	0.007	11
	Satara	2.154	1.000	0.883	0.755	0.875	H > E > I	0.007	14
	Ratnagiri	2.187	0.961	0.953	0.586	0.819	E > H > I	0.017	29
	Sindhudurg	1.330	0.977	0.975	0.687	0.871	E > H > I	0.009	32
	Kolhapur	1.795	0.967	0.958	0.817	0.912	E > I > H	0.003	31
	Sangli	1.369	0.971	0.955	0.813	0.911	E > I > H	0.002	32
	Palghar	1.863	0.988	0.793	0.675	0.811	H > E > I	0.010	22
	Thane	1.874	0.979	0.923	0.896	0.932	I > E > H	0.000	26
	Senapati	2.762	0.986	0.917	0.302	0.674	H > E > I	0.105	24
	Tamenglong	2.579	0.950	0.900	0.189	0.587	E > H > I	0.164	24
	Churachandpur	2.012	0.976	0.870	0.422	0.723	H > E > I	0.044	11
	Bishnupur	1.993	0.972	0.917	0.430	0.740	E > H > I	0.045	29
	Thoubal	2.004	0.971	0.894	0.422	0.728	E > H > I	0.045	29
	Imphal West	2.024	0.991	0.933	0.697	0.867	H > E > I	0.008	14
	Imphal East	1.900	0.957	0.907	0.597	0.808	E > H > I	0.011	29
	Ukhrul	3.012	0.966	0.900	0.164	0.571	E > H > I	0.201	24
	Chandel	2.772	0.962	0.857	0.410	0.710	E > H > I	0.039	11
Manipur	South West Garo Hills	2.071	0.973	0.826	0.301	0.646	H > E > I	0.085	20
	South Garo Hills	3.076	0.994	0.941	0.268	0.662	H > E > I	0.135	24
	North Garo Hills	2.098	0.974	0.923	0.277	0.658	E > H > I	0.113	24
	East Jantia Hills	2.314	0.947	0.696	0.274	0.586	H > E > I	0.061	17
	East Khasi Hills	3.877	1.000	0.780	0.472	0.725	H > E > I	0.046	20
	Ribhoi	1.867	0.964	0.739	0.261	0.595	H > E > I	0.089	20

(continued)

Table 8.5 (continued)

State/ Union Territory	District	TFR	HE	ED	SL	HDS	HDP	HDU	Node
	South West Khasi Hills	1.980	0.943	0.813	0.203	0.573	E > H > I	0.125	18
	East Garo Hills	3.630	0.991	0.867	0.330	0.678	H > E > I	0.089	11
	West Garo Hills	2.095	0.993	0.940	0.480	0.776	H > E > I	0.044	30
Meghalaya	West Jaintia Hills	3.474	0.951	0.587	0.236	0.531	H > E > I	0.088	17
	West Khasi Hills	4.701	0.929	0.652	0.112	0.457	H > E > I	0.180	17
Manipur	Mamit	2.224	0.980	0.857	0.655	0.822	H > E > I	0.006	11
	Kolasib	1.767	0.991	0.900	0.820	0.902	H > E > I	0.002	14
	Aizawl	1.677	0.970	0.896	0.935	0.933	I > E > H	0.002	26
	Champhai	1.702	0.990	0.900	0.813	0.899	H > E > I	0.002	14
	Serchhip	1.537	0.981	0.857	0.840	0.891	H > I > E	0.001	14
	Lunglei	1.380	1.000	0.833	0.754	0.858	H > I > E	0.010	22
	Lawngtlai	2.137	0.956	0.750	0.472	0.704	H > E > I	0.013	20
	Saiha	1.657	0.979	1.000	0.707	0.887	E > H > I	0.010	32
Nagaland	Mon	2.222	0.982	0.800	0.124	0.517	H > E > I	0.252	19
	Mokokchung	1.884	0.984	0.909	0.534	0.789	H > E > I	0.025	30
	Zunheboto	2.714	0.976	0.800	0.240	0.602	H > E > I	0.122	20
	Wokha	2.223	0.958	1.000	0.467	0.777	E > H > I	0.046	24
	Dimapur	1.830	0.986	0.939	0.794	0.904	H > E > I	0.002	14
	Phek	2.545	0.975	0.833	0.225	0.601	H > E > I	0.137	20
	Tuensang	3.015	0.967	0.739	0.187	0.548	H > E > I	0.150	19
	Longleng	3.006	0.953	0.750	0.130	0.503	H > E > I	0.204	19
	Kiphire	2.598	0.917	0.800	0.159	0.532	E > H > I	0.153	18
	Kohima	1.968	0.984	0.933	0.655	0.848	E > H > I	0.010	30
	Peren	2.457	0.900	0.800	0.345	0.644	E > I > H	0.058	18
	Bargarh	1.652	0.972	0.825	0.370	0.682	H > E > I	0.054	20
	Jharsuguda	1.773	0.940	0.911	0.511	0.767	E > H > I	0.024	29
	Sambalpur	1.559	0.966	0.833	0.380	0.689	H > E > I	0.048	20
	Debagarh	1.810	0.958	0.808	0.247	0.604	H > E > I	0.102	20
	Sundargarh	1.666	0.925	0.780	0.411	0.677	E > H > I	0.019	15
	Kendujhar	2.367	0.945	0.747	0.290	0.610	H > E > I	0.057	18
	Mayurbhanj	1.965	0.953	0.698	0.154	0.509	H > E > I	0.162	19
	Baleshwar	2.137	0.971	0.881	0.362	0.694	E > H > I	0.063	24
	Baudh	1.720	0.949	0.717	0.329	0.623	H > E > I	0.041	15
	Kendrapara	2.060	0.939	0.906	0.430	0.727	E > H > I	0.039	29
	Jagatsinghapur	1.740	0.972	0.884	0.515	0.770	E > H > I	0.022	29
	Cuttack	1.485	0.983	0.786	0.583	0.771	H > E > I	0.015	22
	Jajapur	2.143	0.941	0.798	0.412	0.687	E > H > I	0.023	15

(continued)

Table 8.5 (continued)

State/ Union Territory	District	TFR	HE	ED	SL	HDS	HDP	HDU	Node
	Dhenkanal	2.155	0.965	0.807	0.383	0.682	H > E > I	0.043	20
	Anugul	1.966	0.985	0.697	0.418	0.670	H > E > I	0.051	20
	Nayagarh	2.174	0.993	0.887	0.452	0.747	H > E > I	0.048	24
	Khordha	1.469	0.989	0.891	0.699	0.853	H > E > I	0.007	14
	Puri	1.720	0.979	0.926	0.570	0.809	E > H > I	0.019	29
	Ganjam	2.271	0.975	0.802	0.639	0.796	H > E > I	0.006	27
	Gajapati	1.983	0.936	0.636	0.283	0.570	H > E > I	0.040	15
	Kandhamal	2.428	0.952	0.745	0.206	0.559	H > E > I	0.117	19
	Bhadrak	2.032	0.957	0.903	0.343	0.686	E > H > I	0.068	24
	Subarnapur	1.820	0.941	0.814	0.433	0.703	E > H > I	0.021	15
	Balangir	1.993	0.959	0.783	0.362	0.663	H > E > I	0.044	20
	Nuapada	2.251	0.952	0.779	0.220	0.578	H > E > I	0.111	20
	Kalahandi	2.016	0.938	0.645	0.247	0.553	H > E > I	0.060	17
	Rayagada	2.278	0.962	0.604	0.260	0.553	H > E > I	0.090	20
	Nabarangapur	2.732	0.957	0.538	0.132	0.447	H > E > I	0.205	19
	Koraput	1.825	0.965	0.635	0.220	0.539	H > E > I	0.116	20
	Malkangiri	2.150	0.889	0.633	0.189	0.502	E > H > I	0.058	17
Odisha	Yanam	1.671	0.959	0.800	0.972	0.908	I > H > E	0.011	28
	Puducherry	1.615	1.000	0.911	0.914	0.941	H > I > E	0.002	14
	Mahe	1.688	1.000	1.000	1.000	1.000	H=I = E	0.000	14
	Karaikal	1.475	0.986	0.893	0.839	0.904	H > I > E	0.001	14
Punjab	Kapurthala	1.862	0.969	0.905	0.949	0.941	I > E > H	0.003	26
	Jalandhar	1.968	0.968	0.911	0.985	0.954	I > E > H	0.004	26
	Hoshiarpur	1.852	0.989	0.971	0.964	0.975	I > E > H	0.000	14
	Shahid Bhagat Singh Nagar	1.652	0.997	0.949	0.965	0.970	H > I > E	0.000	14
	Fatehgarh Sahib	1.811	0.985	0.895	0.970	0.950	I > H > E	0.002	14
	Ludhiana	2.236	0.960	0.880	0.955	0.931	I > E > H	0.006	26
	Moga	1.825	0.966	0.843	0.929	0.912	I > H > E	0.004	26
	Muktsar	1.991	0.935	0.840	0.894	0.889	I > E > H	0.017	16
	Faridkot	1.969	0.956	0.783	0.901	0.877	I > H > E	0.009	28
	Bathinda	1.964	0.964	0.787	0.907	0.884	I > H > E	0.008	28
	Mansa	2.043	0.960	0.881	0.878	0.906	I > E > H	0.003	26
	Patiala	1.972	0.967	0.879	0.963	0.936	I > E > H	0.004	26
	Amritsar	1.918	0.975	0.867	0.956	0.932	I > H > E	0.003	26
	Tarn Taran	2.048	0.987	0.820	0.873	0.891	H > I > E	0.005	22
	Rupnagar	1.969	0.975	0.916	0.957	0.949	I > E > H	0.002	26

(continued)

Table 8.5 (continued)

State/ Union Territory	District	TFR	HE	ED	SL	HDS	HDP	HDU	Node
	Sahibzada Ajit Singh Nagar	1.767	0.937	0.944	0.967	0.950	I > E > H	0.024	26
	Sangrur	1.934	0.920	0.915	0.960	0.932	I > E > H	0.046	26
	Barnala	1.734	0.950	0.914	0.947	0.937	I > E > H	0.011	26
	Fazilka	2.120	0.951	0.809	0.820	0.858	I > E > H	0.004	28
	Firozpur	1.956	0.960	0.805	0.901	0.887	I > H > E	0.006	28
	Gurdaspur	1.778	0.959	0.940	0.937	0.945	I > E > H	0.006	26
	Pathankot	1.872	0.967	0.982	0.962	0.970	E > I > H	0.005	31
Rajasthan	Ganganagar	2.011	0.971	0.849	0.778	0.863	H > I > E	0.000	26
	Hanumangarh	2.439	0.975	0.858	0.786	0.871	H > I > E	0.000	26
Puducherry	Bikaner	2.520	0.977	0.793	0.754	0.837	H > I > E	0.004	27
	Churu	2.518	0.971	0.875	0.700	0.843	H > E > I	0.002	26
	Jhunjhunun	2.180	0.967	0.905	0.835	0.901	E > I > H	0.001	31
	Alwar	2.781	0.960	0.803	0.675	0.807	H > E > I	0.001	27
	Bharatpur	2.958	0.975	0.777	0.568	0.760	H > E > I	0.013	20
	Dhaulpur	3.025	0.947	0.833	0.479	0.731	E > H > I	0.016	15
	Karauli	3.260	0.973	0.838	0.437	0.720	H > E > I	0.035	20
	Sawai Madhopur	2.977	0.928	0.826	0.525	0.744	E > I > H	0.013	15
	Dausa	2.749	0.968	0.915	0.567	0.801	E > H > I	0.016	29
	Jaipur	2.278	0.966	0.925	0.836	0.908	E > I > H	0.002	31
	Sikar	2.402	0.964	0.880	0.804	0.881	E > I > H	0.000	31
	Nagaur	2.312	0.959	0.852	0.780	0.861	I > E > H	0.001	26
	Jodhpur	2.260	0.974	0.793	0.743	0.832	H > I > E	0.003	27
	Jaisalmer	2.618	0.972	0.715	0.637	0.765	H > I > E	0.012	27
	Barmer	2.454	0.992	0.799	0.588	0.780	H > E > I	0.020	22
	Jalor	2.571	0.977	0.806	0.652	0.804	H > E > I	0.006	27
	Sirohi	2.927	0.960	0.759	0.560	0.746	H > E > I	0.006	20
	Pali	2.346	0.976	0.900	0.839	0.904	E > H > I	0.000	32
	Ajmer	2.121	0.939	0.886	0.821	0.881	E > I > H	0.012	31
	Tonk	2.558	0.953	0.851	0.532	0.762	E > H > I	0.012	11
	Bundi	2.467	0.945	0.854	0.513	0.752	E > H > I	0.014	11
	Bhilwara	2.709	0.960	0.862	0.605	0.798	E > H > I	0.006	11
	Rajsamand	2.137	0.969	0.862	0.652	0.820	H > E > I	0.004	11
	Dungarpur	2.340	0.988	0.886	0.361	0.699	H > E > I	0.075	24
	Banswara	2.266	0.952	0.774	0.282	0.615	H > E > I	0.071	20
	Chittaurgarh	2.358	0.939	0.802	0.592	0.767	E > H > I	0.003	16
	Kota	2.026	0.956	0.902	0.835	0.897	E > I > H	0.003	31
	Baran	2.264	0.959	0.836	0.544	0.764	E > H > I	0.010	20

(continued)

Table 8.5 (continued)

State/ Union Territory	District	TFR	HE	ED	SL	HDS	HDP	HDU	Node
	Jhalawar	2.314	0.952	0.811	0.441	0.709	E > H > I	0.022	20
	Udaipur	2.430	0.980	0.908	0.507	0.776	H > E > I	0.029	29
	Pratapgarh	2.631	0.953	0.810	0.286	0.628	E > H > I	0.076	20
Sikkim	North District	1.604	0.955	1.000	0.667	0.864	E > H > I	0.017	29
	West District	1.474	0.985	1.000	0.635	0.860	E > H > I	0.018	30
	South District	1.392	1.000	0.917	0.739	0.880	H > E > I	0.008	14
	East District	1.418	0.992	0.900	0.869	0.919	H > I > E	0.001	14
Tamil Nadu	Thiruvallur	1.685	0.981	0.931	0.908	0.940	I > E > H	0.000	14
	Chennai	1.400	1.000	0.974	0.976	0.983	H > I > E	0.000	14
	Kancheepuram	1.719	0.988	0.889	0.862	0.912	H > I > E	0.001	14
	Vellore	2.002	0.980	0.976	0.854	0.935	E > H > I	0.001	32
	Tiruvannamalai	1.647	0.958	0.932	0.698	0.856	E > H > I	0.006	31
	Viluppuram	2.103	0.958	0.922	0.644	0.832	E > H > I	0.008	29
	Salem	1.895	0.970	0.944	0.778	0.894	E > H > I	0.002	32
	Namakkal	1.462	0.971	0.938	0.846	0.917	E > I > H	0.001	32
	Erode	1.621	0.994	0.842	0.852	0.894	H > I > E	0.005	14
	The Nilgiris	1.578	0.985	0.974	0.849	0.935	E > H > I	0.001	14
	Dindigul	1.758	0.960	0.907	0.666	0.837	E > H > I	0.005	29
	Karur	1.523	0.978	0.934	0.713	0.869	E > H > I	0.005	32
	Tiruchirappalli	1.669	0.991	0.944	0.739	0.886	H > E > I	0.006	14
	Perambalur	2.027	0.978	0.939	0.734	0.879	E > H > I	0.004	32
	Ariyalur	1.910	0.958	0.937	0.539	0.792	E > H > I	0.022	29
	Cuddalore	1.540	0.955	0.925	0.687	0.849	E > H > I	0.007	31
	Nagapattinam	1.630	0.985	0.901	0.581	0.808	H > E > I	0.018	30
	Thiruvarur	1.725	1.000	0.949	0.616	0.841	H > E > I	0.021	30
	Thanjavur	1.582	0.971	0.946	0.741	0.882	E > H > I	0.004	32
	Pudukkottai	2.005	0.953	0.884	0.560	0.784	E > H > I	0.012	29
	Sivaganga	1.559	0.966	0.923	0.756	0.878	E > H > I	0.002	31
	Madurai	1.513	1.000	0.877	0.876	0.916	H > I > E	0.004	14
	Theni	1.870	0.960	0.940	0.853	0.917	E > I > H	0.004	31
	Virudhunagar	1.557	1.000	0.820	0.763	0.857	H > I > E	0.010	22
	Ramanathapuram	1.723	0.991	0.943	0.751	0.890	H > E > I	0.005	14
	Thoothukkudi	1.900	0.973	0.907	0.871	0.916	I > E > H	0.000	26
	Tirunelveli	1.580	0.993	0.920	0.831	0.913	H > E > I	0.002	14
	Kanniyakumari	1.523	1.000	0.953	0.938	0.963	H > I > E	0.001	14
	Dharmapuri	1.712	0.972	0.965	0.759	0.894	E > H > I	0.004	32
	Krishnagiri	2.038	0.968	0.894	0.808	0.888	E > I > H	0.000	31

(continued)

Table 8.5 (continued)

State/ Union Territory	District	TFR	HE	ED	SL	HDS	HDP	HDU	Node
	Coimbatore	1.405	0.957	0.927	0.896	0.926	I > E > H	0.005	26
	Tiruppur	1.611	0.994	0.917	0.880	0.930	H > I > E	0.001	14
	Bhadradri Kothagudem	1.591	0.939	0.808	0.732	0.823	I > E > H	0.004	16
	Adilabad	2.433	0.971	0.662	0.579	0.724	H > I > E	0.023	20
	Hyderabad	1.782	0.975	0.904	0.981	0.953	I > E > H	0.003	26
	Jagitial	1.757	0.961	0.859	0.803	0.872	I > E > H	0.001	26
	Jangoan	1.821	0.977	0.976	0.812	0.919	E > H > I	0.003	32
	Jayashankar Bhupalapally	1.743	0.948	0.895	0.610	0.807	E > H > I	0.009	29
	Jogulamba Gadwal	2.150	0.978	0.745	0.701	0.801	H > I > E	0.010	27
	Kamareddy	1.766	0.971	0.864	0.576	0.790	H > E > I	0.011	11
	Karimnagar	1.469	0.992	0.891	0.859	0.913	H > I > E	0.001	14
	Khammam	1.643	0.989	0.853	0.815	0.884	H > I > E	0.003	14
	Komaram Bheem Asifabad	1.761	0.966	0.702	0.444	0.679	H > E > I	0.026	20
	Mahabubabad	1.631	0.992	0.902	0.634	0.832	H > E > I	0.014	30
	Mahabubnagar	2.098	0.982	0.887	0.719	0.857	H > E > I	0.003	14
	Mancherial	1.495	1.000	0.923	0.717	0.874	H > E > I	0.009	14
	Medak	1.984	0.954	0.925	0.512	0.776	E > H > I	0.024	29
	Medchal-Malkajgiri	1.866	0.981	0.909	0.963	0.951	I > H > E	0.001	14
	Nagarkurnool	1.809	0.971	0.900	0.724	0.860	E > H > I	0.002	32
	Nalgonda	1.784	0.991	0.963	0.816	0.921	E > H > I	0.002	14
	Nirmal	1.600	0.976	0.849	0.648	0.816	H > E > I	0.006	11
	Nizamabad	1.643	0.991	0.825	0.765	0.856	H > I > E	0.006	22
	Peddapalli	1.563	0.989	0.875	0.831	0.897	H > I > E	0.002	14
	Rajanna Sircilla	1.970	0.955	0.953	0.839	0.915	E > I > H	0.007	31
	Ranga Reddy	1.999	0.942	0.891	0.913	0.915	I > E > H	0.014	26
	Sangareddy	2.298	0.966	0.938	0.620	0.829	E > H > I	0.012	29
	Siddipet	1.827	0.974	0.929	0.775	0.890	E > H > I	0.002	32
	Suryapet	1.857	0.955	0.969	0.790	0.902	E > I > H	0.008	31
	Vikarabad	2.120	0.935	0.857	0.609	0.791	E > I > H	0.009	11
	Wanaparthy	1.930	0.936	0.951	0.767	0.881	E > I > H	0.019	31
	Warangal Rural	1.808	0.992	0.933	0.629	0.839	H > E > I	0.016	30
	Warangal Urban	1.778	0.973	0.917	0.862	0.916	E > I > H	0.000	32
	Yadadri Bhuvanagiri	1.798	0.948	0.938	0.862	0.915	E > I > H	0.010	31
Telangana	Dhalai	2.277	0.979	0.905	0.217	0.615	H > E > I	0.158	24
	Gomati	1.803	0.933	0.872	0.340	0.670	E > H > I	0.058	24

(continued)

Table 8.5 (continued)

State/ Union Territory	District	TFR	HE	ED	SL	HDS	HDP	HDU	Node
	Khowai	1.669	0.945	0.857	0.239	0.610	E > H > I	0.109	11
	North Tripura	1.919	0.932	0.833	0.375	0.677	E > H > I	0.038	15
	Sepahijala	2.220	0.964	0.879	0.328	0.673	E > H > I	0.074	24
	South Tripura	1.793	0.953	0.860	0.225	0.603	E > H > I	0.125	11
	Unakoti	2.465	0.958	0.778	0.219	0.578	H > E > I	0.117	19
	West Tripura	1.701	0.970	0.886	0.547	0.784	E > H > I	0.016	29
Uttar Pradesh	Saharanpur	2.210	0.938	0.677	0.783	0.794	I > H > E	0.017	16
	Bijnor	2.487	0.950	0.768	0.732	0.813	I > H > E	0.002	16
	Rampur	3.499	0.935	0.562	0.683	0.715	I > H > E	0.045	17
	Jyotiba Phule Nagar	3.044	0.943	0.706	0.709	0.780	I > H > E	0.006	16
	Meerut	2.932	0.943	0.739	0.886	0.853	I > H > E	0.016	16
	Baghpat	2.559	0.951	0.814	0.824	0.861	I > E > H	0.004	16
	Gautam Buddha Nagar	2.201	0.947	0.850	0.929	0.908	I > E > H	0.011	26
	Bulandshahr	2.612	0.939	0.786	0.765	0.827	I > E > H	0.005	16
	Aligarh	2.476	0.924	0.735	0.690	0.778	I > E > H	0.009	16
	Mahamaya Nagar	2.532	0.923	0.778	0.567	0.745	E > I > H	0.008	16
	Mathura	2.999	0.943	0.709	0.729	0.789	I > H > E	0.007	16
	Agra	2.942	0.939	0.718	0.812	0.819	I > H > E	0.013	16
	Firozabad	2.953	0.912	0.768	0.634	0.765	E > I > H	0.020	16
	Mainpuri	2.955	0.889	0.839	0.487	0.720	E > I > H	0.088	18
	Bareilly	2.935	0.961	0.565	0.679	0.722	H > I > E	0.057	27
	Pilibhit	2.621	0.930	0.596	0.472	0.646	H > I > E	0.010	15
	Shahjahanpur	3.670	0.900	0.669	0.481	0.667	E > I > H	0.014	15
	Kheri	2.738	0.904	0.643	0.288	0.567	E > H > I	0.017	15
	Sitapur	2.626	0.917	0.529	0.240	0.508	H > E > I	0.036	15
	Hardoi	3.093	0.931	0.579	0.252	0.534	H > E > I	0.047	17
	Unnao	2.396	0.945	0.659	0.382	0.631	H > E > I	0.021	15
	Lucknow	2.062	0.962	0.794	0.748	0.831	H > I > E	0.001	27
Tripura	Farrukhabad	2.969	0.927	0.726	0.524	0.712	E > I > H	0.001	15
	Kannauj	2.961	0.923	0.726	0.382	0.646	E > H > I	0.014	15
	Etawah	2.604	0.941	0.844	0.641	0.801	E > I > H	0.005	11
	Auraiya	2.946	0.931	0.796	0.410	0.683	E > H > I	0.022	15
	Kanpur Dehat	2.500	0.948	0.761	0.379	0.662	H > E > I	0.029	15
	Kanpur Nagar	2.282	0.946	0.827	0.714	0.825	E > I > H	0.002	16
	Jalaun	2.464	0.956	0.785	0.466	0.713	H > E > I	0.016	20
	Jhansi	2.314	0.966	0.872	0.574	0.790	E > H > I	0.011	29
	Lalitpur	3.171	0.929	0.790	0.323	0.636	E > H > I	0.044	18

(continued)

Table 8.5 (continued)

State/ Union Territory	District	TFR	HE	ED	SL	HDS	HDP	HDU	Node
	Hamirpur	2.501	0.946	0.827	0.393	0.688	E > H > I	0.033	15
	Mahoba	2.433	0.952	0.822	0.386	0.685	E > H > I	0.036	20
	Banda	2.941	0.932	0.669	0.234	0.551	H > E > I	0.062	17
	Chitrakoot	2.491	0.952	0.679	0.261	0.575	H > E > I	0.072	20
	Fatehpur	2.817	0.931	0.687	0.320	0.605	H > E > I	0.026	15
	Pratapgarh	2.816	0.956	0.834	0.430	0.711	E > H > I	0.028	20
	Kaushambi	3.767	0.933	0.677	0.305	0.594	H > E > I	0.031	15
	Prayagraj	3.186	0.935	0.771	0.468	0.704	E > H > I	0.009	15
	Bara Banki	3.025	0.930	0.557	0.333	0.569	H > I > E	0.028	15
	Faizabad	2.333	0.967	0.806	0.482	0.729	H > E > I	0.020	20
	Ambedkar Nagar	2.911	0.963	0.827	0.347	0.668	H > E > I	0.057	20
	Bahraich	3.110	0.936	0.395	0.192	0.436	H > I > E	0.294	17
	Shrawasti	3.452	0.955	0.449	0.195	0.461	H > I > E	0.211	19
	Balrampur	2.912	0.933	0.522	0.325	0.553	H > I > E	0.048	17
	Gonda	2.801	0.951	0.683	0.397	0.647	H > E > I	0.023	15
	Siddharthnagar	3.140	0.986	0.589	0.370	0.611	H > I > E	0.091	20
	Basti	2.318	0.949	0.754	0.457	0.697	H > E > I	0.012	15
	Sant Kabir Nagar	2.330	0.965	0.738	0.369	0.654	H > E > I	0.043	20
	Maharajganj	2.783	0.951	0.763	0.472	0.708	H > E > I	0.012	20
	Gorakhpur	2.612	0.942	0.813	0.550	0.755	E > H > I	0.006	15
	Kushinagar	2.364	0.965	0.746	0.435	0.689	H > E > I	0.026	20
	Deoria	2.124	0.985	0.864	0.570	0.791	H > E > I	0.018	11
	Azamgarh	2.296	0.933	0.875	0.448	0.725	E > H > I	0.030	29
	Mau	2.516	0.970	0.784	0.516	0.739	H > E > I	0.016	20
	Ballia	2.361	0.965	0.842	0.471	0.735	H > E > I	0.024	11
	Jaunpur	2.263	0.976	0.904	0.502	0.771	E > H > I	0.028	29
	Ghazipur	2.667	0.972	0.867	0.403	0.711	H > E > I	0.047	11
	Chandauli	2.447	0.944	0.852	0.468	0.731	E > H > I	0.020	11
	Varanasi	2.298	0.989	0.791	0.679	0.812	H > I > E	0.011	22
	Sant Ravidas Nagar (Bhadohi)	2.564	0.927	0.871	0.488	0.741	E > H > I	0.026	11
	Mirzapur	2.466	0.947	0.851	0.466	0.730	E > H > I	0.021	11
	Sonbhadra	2.569	0.960	0.706	0.254	0.581	H > E > I	0.087	20
	Etah	3.440	0.942	0.747	0.439	0.685	H > E > I	0.012	15
	Kanshiram Nagar	3.771	0.921	0.597	0.463	0.640	H > I > E	0.005	15
	Amethi	3.167	0.885	0.713	0.353	0.618	E > I > H	0.060	17
	Budaun	3.386	0.947	0.499	0.401	0.584	H > I > E	0.083	17
	Ghaziabad	2.529	0.970	0.844	0.968	0.926	I > H > E	0.005	26

(continued)

Table 8.5 (continued)

State/ Union Territory	District	TFR	HE	ED	SL	HDS	HDP	HDU	Node
	Hapur	2.593	0.930	0.777	0.856	0.852	I > E > H	0.019	16
	Moradabad	2.547	0.933	0.728	0.709	0.785	I > E > H	0.006	16
	Muzaffarnagar	2.479	0.981	0.695	0.822	0.826	H > I > E	0.023	27
	Rae Bareli	2.824	0.954	0.758	0.347	0.646	H > E > I	0.043	20
	Sambhal	2.814	0.931	0.598	0.540	0.674	I > H > E	0.013	15
	Shamli	2.582	0.944	0.594	0.779	0.762	I > H > E	0.043	17
	Sultanpur	3.064	0.950	0.817	0.362	0.671	E > H > I	0.042	18
Uttarakhand	Uttarkashi	2.125	0.952	0.972	0.604	0.829	E > H > I	0.019	29
	Chamoli	1.934	0.942	0.980	0.659	0.851	E > I > H	0.020	29
	Rudraprayag	2.317	0.969	0.964	0.654	0.852	E > H > I	0.011	29
	Tehri Garhwal	2.243	0.987	0.986	0.698	0.882	E > H > I	0.010	14
	Dehradun	1.928	0.967	0.892	0.950	0.936	I > E > H	0.003	26
	Garhwal	3.217	0.980	0.979	0.652	0.859	E > H > I	0.013	29
	Pithoragarh	2.106	0.939	0.971	0.624	0.833	E > I > H	0.024	29
	Bageshwar	2.020	0.983	0.969	0.564	0.820	E > H > I	0.025	30
	Almora	1.672	0.979	0.974	0.481	0.783	E > H > I	0.041	29
	Champawat	2.616	0.956	0.950	0.562	0.806	E > H > I	0.020	29
	Nainital	2.021	0.979	0.865	0.793	0.876	H > I > E	0.001	26
	Udham Singh Nagar	2.541	0.940	0.795	0.803	0.844	I > E > H	0.007	16
	Hardwar	2.227	1.000	0.656	0.812	0.814	H > I > E	0.046	22
West Bengal	Dakshin Dinajpur	1.681	0.986	0.850	0.252	0.625		0.128	11
	Hugli	2.288	0.976	0.880	0.536	0.779	H > E > I	0.020	29
	Jalpaiguri	2.334	0.966	0.852	0.405	0.707	H > E > I	0.042	11
	Uttar Dinajpur	2.537	0.961	0.783	0.239	0.593	H > E > I	0.106	20
	Birbhum	1.954	0.974	0.770	0.287	0.622	H > E > I	0.088	20
	Maldah	2.357	0.970	0.799	0.306	0.640	H > E > I	0.076	20
	Murshidabad	2.138	0.955	0.826	0.287	0.633	H > E > I	0.079	20
	Bankura	2.198	0.965	0.903	0.238	0.626	E > H > I	0.131	24
	Nadia	1.901	0.977	0.911	0.386	0.717	E > H > I	0.061	24
	North Twenty Four Parganas	1.772	0.990	0.892	0.653	0.836	E > H > I	0.011	30
	Haora	1.444	0.990	0.821	0.655	0.814	H > E > I	0.011	22
	Kolkata	2.040	0.987	0.884	0.882	0.917	H > E > I	0.001	14
	Puruliya	2.464	0.969	0.865	0.227	0.609	H > I > E	0.135	11
	Darjiling	1.507	0.991	0.908	0.589	0.815	H > E > I	0.019	30
	Koch Bihar	1.285	0.956	0.875	0.201	0.591	H > E > I	0.150	24

(continued)

Table 8.5 (continued)

State/ Union Territory	District	TFR	HE	ED	SL	HDS	HDP	HDU	Node
	South Twenty Four Parganas	2.108	0.973	0.811	0.383	0.685	E > H > I	0.049	20
	Paschim Medinipur	2.010	0.978	0.874	0.210	0.602	H > E > I	0.158	24
	Purba Medinipur	1.639	0.984	0.882	0.209	0.605	H > E > I	0.165	24
	Paschim Barddhaman	1.867	0.945	0.809	0.636	0.789	H > E > I	0.001	16
	Purba Barddhaman	2.000	0.975	0.866	0.377	0.699	E > I > H	0.057	11

Source Author

References

Bongaarts J (1972) A framework for analyzing the proximate determinants of fertility. Popul Dev Rev 4(1):105–132

Bongaarts J (2015) Modeling the fertility impact of the proximate determinants: Time for a tune-up. Demogr Res 33(19):535–560

Bongaarts J (1978) The fertility-inhibiting effects of the intermediate fertility variables. Stud Fam Plan 13(6/7):178–189

Breiman L, Friedman JH, Olshen RA, Stone CJ (1984) Classification and regression trees. Wadsworth, Inc., Belmont

Bryant J (2007) Theories of fertility decline and the evidence from development indicators. Popul Dev Rev 33(1):101–127

Davis K (1945) The world demographic transition. Ann Am Acad Polit Soc Sci 237:1–11

Davis K, Blake J (1956) Social structure and fertility: an analytical framework. Econ Dev Cult Chang 4(3):211–235

Esping-Andersen G, Billari FC (2015) Re-theorizing family demographics. Popul Dev Rev 41(1):1–31

Fox J, Klüsener S, Myrskylä M (2019) Is a positive relationship between fertility and economic development emerging at the sub-national regional level? Theoretical considerations and evidence from Europe. Eur J Popul 35(3):487–518

Furuoka F (2009) Looking for a J-shaped development-fertility relationship: do advances in development really reverse fertility declines. Econ Bull 29(4):3067–3074

Goldscheider F, Bernhardt E, Lappegård T (2015) The gender revolution: a framework for understanding changing family and demographic behavior. Popul Dev Rev 41(2):207–239

Han J, Kamber M, Pei J (2012) Data mining: concepts and techniques, 3rd edn. Elsevier, New York

Harttgen K, Vollmer S (2014) A reversal in the relationship of human development with fertility? Demography 51(1):173–184

Ho TK (1998) The random subspace method for constructing decision forests. IEEE Trans Pattern Anal Mach Intell 20(8):832–844

Kass GV (1980) An exploratory technique for investigating large quantities of categorical data. Appl Stat 29:119–127

Leon C (1998) Descriptive and inferential statistics. In: Bellack AS, Hersen M (eds) Comprehensive Clinical Psychology. Pergamon

Liu S, McGree J, Ge Z, Xie Y (2016) Classification methods. In: Liu S, McGree J, Ge Z, Xie Y (eds) Computational and Statistical Methods for Analysing Big Data with Applications. New York

Loh W, Shih Y (1997) Split selection methods for classification trees. Stat Sin 7:815–840
Luci A, Thévenon O (2010) Does economic development drive the fertility rebound in OECD countries? Paper presented in the European Population Conference 2010 EPC2010, Vienna, Austria, 1-4
Muratbek BA, Bektemisova BU (2024) Using neural networks for demographic prediction. J Probl Comput Sci Inf Technol 1(2):25–33
Myrskylä M, Kohler HP, Billari FC (2009) Advances in development reverse fertility declines. Nature 460(7256):741
Myrskylä M, Kohler HP, Billari FC (2011) High development and fertility: fertility at older reproductive ages and gender equality explain the positive link. MPIDR Working Paper WP 2011–017
Notestein FW (1945) Population-The long view. In: Schultz, Theodore W. (ed.) Food for the World. 36–57
Parihari S, Chatterjee ND, Das K, Bhattacharya RK (2023) Role of groundwater potentiality and soil nutrient status on agricultural productivity: a case study in Paschim Medinipur District, West Bengal. In: Shit P, Bhunia G, Adhikary P (eds) case studies in geospatial applications to groundwater resources. Elsevier, pp 39–66
Song Y, Lu Y (2015) Decision tree methods: applications for classification and prediction. Shanghai Arch Psychiatry 27(2):130–135
Sun Y, Wong AKC, Kamel MS (2009) Classification of imbalanced data: a review. Int J Pattern Recognit Artif Intell:687–719
Tan PN, Steinbach M, Karpatne A, Kumar V (2014) Introduction to data mining. Essex, Pearson
Urso A, Fiannaca A, La Rosa M, Ravì V, Rizzo R (2019) Data mining: classification and prediction. In: Ranganathan S, Gribskov M, Nakai K, Schönbach S (eds) Encyclopedia of bioinformatics and computational biology. New York
Venkatesan R, Meng J, Er (2016) A novel progressive learning technique for multi-class classification. Neurocomputing 207: 310–321
Wilson C, Airey P (1999) How can a homeostatic perspective enhance demographic transition theory? Popul Stud 53(2):117–128

The manufacturer's authorised representative in the EU is Springer Nature Customer Service Centre GmbH, Europaplatz 3, 69115 Heidelberg, Germany. If you have any concerns regarding our products, please contact ProductSafety@springernature.com

Printed and bound by CPI Group (UK) Ltd, Croydon, CR0 4YY
21/09/2026
02220012-0017